Powder and Propellants:

ENERGETIC MATERIALS AT INDIAN HEAD, MARYLAND, 1890–2001

Second Edition

RODNEY CARLISLE

University of North Texas Press
Denton, Texas

First edition published in 1990.
Second edition published in 2002 by University of North Texas Press.

10 9 8 7 6 5 4 3 2 1

Permissions:
University of North Texas Press
P.O. Box 311336
Denton, TX 76203-1336

The paper used in this book meets the minimum requirements of the American
National Standard for Permanence of Paper for Printed Library Materials, z39.48.1984.
Binding materials have been chosen for durability.

Library of Congress Cataloging-in-Publication Data

Carlisle, Rodney P.
    Powder and propellants : energetic materials at Indian Head,
  Maryland, 1890–2001 / Rodney Carlisle.—2nd ed.
       p.   cm.
    Includes bibliographical references and index.
    ISBN 1-57441-149-7 (cloth : alk. paper)
       1. United States Naval Ordnance Station (Indian Head, Md.)—
  History. 2. United States. Navy—Ordnance and ordnance stores—
  History. 3. Ordnance—Research—Maryland—Indian Head. I. Title

VF384.U6 C3823 2002
359.7'5'0975247—dc21

                                                                  2002018098

This book was supported under Contract NOO174-98-D-0004

Table of Contents

List of Tables   vi
Preface to the Second Edition   vii
Preface   ix
About the Author   xii
Acknowledgments   xiii
List of Abbreviations   xv

CHAPTER 1: Origins of the Naval Ordnance Station   1

A Patented Breech—The Naval Inventor at Work   5
The Proving Ground—Structured Innovation in Ordnance   8
Structured Innovation Under Way: Testing of Guns, Armor, Shells   10
1889–1896: Ordnance Issues and the Place of Indian Head   16

CHAPTER 2: Early Powder Manufacturing and Ordnance Testing   19

Powder Under Development   20
Powder Factory: Industrial Management   24
Doubled Capacity   26
The Beginnings of Civilian Management   27
Management Problems: Information, Staff, Costs   32
Cost Accounting and Chemical Engineering   34
Proving Ground: Guns, Armor, Shells   35
Experimental Work   37
Zealots at Work   40

CHAPTER 3: Ordnance Technology in the World War I Era   43

Range Safety   45
Standardization of Shells and Powder   47
Powder Expert at Work   50
Relationship with Private Industry—The Roles Redefined   51
Powder Production   55
Indian Head Goes to War   57
Evolution of Mission   62

CHAPTER 4: Indian Head in Ordnance Research, 1921–1941   63

Factors Shaping the Research Agenda   66
Specific Problems: Lessons Applied   69
The Variable Delay Fuze—A Case of Incremental Innovation   73
Policies, Budgets, Treaty, Foreign Progress   75
The Agenda Juggled: Ships Challenged From Above and Below   78
Indian Head—A Place in the Network   81

The 1930s—Experiments, Personnel, Technology Transfer   83
The Interwar Years—Quiet Progress and Cross-Currents   86

CHAPTER 5: Indian Head: The Navy and the Community, 1890–1940   89

Indian Head's Earliest Years   91
World War I and Postwar Growth   95
Indian Head Between the Wars   99

CHAPTER 6: World War II   105

Indian Head in Rockets, World War II   107
The JPL at Indian Head   113
Explosives Investigation Laboratory   114
Powder Factory Research and Development During the War   118

CHAPTER 7: Postwar Patterns   121

Early Work on JATOs   122
The Postwar Search for a Research and Development Mission   123
Pilot Plants and Manufacturing Expansion   127
Research and Development: Problems with Implementing the Mission   131

CHAPTER 8: Years of Transition, 1955–1963   137

Continuing Problems: Personnel, Safety, and Equipment   139
The Search for a Mission   141
Internal Divisions   143
Ad Hoc Mission Add-Ons: CAD, Terrier, Weapon A   146
Mission Redefinition by Planned Remodelling   149
The Shaping of a Dominant Personality: Joe L. Browning   151
Management Problems Addressed   152
Browning's Management Ideas   154
Polaris Work   156
Challenge and Response   159

CHAPTER 9: From Naval Propellant Plant to Naval Ordnance Station   161

Mission Visions, Market Realities   162
Products, Plants, and Processes in 1960   164
The Propellant Plant in 1962   166
Management Concepts   168
Workload Development: NASA Efforts and Inert Diluent   169
From Research to Production: Otto II   170
The Bootstrap Problem Mastered   172
Shots in the Arm: 2.75 Revived, Zuni, Line-Thrower, C-3 Plastic   174
Guns: Old Weapons Redesigned   176
A Crisis in Workload—Pessimists' View   177

The Optimists' View   179
Facing the Seventies   182

CHAPTER 10: The Post-Vietnam Years   185

Lee Aboard—Reorganization and Public Relations   188
Adding Tasks Through the Seventies   190
    Ordnance Devices   190
    Standard Anti-Radar Missile   192
    Simulators   192
    JATOs   193
    Work for Others   193
    Guns   194
Reorganizations   194
Problems and Solutions   195
Production and Engineering: Striking the Balance   196
New Assignments   201
Planning Conferences, Review Panels, Training   203
Successes and Adjustments   206
A New Technical Director   206
Glassware to Stainless Steel; Laboratory to Production   211

CHAPTER 11: The Second Century Begins   213

A Decade of Turbulence   214
Navigating into a New Century   217
Buffeted by BRAC   220
Growing Role in a Shrinking Business   222
Responding to Requirements   228
Integrated Circuits and the Internet   231
Energetics and the Environment   232
Continuities and Changes   235

EPILOGUE: Retrospectives and Prospects   237

APPENDIXES

1. Tours of Duty, Inspectors of Ordnance in Charge
   and Commanding Officers, Indian Head   245
2. Biographical Information, Officers in Charge and Commanding
   Officers, Indian Head   247
3. Oral History Interviews Conducted for this Work   258
4. Milestones in Technology and Management   259
5. Chiefs of the Navy Bureau of Ordnance and its Successors, 1881–2001   263

NOTES   265

GLOSSARY OF TECHNICAL TERMS   303

BIBLIOGRAPHY   311

INDEX   317

List of Tables

Table 1   Tenure, Early Officers in Charge
Table 2   Continuity of Civilian Personnel 1901–1906
Table 3   Navy Smokeless Powder Deliveries, July 1913–June 1914
Table 4   Powder Production 1910–1919, Indian Head
Table 5   Testing Program, Indian Head, 1915–1919
Table 6   Experimental Ammunition Unit Problem List, 1919
Table 7   Experimental Ammunition Unit, Problems Added, 1920–1926
Table 8   Indian Head Personnel, December 1939
Table 9   Indian Head JPL Projects, July 1944
Table 10   Categories of R and D Projects, December 1947
Table 11   Postwar Commanding Officers
Table 12   Civilian Employment, Indian Head, 1950–1962
Table 13   Partial List of Rocket and Missile Systems, circa 1960
Table 14   Wage and Professional Employees, NOS, 1973–1978
Table 15   Program Funding, NOS, Fiscal Year 1979
Table 16   Evolution of Sea Systems Management Information Activity

Preface to the Second Edition

A decade after the first edition of this work was printed, staff at Indian Head found themselves confronted by two problems. The first press run of the book had been exhausted through sales and gifts, and visitors, new employees, and friends of the institution throughout the Navy, southern Maryland communities, and in the federal government kept asking for copies. Secondly, the turbulent decade of the 1990s seemed one of the most dramatic in the history of Indian Head and an update to the history seemed appropriate.

Every generation thinks that its times are more challenging than most prior eras, probably because history lacks the immediacy of first hand experience. However, in the case of the adjustment of naval technology to the changes of the 1990s, there is a measure of objective truth in the sensation that this decade at Indian Head was more turbulent than those that preceded it. The end of the Cold War and the consequent downsizing of facilities and of military production simply spelled the end of many military bases, naval shore establishments, and service-managed research and development detachments and field activities. Other changes that had been underway in the 1980s came to fruition in the 1990s, with the change of defense procurement to an industrial model, in which government facilities competed with each other and with the private sector for contracts. This change had mixed consequences. A facility like Indian Head, that had production at its core, found itself in competition with the private sector. Yet in times of declining military purchases and the emergence of mega-corporations, the business community found re-supply of the specialized weapons of the Navy less attractive. Government re-invention, reductions-in-force, the emergence of the Internet, a national consciousness of the polluting effects of chemicals essential in explosives and propellants, all put Indian Head through crisis after crisis in the 1990s.

Morale suffered, despite the energetic efforts of the new Technical Director, Roger Smith, who took office in 1989. A member of the Senior Executive Service with a rich technical and acquisition background, Smith was well aware of the new management style of the 1990s. His charm and good humor, as well as his efforts to instill a sense of shared responsibility and Total Quality Management won him a strong personal following. Yet fundamental problems persisted. When Smith took charge in 1989, the station employed about 3000 personnel; a decade later, the number was down to about 2000 and still declining. Roger Smith passed away January 31, 1999. The Indian Head community paid emotional tribute to him, mourning their rather sudden loss. A week later, the Naval Sea Systems Command announced the appointment of Mary Lacey as Technical Director at Indian Head. She had served as a department head at the Dahlgren Weapons Center and had prior experience as a scientist at White Oak.

Lacey faced continued crises, among the first of which was a plan to consolidate the Navy's energetics work with those of the Naval Air Systems Command facility at China Lake, California. Although the plans never came to fruition, some at Indian

Head feared their mission might diminish or even vanish in such a realignment. The thought that many Indian Head personnel could face the choice of relocation or resignation sent shudders through Charles County, where Indian Head had become part of the local economy as well as the local tradition and culture. Although that plan did not work out, the Explosives Ordnance Disposal school, that had been a tenant for decades on the Indian Head peninsula, was relocated to Elgin Air Force Base in Florida a few months later.

Then, in 2001, as the second edition was going to press, the terrorist attack of September 11, 2001, gave new urgency and a need for re-dedicated effort on many of the missions of the Indian Head facility.

The last decade of the twentieth century, and the first decade of Indian Head's second century had been one of its most turbulent times. A new chapter in "The Book," as the first edition came to be known, would be fully justified.

Over the year 2000-2001, the author interviewed a number of the veterans of the decade, including some he had interviewed during the preparation of the first edition. The public affairs office maintained excellent files of the base newspaper, together with materials from the ceremonies installing new commanding officers and from a variety of award ceremonies and events. Other documentation included copies of technical reports at the technical library, and packets of materials prepared to deal with "data calls" when the facility had been faced with the Base Realignment and Closure exercises of the 1990s and retained in the Corporate Operations Office. Soon the picture emerged, one that could merit a book of its own. However, facing a mandate to keep within the bounds and the dimensions established in the first ten chapters, the author told the story in a new concluding chapter.

With additions to the appendices to bring them up to date and a few changes to the bibliography and notes to reflect new materials, this new work tells the story of energetic materials at Indian Head, from its origins in 1890 through the year 2001.

Rodney Carlisle
History Associates Incorporated
Rockville, Maryland
November 5, 2001

Preface

The modern American Navy was born in the 1880s, with the transition from wood and sails to steel and steam. With that change came others in weapons and armor. By the end of the nineteenth century the U.S. Navy had already taken an interest in submarines and aircraft. In a less spectacular fashion, it also financed research and development in smokeless powder and in the building of establishments to bring the methods of engineering and science to improve steel, weaponry, and ships.

A century later, several of the institutions created in that period survive. One, the focus of this book, is the Naval Ordnance Station at Indian Head, Maryland, founded in 1890. Twenty-six miles from the southeast edge of the District of Columbia, on a three thousand acre site on the Potomac River, the Navy built and maintained a facility devoted to the modernization of naval ordnance. Ordnance research included all the arts and sciences surrounding the propelling, by chemical charge, of steel shell and explosive warhead against steel armor. The century at Indian Head has reflected the evolution of that science and art, that juxtaposition of energetic chemicals and cold steel. At the turn of the century, the concentration was on guns, gun mechanisms, and gun powder improvements, reflecting the advances of metallurgy, mechanical arts, and chemistry.

As the Navy adapted to the exigencies of modern warfare in the twentieth century, the naval command developed new vocabularies of research and development. Officers struggled with questions of exactly how to manage and organize the skills of innovative people so that modernization of complex systems could be planned and accomplished. How to fit the Navy's needs into a market economy remained an issue. Mechanisms to demand consistent quality while purchasing from the lowest bidder in a competitive marketplace evolved in various forms over the decades. For more than twenty years, during the heyday of the large rifled naval gun as the greatest weapon on earth, Indian Head served as a proving ground for testing guns, propellant powder, shells, mounts, and armor. Its location on the river provided an open range, yet boat traffic increasingly interfered with experimental shots.

While steel was being tested to resist impact, a new propellant—smokeless powder—was produced in the first Navy-owned and Navy-operated powder factory, also built at the same location. Convenient to Washington, yet isolated in a wooded and remote section of agricultural Charles County, Indian Head repeatedly proved to be an ideal spot. The locale would host an assortment of research, development, production, testing, and maintenance tasks related to energetic materials: gun propellants, warhead explosives, pyrotechnic mixtures, and rocket fuels.

During and after World War II, the Navy developed and adopted rocket-propelled missiles and large Jet Assist Takeoff, or JATO, rockets for heavy bombers aboard aircraft carriers and on short runways. Indian Head provided a site close to Washington for the manufacture and testing of such rockets. The rockets ranged from 2-inch barrage and air-to-ground weapons through the large JATOs developed at the California Institute of Technology. That work would set precedents and establish both skills and equipment that would prove useful when the facility moved into work on Zuni, Talos, Polaris, and later generations of naval missiles. The propellants for some of the large-diameter rockets would be cast, and poured as a liquid to harden as molds in the individual rockets.

The armed services found more and more applications for propellant and cartridge actuated devices, beginning with aircraft pilot ejection devices that required rapid application of high energy to blow off the canopy, eject the pilot with seat and parachute, and then separate the pilot from the seat. Separation of stages, deployment of fins, and other steps in staging missiles required similar instantaneous applications of energy. For all such uses, the chemistry of energetic materials and their engineering into delicate machines required an unusual mix of skills, facilities, and equipment. The job had to be performed with precision under strict safety conditions, yet management had to find and encourage an innovative style of thought.

The tension between innovation and risk-taking on the one hand, and careful procedures for both safety and reliable quality on the other hand, always haunted Indian Head and other facilities that dealt with such powerful and hazardous materials. Forethought in laying out the facilities, use of bunkers, blowout walls, judicious spacing of buildings, escape slides and water-flooding, limited-quantity storage, and strict adherence to fire and safety regulations prevented a major catastrophic fire or explosion such as occurred at some other manufacturing plants and ammunition depots. Yet, over the course of the decades, a few fatal accidents provided tragic reminders of the hazards of the work.

For a century Indian Head survived not only fire hazards but the hazards of changing technology and naval organization as well. As guns and smokeless powder gave way to rockets, and as the Navy's procurement bureaus reorganized to conform with systems theory between the 1950s and the 1970s, Indian Head could easily have been outmoded or closed. In weaponry, as in other engineering fields, one generation's "high technology" soon becomes the "obsolete junk" of the next. The struggle to readjust the mission and to find the best use for talent and equipment absorbed the energies of the Station's leaders.

As Indian Head prepares to celebrate its centennial in 1990, we can look back on a history of much more than a single institution. Here, at one location, is the story of ordnance evolution over a century, the development of a community to support a naval station, and the difficulties and successes of creating a large civilian-manned operation under the control of naval officers.

Perhaps most central to the survival and flourishing of the institution was its adjustment to the demands of changing military technology and changing naval organization. That adjustment came despite a natural inertia built in by personnel and practice. Skills, equipment, and facilities represented a set of assets that could be used for different tasks. Yet those very assets represented a limitation and a definition of capability which could restrict adaptability. The established institutional assets represented a force for stability, a matrix that would channel the choice of tasks.

While these forces for institutional inertia made change difficult, each new addition of personnel, machinery, or laboratory equipment, and each refinement of capability would create new opportunities. It was precisely because the institution had to adjust to survive, to alter its resources in order to adapt, that the role of the individual leaders—some of them naval officers, some civilian chemists or engineers, some specialists in management—became so important.

The force of individual personalities would shape the nature of the institution. The commanding officers, usually naval captains, ordinarily stayed for a tour of duty that lasted two to three years. Some of them would go on to achieve greater fame and prominence in their careers. Some stood out as well-known members of the "gun club" which dominated the Navy in the first part of the twentieth century. Captain Joseph Strauss, who served as commanding officer for two terms in the first decade of the twentieth century, went on to develop the North Sea Mine Barrage in World War I. Captain Harold R. Stark later gained renown in World War II as Chief of Naval Operations.

But the generally short tours left the opportunity to shape and lead to a handful of leading civilians, managers of the production and research facilities at Indian Head. Their effort to find and hold a consistent mission in the face of larger forces such as technology, warfare, national policies, and the changing manner of conducting naval business made the drama one of personal struggle, not a mindless conflict of impersonal forces.

Like ruling heads of state, the commanding officers, plant managers, and technical directors at Indian Head would shape policy, make choices, and establish their imprint on the Station for a decade or more. The local officials would make their choices, taking blame and credit for their decisions on how to play the role; but the choices they made were usually narrowly defined and limited at headquarters in Washington.

As weapons became more complex and the manufacturing establishments supporting them spread across the country, Indian Head found itself joined by other institutions in the businesses of powder, propellant, and explosive against steel. For the shifting procurement bureaus and commands involved in ordnance, Indian Head remained the only Navy-owned and Navy-operated production facility, joined by a burgeoning array of government-owned, contractor-operated laboratories and factories, universities, and private enterprises. Several of the new facilities were direct offspring of Indian Head—carved like Eve out of Adam's rib—and specialized in one or another aspect of research and testing. The weapons test facility at Dahlgren and the Naval Ordnance Laboratory at White Oak later merged into the Naval Surface Weapons Center. Indian Head operated the facility at Dahlgren in its early years and also hosted one of the groups that joined others to form the Naval Ordnance Laboratory at White Oak. The rocket testing program that formed the core of the China Lake Naval Ordnance Test Station drew one cluster of its personnel from Indian Head. The oldest of these ordnance facilities, Indian Head, continued to adapt to changes in both technology and management.

This book is written to tell the story of the adapting institution, the individuals who shaped it, and the heritage they built. Through the story of powder and propellants at Indian Head we can see the effects upon ordnance of larger transformations in technology, warfare, and naval organization over the century since the beginning of the modern Navy.

About the Author

Rodney Carlisle is vice president and a senior associate of History Associates Incorporated, a historical services firm located in Rockville, Maryland. He is professor emeritus of history at Rutgers University, The State University of New Jersey, at the Camden campus. He holds an A.B. in History from Harvard College and an M.A. and Ph.D. in History from the University of California at Berkeley.

Dr. Carlisle is the author of numerous books in the history of military and naval technology. Recent works include *Where the Fleet Begins: a History of the David Taylor Research Center* (Naval Historical Center, 1998), and *Supplying the Nuclear Arsenal: American Production Reactors, 1942–1992* (Johns Hopkins University Press, 1996). He co-authored *Brandy: Our Man in Acapulco* (University of North Texas Press, 1999). His *Jack Tar: A Sailor's Life, 1750–1910* (Antique Collector's Club, 1999), co-authored with J. Welles Henderson, won the Outstanding Work of Non Fiction Award by the Philadelphia Athenaeum. He edited *Encyclopedia of the Atomic Age* (Facts on File, 2001), and he has completed a volume for the Facts on File America at War series, *The Persian Gulf War*, to be published in 2003. He and his wife, Loretta, make their home in Cherry Hill, New Jersey.

Acknowledgments

A great many people assisted in the production of this work, which is a corporate product in more ways than one. At History Associates Incorporated, we strive to meet the standards of historical scholarship and, at the same time, to produce research and writing through a team effort. For that reason, these acknowledgments constitute recognition of collective effort, in addition to personal thanks by the principal author.

At Indian Head, a regularly constituted advisory committee, chaired by Diane A. Santiago, provided many specific suggestions. Ms. Santiago coordinated the efforts of the committee and provided her own helpful editorial suggestions which sharpened the focus of the work at many points. The committee included Drs. Dean Allard and Ronald Spector from the Naval Historical Center, and Naval Ordnance Station personnel including Dr. Dominic J. Monetta, John W. (Bill) Jenkins, Vincent C. Hungerford, Stephen E. Mitchell, Robert H. Spotz, Jean McGovern, Edith A. Chalk, John W. Murrin, and John P. McDevitt. Readings and suggestions from Wayne Vreatt also helped clarify many points. In addition, the individuals listed in Appendix 3 gave freely of their time for oral history interviews. The work was performed under an omnibus contract through Richard Carson and Associates, and the staff there helped arrange fiscal details.

Archival and library support from Charles F. Gallagher at the Naval Ordnance Station Technical Library and from his staff members, Meriam L. Washington, Mary Jane Edmonds, and Janet S. Ferrell, proved invaluable. At the National Archives, Richard Von Dornhoff and Barry Zerbi assisted in locating materials, while at the Washington Navy Library, the assistance and hospitality of John Vajda, Paula Murphy, and Janis Beatty were very much appreciated. At the Washington National Records Center in Suitland, Maryland, the assistance of Ernest Byrd, Ben Cooper, Helen Embleton, and Victoria Washington was extremely helpful. Mark Weber at the Naval Photographic Center assisted in gathering illustrations. Dorothy Artes provided generous access to her private collection of local history materials and suggested several leads to other sources. Norma Hurley's assistance at the Charles County Board of Education was of great help in deciphering early county school records.

At History Associates Incorporated, Jim Gilchrist provided consistent and valuable assistance through the project, conducting research, checking citations, preparing an excellent draft of Chapter Five, conducting interviews, and keeping the project on track. In the later phases of the project, Jeff Covell and James Lide assisted in picture research and in helping to meet deadlines with research, citation checking and editing, and gathering material for appendixes. Gail Mathews provided excellent professional assistance in word processing, editing, and layout. DyAnn Smith, Renee Sabin, Katharine Jewler, Kim Kilpatrick, Barbara Hunt, and Margaret Belke made sure the billing and progress report procedures were properly handled. Support, suggestions, and encouragement from Drs. Philip Cantelon, Richard Hewlett, Ruth Harris, and Ruth Dudgeon eased the path to particular sources and through the thickets of naval acronyms and prose.

Several excellent works in the history of naval ordnance and in the history of management, noted in the bibliography, provided guidance. For the most part, however, this work navigated through uncharted waters, deriving conclusions from masses of archived or stored federal records, whose quantity was measured not by folder but by hundreds of cubic feet. For these reasons, the assistance from participants, both on the advisory committee and through the oral history interviews, was a great boon. Any mistakes, omissions, or erroneous conclusions remain the responsibility of the author.

Rodney Carlisle
January 1990

Rockville, Maryland

List of Abbreviations

ABL	Allegany Ballistics Laboratory
Acc.	Accession
ADAMS	Automatic Dynamic Analysis of Mechanical Systems
ADM	Admiral
AMB	Assistant Management Board
ARM	Anti-Radar Missile
ASROC	Anti-Submarine Rocket
BRAC	Base Realignment and Closure
BuOrd	Bureau of Ordnance
BuWeps	Bureau of Weapons
C-3	Composite-3 Plastic Explosive
CAD	Cartridge Actuated Device
Cal Tech	California Institute of Technology
CAPT	Captain
CASDO	Computer Applications Support and Development Office
CBIRF	Chemical Biological Incident Response Team
CCC	Civilian Conservation Corps
CCCC	Charles County Community College
CDR	Commander
CEE	Captured Enemy Equipment
CENO	Central Naval Ordnance Management Information System Office
CIT	California Institute of Technology
CIWS	Close-In Weapon System
CO	Commanding Officer
CO_2	Carbon Dioxide
Como	Commodore
Cong.	Congress
CPIA	Chemical Propulsion Information Agency
DBX	Depth Bomb Explosive
D. E. L.	David E. Lee
DIM	Demolition Investigation Memoranda
DINA	Diethanol Nitramine Dinitrate
DNPOH	2-2-Dinitroproponal
DOD	Department of Defense
DTRC	David Taylor Research Center
EAU	Experimental Ammunition Unit
EIL	Explosives Investigation Laboratory
ELBA	Ethyl Lactate-Butyl Acetate
ENS	Ensign
EODTC	Explosive Ordnance Disposal Technical Center

ERDA	Energy Research and Development Administration
FFAR	Folding-Fin Aircraft Rocket
GALCIT	Guggenheim Aeronautical Laboratory at the California Institute of Technology
GOCO	Government-Owned Contractor-Operated
GPO	Government Printing Office
GRI	Gas Research Institute
GS	Government Service
HARP	High Altitude Research Projectile
HBNQ	High-Bulk Density Nitroguanidine
HC	High Capacity
HMX	Cyclotetramethylenetetranitramine
H. R.	House Resolution
HVAR	High Velocity Antiaircraft Rocket
IHDiv-NSWC	Indian Head Division-Naval Surface Warfare Center
IHSP	Indian Head Smokeless Powder (in text only)
IHSP	Indian Head Special Project (in endnotes only)
IHTL	Indian Head Technical Library
IHTR	Indian Head Technical Report
IPT	Integrated Process Team
JAN	Joint Army Navy
JANAF	Joint Army Navy Air Force
JASNE	Journal of the American Society of Naval Engineers
JATO	Jet Assist Takeoff Unit
JLB	Joe L. Browning
JP	Jet Propellant
JPL	Jet Propulsion Laboratory
JPN	Jet Propellant, Navy
LCDR	Lieutenant Commander
LORAP TAFT FOGS	Long Range Planning Task Force Team for Gun Systems
LOVA	Low-Vulnerability Ammunition
LRBA	Long Range Bombardment Ammunition
Mark	Model
MEMS	Microelectricalmechanical Systems
Mod	Modification
NACO	Navy Cool
NARA	National Archives and Records Administration
NASA	National Aeronautics and Space Administration
NATO	North Atlantic Treaty Organization
NAVAIR	Naval Air Systems Command
NAVORD	Naval Ordnance Systems Command
NAVSEA	Naval Sea Systems Command
NAVSHIPS	Naval Ship Systems Command
NC	Nitrocellulose
NDRC	National Defense Research Council
NG	Nitroglycerin
NOL	Naval Ordnance Laboratory

NOMIS	Naval Ordnance Management Information System
NOS	Naval Ordnance Station
NOSOL	No-Solvent Propellant
NOTS	Naval Ordnance Test Station
NPF	Naval Powder Factory
NPG	Naval Proving Ground
NPP	Naval Propellant Plant
NQ	Nitroguanidine
NRL	Naval Research Laboratory
NSWC	Naval Surface Warfare Center
NUMIS	Naval Uniform Management Information System
OD	Ordnance Data
OIM	Ordnance Investigation Memoranda
ONR	Office of Naval Research
OP	Operating Procedures
OPNAV	Office of the Chief of Naval Operations
OSHA	Occupational Safety and Health Administration
PAD	Propellant Actuated Device
PBX	Plastic Bonded Explosive
PBXN	Professional Development Council
PECC	Pilot Ejection Catapult Cartridge
PERT	Program Evaluation Review Technique
PETN	Pentaerythritol Tetranitrate
PGDN	Propylene Glycol Dinitrate
PM	Program Management
PNAB	Philadelphia National Archives Branch
PTA	Parent-Teacher's Association
PT Boat	Power Torpedo Boat
PWA	Public Works Administration
R and D	Research and Development
RAPEC	Rocket Assisted Pilot Ejection Catapult
RAT	Rocket Assisted Torpedo
RDX	Cyclonite or Cyclo Tri-methylenenitramine
RG	Record Group
RIF	Reduction in Force
RMMP	Research-Missiles, Missile-Propulsion
SAE	Society of Automotive Engineers
SCAMP	Sub-Calibre Anti-Material Projectile
SDA	Sea Systems Design Activity
S. Doc.	Senate Document
SEAADSA	Sea Automated Data Systems Activity
SEAADSO	Sea Automated Data Support Office
SES	Senior Executive Service Sess. Session
SJP	Standard Job Procedure
SOP	Standard Operating Procedure
SP	Smokeless Powder
SPAG	Solid Propellant Advisory Group

S.P.D.	Smokeless Powder Rosaniline
SPU	Sidewinder Propulsion Unit
TD	Technical Director
TNT	Trinitrotoluene
USNA	United States Naval Academy
VD7F	Variable Delay, Mark 7 Fuze
WAVES	Women's Reserve for the U.S. Naval Reserve
WHIPS	White Papers Produced by TAFT FOGS
WNRC	Washington National Records Center
WQEC	Weapons Quality Engineering Center
WPI	Worcester Polytechnic Institute

POWDER AND PROPELLANTS

CHAPTER 1

Origins of the Naval Ordnance Station

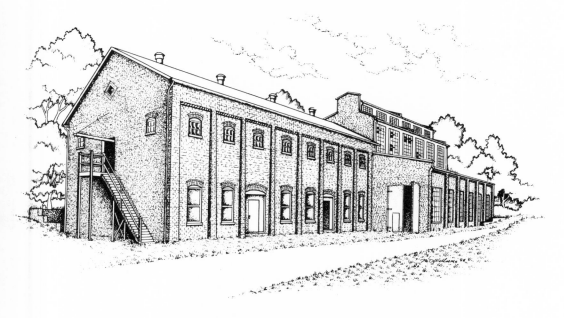

Ether House (l) and Distillation House (r)
Buildings 164 and 164A
1893 and 1919

Robert Brooke Dashiell, ensign, U.S. Navy, worked late into the night, perfecting his design sketches for an improved breech for rapid-firing 5-inch naval ordnance. The crucial element was the mounting of a double hinge; the idea was simple, but the sketch must be precise enough for the draftsmen and machinists at the Washington Navy Yard, who would produce the blueprints and the working model, to execute. A sputtering candle provided the only light as he sat at a rude desk in an abandoned fishing shack. The candlelight illuminated his square-featured, blond head, and his firm mouth which was framed by a determined moustache. Occasionally, he slapped at besieging mosquitoes—laden with infectious malaria and yellow fever—which bred in the swamps surrounding the shack.[1]

It was September 1890, still summer along these desolate Maryland shores of the Potomac. Only twenty-one miles by river from the nation's capital, Dashiell knew he was far out of touch with the bustling activity at Bureau of Ordnance

1

headquarters located there. Here, where the Mattawoman Creek flowed into the Potomac, the landing facility was a broken-down, long-abandoned wharf. Under his supervision, it was to be replaced by a heavy-duty dock to receive naval guns and ammunition for proof testing.

Ensign Dashiell had been ordered to Indian Head that summer to take over construction and to supervise the opening of a new proving ground for the U.S. Navy. The mission was clear: get the proving ground into operation as soon as possible in order to proceed with testing of guns and armor, shells and mounts, for a rapidly expanding fleet. Without direct orders, but on the basis of encouragement from his commanding officer, Dashiell had also made it his personal mission to invent a new mechanism for the breech loading of large naval guns, one that would allow for fewer motions by the gunner and therefore more rapid firing. These two missions, one official, the other less formal, fully engaged the energies of this determined and ambitious young officer. The missions themselves grew out of the interaction of an aggressive naval construction policy and the almost daily progress of technology in the 1880s and 1890s.

The Navy was in the midst of a mounting technological revolution, and despite his geographic isolation Dashiell was determined to be at the center of that revolution. Since the launching of the first steel-hulled "ABCD" ships, *Atlanta*, *Boston*, *Chicago*, and *Dolphin*, in the mid-1880s, the Navy had expanded in every direction: in training, in science, in engineering as well as in the planning of eighteen more new ships. In 1890 Secretary of the Navy Benjamin Tracy announced a policy of capital ship construction, apparently stimulated by the writings of Alfred Thayer Mahan. The resulting second generation of steel-hulled, steam-powered ships would require the most modern and reliable guns and armor. It was a "New Navy."[2]

The particular invention that Dashiell worked on by candlelight in his shanty would, if it succeeded, speed up the firing cycle of hand-loaded 5-inch guns, tactically used for clearing the decks and unarmored sections of enemy ships. The existing standard breech mechanism was slow to operate—it took at least three motions by the gunner to twist open the heavy slotted-screw breech, turn it to the side out of the way and pop the extractor so the spent casing would flip out. After the new shell was inserted, another set of motions was required to close and lock the breech. With black powder and brown or "cocoa" powder, firing guns created such a cloud of smoke in windless conditions that it could take several minutes to achieve a clear sight of the target for a second shot. Consequently, behind the cloud of smoke, the slow steps of opening, turning, extracting, loading, turning, and locking would not matter, since no clear aim was possible.[3]

But a new development loomed that could change all that. A form of shell propellant that was nearly smokeless—what the British called cordite—was under experimentation by all the modern navies. France, Britain, Russia, and Austria-Hungary, each with a navy far stronger than the fledgling new steel and steam fleet of the United States, vied to improve the sportsmen's smokeless powder for use in naval ordnance. The anticipated development of smokeless powder would work as an incentive to improve the speed of breech-loaded guns. Not only would use of the new powder allow an immediate clear shot, but also it would not leave a residue in the gun requiring a delay for sponging-out, as had black and brown powder. It would be only a matter of years, perhaps even months, Dashiell expected, before some form of smokeless powder entered U.S. naval magazines. When that happened, gunnery had to be ready. The rapid-fire mechanism or "R.F. gun" that he and half a dozen other European and American inventors searched for would take advantage of the clean and smokeless qualities of the new powder. If a rapid-fire mechanism could be introduced before smokeless powder, it would still offer advantages over the earlier awkward and slower breeches.[4]

The new proving ground facility that Dashiell planned reflected the institutional side of the Navy's technological revolution. That revolution required not only new inventions but also a range of new and improved shore facilities. Older shops, yards, and testing grounds were pushed to their capacity, not only to build and test but also to innovate. During the 1880s Secretary of the Navy William C. Whitney had concentrated on improving existing shore facilities to keep pace. Under Secretary Tracy, the private steel firms of Carnegie and Bethlehem produced heavy forgings for guns and slowly adjusted to Navy pressure for better armor.[5]

As the United States moved to the high seas as a world power, the Navy sought ways to harness the creative energies of its own officers and inventors in the civilian world. At first those efforts would be halting because nineteenth century planners usually visualized invention as the work of inspired individuals, expecting the new devices and weapons to come as the result of hard work and genius, not the product of planned or structured corporate effort. Hence Dashiell and many other officers worked in the heroic tradition of such famous inventors as ship designers Robert Fulton and John Ericcson, and ordnance innovators like John Dahlgren and Robert Parrot. The organizations that in fact had backed those men, such as shops, drafting rooms, foundries, and shipyards, were not remembered; the contributions were perceived as personal acts of brilliance, not as team or corporate products. Even when an invention came out of an organization, the prevailing individualism of the times required that innovation be viewed as personal; hence weapons, like other nineteenth century inventions, were named for individuals. Men went to war firing Gatlings, Parrots, Dahlgrens, and Maxims.

Despite the prevalence of the old methods of personal invention, court battles over private patents, favoritism in the selection of contractors, and inside information for rigging bids that sometimes characterized the nineteenth century way of procuring military goods, Secretaries of the Navy and bureau chiefs through the 1880s and 1890s worked to put in place a more organized and scientific set of procedures necessary to sort out complex technical issues and to stimulate progress. The beginnings of that structured and less individualistic approach to innovation could be found in the era. The Navy made plans to institute new laboratories, experiment stations, and factories that would apply the findings and methods of science and technology to new and long-standing naval problems.[6]

Dashiell was at once the representative of the older mode of invention by individual effort and a builder of a testing facility that would later become one of the Navy's large collection of research and development institutions, part of its corporate approach. In this regard, he was an early member of a transitional elite of innovative officers, which would include the more famous line officer Bradley A. Fiske, Naval Constructor David W. Taylor, Bureau of Steam Engineering Chief George Melville, and others whose work helped change Navy hardware.

Dashiell served at the time when the change from individual to organizational approach to innovation began in the Navy, and like Taylor and Melville after him he made contributions in both modes. In the older style they earned fame for their personal contributions. Taylor published definitive tables of the hydrodynamic resistance of various hull-shapes of ships; Melville designed a transmission for steam turbines. At the organizational level, Taylor would personally oversee the construction of the Navy's first Experimental Model Basin, which would open in 1899, and Melville would urge the establishment of a Naval Engineering Experiment Station, begun in 1904. Dashiell, despite a career cut short by an untimely death in 1899, had a similar dual career of personal and corporate achievement.[7]

This book will detail the Navy's evolution from individual innovation to institutionalized development in ordnance, using as its focus the facility at Indian Head, from the small beginnings in Dashiell's isolated fishing shack, through the growth of a testing facility, to the emergence of a large-scale institution devoted to constant modernization. The Bureau of Ordnance had begun the move to a new testing facility a full decade before the other technical bureaus, Construction and Repair, which supervised the design and construction of ships, and Steam Engineering, which concentrated on the ship's primary and auxiliary engines.

In his search for advancement in the "New Navy," Dashiell had pursued a line career with emphasis in ordnance, rather than an engineering or construction corps career. This meant that he would alternate sea service with service ashore. Typically each tour of duty would be about three years. Ordnance officers remained in the "line," commanding ships and men, unlike the engineering and construction corps specialists who would stay ashore and concentrate for years in shop, factory, or experimental laboratory. Few ordnance officers, working on their two- and three-year shore assignments, could find the sustained time required for extensive technical work. Dashiell, with his determination and concentration, made the time. After service aboard the cruisers *Essex* and *Pensacola* to the Mediterranean and back, Dashiell had reported to Commander J. H. Dayton at the old proving ground at Annapolis in April 1888, still an ensign at age twenty-eight, his apparently junior rank due to the excess of officers over available billets in the period. Working under Dayton, Dashiell learned the principles of proof-testing of guns.[8]

Across the Severn River from the academy, Dayton ranged newly forged guns from the Naval Gun Factory, which had opened at the Washington Naval Yard the year before under Secretary Whitney's expansion program. The new gun barrels were shipped by barge and followed a circuitous course down the Potomac, up the Chesapeake and the Severn to be unloaded by crane at the dock across from the Naval Academy. The Severn, however, was cluttered with the white sails of pleasure craft, and with lumbering steam traffic into the academy and to the port of Annapolis, the state capital. The occasional wild shot or ricochet from experimental shell or guns endangered individuals strolling the Severn's shores or fishing from its banks. Finding a gap in the traffic for a clear shot became ever harder. The Annapolis Proving Ground, like other naval facilities, was hard pressed to keep up with the pace of expansion.[9]

As early as 1887, Congress had set aside funds for the purchase of land to be used for a new proving ground at a more isolated spot, yet one closer by barge to the gun factory at Washington. Planned construction of a lighthouse at Greensbury's Point near Annapolis would close off even more of the Severn over-water range, adding urgency to the plans.[10]

In 1890 the Bureau of Ordnance acquired the Potomac River site on Cornwallis Neck known as Indian Head and ordered Dashiell to oversee the construction and begin the operation of its new proving ground. Not until Dashiell got the construction at Indian Head completed could Commander Dayton shut down the increasingly dangerous Annapolis facility. Then, all proof-testing of new guns could shift to Indian Head. Meanwhile, Dashiell would spend his evenings designing the rapid-firing breech-loading mechanism for the 5-inch gun.

A PATENTED BREECH—THE NAVAL INVENTOR AT WORK

The story of the invention by Dashiell of the rapid-firing breech that would bear his name reflects the older method of invention by an individual. Through the account we can see how cumbersome that method had become by the early 1890s, how the informal system began to take on aspects of a formal system, and why the new breech can also be viewed as the Navy's first major product to come out of Indian Head, the institution.

Before Dashiell received his appointment to take up the Indian Head task, he joined Capt. William Folger, the naval officer in charge of the gun factory, on the trolley car from the Washington Navy Yard. Dashiell remarked, "Captain, I have a rapid fire mechanism for you."

"That is good," Folger replied. "I would like to have you get up a gun like that. . . . Shove along with it." That was all the encouragement he needed, Dashiell would later recall.

Later that year Folger would be appointed chief of the Bureau of Ordnance. A strong-willed, domineering officer devoted to developing American ordnance products, he fitted the needs of the new Secretary of the Navy, Benjamin Tracy, who had announced the plan to construct a whole fleet of American-produced ships, equipped with American-made armor and weapons. The battleships for a main battle fleet, monitors for coastal defense, and cruisers for commerce-raiding would all need new guns. As chief of the Bureau of Ordnance, Folger oversaw not only the old and new proving grounds but also the gun factory at the navy yard and the torpedo station in Newport, Rhode Island. Through a corps of "inspectors," Folger procured all the Navy's powder, shells, guns, and armor plate from the private sector. Together with Tracy he was eager to force the pace, and appeared to take the young Dashiell under his wing as one of the new breed of ordnance specialists he would like to advance.[11]

Following up on the streetcar discussion and the encouragement from Folger,

Dashiell studied the breech problem. While still at Annapolis, he read everything he could get his hands on, at the St. Johns College library, in journals, in military manuals and reports, and in the *Official Gazette* of the U. S. Patent Office, on the various breech mechanisms under development.[12]

In April 1890 Folger ordered Dashiell to Washington to work on the plans for the new proving ground. Over the summer, in regular working hours, Dashiell planned the proving ground docks and firing ranges. In evenings and on weekends, he returned to the breech and developed a working model in cooperation with a Baltimore model maker.[13]

On the last Sunday in August 1890, before taking the launch down to the new site, Dashiell and Folger chatted again about the breech design. Dashiell showed Folger the preliminary model. Although Folger liked it, Dashiell felt the model needed further work. Later, Folger told Dashiell to be careful that his design not infringe the existing patents of others including a design by Navy Lt. Samuel Seabury. Dashiell had already reviewed the patent drawings of the Seabury breech, and he believed he had a superior idea. The Seabury breech was simply not strong enough to stand up in service, he said, and it required an almost impossible adjustment to one two-hundredth of an inch to seat properly. As far as Dashiell was concerned, Seabury had invented an "utterly worthless contrivance called a rapid fire gun."[14]

Folger and Dashiell also discussed the other breeches invented by Canet, Farcot, Hotchkiss, and Nordenfelt, and a design by Count Baronowski, which Dashiell had noted while visiting a Russian ship on his cruise to the Mediterranean. Dashiell set himself several goals in regard to the breech, reflecting a mix of technical, patriotic, and policy objectives. As he stated them, his personal goals reflected the official orientation of Folger and Tracy: "First, to get an American invention; second, to get a mechanism that was as short to the rear as possible; third, one that could be made cheaply; fourth, one that was strong [enough to] pull out a fired cartridge case." And, he insisted, it had to be one that could be made at the Washington Navy Yard. Nevertheless, Seabury had a recent patent on his breech and had already submitted his plans to Folger for possible adoption by the Navy. Dashiell's invention had to be unique not only to be effective but also to avoid legal claims from the other inventors.[15]

For reasons that derived from patent law, Dashiell construed his work as self-motivated, encouraged, but not "ordered" by Folger. But even from his own account, it is clear that his invention met the needs of the Bureau, conformed in its design to Bureau requirements and policies, and was to be made in Bureau shops, and also that the rights would be sold only to the Bureau. Was this a case of the independent inventor, in nineteenth century style, or the beginnings of developmental technology, in twentieth century style? Dashiell was at the cusp of the transition.

Dashiell knew that naval career advancement and perhaps some royalties could flow if he could simultaneously get the new proving ground into operation and solve the breech problem, with a simple, efficient, and rugged design. He was always careful to work on the breech design on his own time and to make it clear that he did not produce the device on direct orders, thus adhering to the ethical code that could provide royalties to government employees in that era.

Under these circumstances, the Navy hoped to encourage its officers to invent weapons and instruments that they would patent and market to the government, or write technical works on which they would receive royalties. The intellectual model was the individual inventor, the motivation structure was typically nineteenth century—face-to-face encouragement and personal, individual repute and profit. By working on one's own time, one could cleanly maintain the distinction between Navy work and personal work. Yet if the product was made by an officer, his identification with a naval career would tend to prevent him from separately exploiting the situation by demanding excessive royalties. It would keep the invention in the family and in accord with Bureau needs, and it was in that spirit that Dashiell continued his design study.

The expectation that smokeless powder would soon enter the fleet appeared well grounded, but that work took far longer to accomplish than the redesign of guns to utilize fully the anticipated benefits of the new powder. At the torpedo station in Newport, Professor Charles Munroe, chemist from the Naval Academy faculty, headed a laboratory that experimented with several formulas for smokeless powder. Technicians analyzed the European samples collected by the Office of Naval Intelligence through embassy naval attaches, and carefully reviewed the manufacturing literature to learn secrets of safety, drying, storage, and the means of producing a powder that would be stable and which would release little or no smoke or residue as it burned. Despite the structured laboratory environment, it would take over two decades to perfect the powder. The full story of that development will be detailed in later chapters, as its production and perfection became part of the mission of Indian Head.

After receiving Dashiell's revised plans for the rapid-fire breech mechanism in the fall of 1890, Folger quickly put the new design into production at the gun factory, antagonizing Lieutenant Seabury who believed his earlier application and earlier patent took precedence. Folger agreed to pay Dashiell $125 royalty for every breech manufactured; with over 150 rapid-fire guns being planned for acquisition in the next five years, Dashiell stood to gain a tidy sum from his invention. Seabury lodged a suit in the courts against Dashiell for patent infringement. While the suit was pending, the Navy continued to manufacture and install the Dashiell breech on the new ships but held off on royalty payments. A later generation might find the competition between two naval officers to perfect the same device—and the resort to the courts over questions of precedent and royalty—an awkward mechanism to achieve modernization. But the legal and institutional process that fostered that older, competitive process of innovation existed side by side, in the 1890s, with the laboratory approach employed at Newport, under Munroe's direction, in the search for improved propellants. Closer to home at Indian Head, the personal invention style ran simultaneously with Dashiell's efforts to set up and run the Proving Ground as part of the institutional apparatus for innovation.[16]

While Dashiell at first had to handle the Seabury patent suit on his own resources, Folger cooperated to some extent by providing Dashiell with orders to travel to Sandy Hook Proving Ground to observe some Army tests on the Seabury design. Seabury suspected a degree of favoritism on Folger's part and eventually named the bureau chief as involved in collusion to infringe his patent. Seabury's

charge provided a legal opening for even more direct help from the Bureau to
Dashiell. Because he had been named, Folger was then able to obtain assistance
from government attorneys. The case moved rapidly through the Circuit Court
of Maryland, which found for Seabury, then through the Circuit Court of Appeals
and the Supreme Court. On 13 April 1896, Dashiell won the case against Seabury
in the U.S. Supreme Court, and the Navy agreed to pay him accrued royalties
of over $17,000. Even though Dashiell won this particular case, Seabury's charge
of favoritism against Folger was part of a larger pattern. The Senate investigated
charges that Folger had similarly decided upon patents by naval officers in several
other cases involving steel processing and breeches. The investigation led to
stricter controls on the naval officers seeking patents.[17]

The development of Dashiell's new breech smacked of both the old and the
new style of invention. On the one hand, it clearly reflected the nineteenth century
individualistic effort. On the other hand, because of the quiet support and encouragement from Bureau Chief Folger and the use of the Naval Gun Factory to
convert the device from model into a service item, the development reflected in
a rudimentary way an effort to produce innovation through a structure, on order
for the Navy. Later, Dashiell would test and improve the navy yard produced
models at the Proving Ground. Even though the breech was Dashiell's own
invention, it could be regarded as the first major innovation to come out of the
new setting at Indian Head, the first of a sequence that followed over the next
hundred years.

Dashiell's breech became one of several Navy standard rapid-fire mechanisms, and gunnery textbooks for the next two decades instructed academy mid-
shipmen on its design and operation. Dashiell himself had the pleasure of conducting
proof tests on his own gun, reporting that gun #11 fired 248 times and the
mechanism worked over 8,000 times with no jamming or "failure of the action
of any part."[18]

He predicted that the tactical function of rapid-fire 4- and 5-inch guns would
be to clear the "unprotected" deck guns and personnel from enemy ships. Dashiell
recognized that improvements in armor generally were ahead of the penetrating
power of projectiles, so that battleships would depend on lighter rapid-fire guns
for destruction of the lightly armored or unprotected parts of enemy ships, rather
than on the main batteries against armored portions. These considerations proved
prophetic. In 1898 when Admiral Dewey, according to legend, ordered, "Fire
when ready, Gridley," the guns that Captain Gridley fired with deadly effect
from the *Olympia* were Dashiell 5-inch RF guns.[19]

THE PROVING GROUND—STRUCTURED INNOVATION IN ORDNANCE

The Indian Head site that the Bureau had purchased on Cornwallis Neck,
below the small settlement and post office of Glymont, was well chosen in several
regards. Although a quiet backwater, it was conveniently close to the nation's
capital. Gun barrels could be loaded on the Anacostia River in Washington in the

morning and arrive at the Indian Head dock by the evening of the same day. At the foot of the dock, a stream meandered through a narrow, steeply walled grassy valley which, with a little improvement, could be drained of its marshes. This provided an ideal spot for testing both armor and guns, since the hills on each side would absorb shots and potential explosions of untried gun barrels. The distance from one side of the valley to the other was about four hundred feet, a short range for a naval gun, allowing for both precision and accuracy. The shell blasted directly into the steel plate, testing at once shell, gun, gun mount, powder charge, and the armor itself.

Range testing was another matter, but the light river traffic allowed for down-river shelling as long as lookouts could be posted to signal all-clear. Unlike the older Annapolis range, the wide reach down the Potomac at Indian Head could provide over eight thousand yards of open water for the required tests. Prospective work for the new range accumulated even as the construction was still under way. In accord with Tracy's expansion program, the Navy ordered literally hundreds of 10- and 12-inch main batteries planned for the ocean-going monitors *Monterey, Miantanomah,* and *Monadnock,* and for the battleship *Maine,* and 8- and 6-inch guns for the *Lancaster* and other smaller ships. Folger reported to Congress that in 1890, 102 gun forgings, out of a total order of 247, had been worked into completed guns; all would have to be tested as received.[20]

After months of constant bickering with the local construction contractors, Dashiell made progress, and by the winter the dock was in operation. Early in 1891 equipment and gun barrels began to flow to the new proving ground. On 24 January 1891 Dashiell could announce the firing for proof-test of the first gun, a 6-inch breech loading "rifle," as the rifled-bore artillery of the period was called. The old Annapolis Proving Ground closed down later in 1891, and henceforth, in the Navy's expanding glossary of acronyms, "NPG" stood for the Naval Proving Ground that Dashiell established at Indian Head.[21]

For Dashiell, every step was a struggle; every bit of progress was part of a campaign. No satisfactory road existed from Washington to the Indian Head outpost, and the nearest point on the railroad line was about twelve miles away through thick woods and farm country. Dashiell wrote out in neat ordnance engineer's penmanship his early reports, hand-carrying them to Glymont for the U.S. Postal pickup there. He finally obtained a typewriter in the first weeks of 1891. With no telegraph or telephone line, all dispatches took forty-eight hours to reach headquarters twenty-one miles up river, if no return launch to the navy yard was available. Equipment had to be scrounged from surplus Dashiell had spotted at the Naval Gun Factory and at the Annapolis Proving Ground. Hesitating to insult his old superior Dayton at Annapolis, he requested that Folger order the transfer of various goods, including a horse, carriage, harness, and feed, as well as gun mounts and heavy equipment. Dayton voluntarily shipped off measuring instruments, powder sieves for measuring powder grains, and other ordnance equipment, assuming they would be of "some use" to his protégé at Indian Head.[22]

Even by the harsh standards of the 1890s Dashiell stood out as a demanding and domineering supervisor. A creative and abrasive individualist, he lacked some of the skills of a good organization man. A civilian employee who sought pay for holiday time off was fired, and Dashiell refused the appeal. Frequent minor injuries, including smashed fingers, powder burns, cuts, and bruises, soon consumed Dashiell's small store of first-aid equipment, but his response was to order more carbolic acid and an ever-faster tempo. He personally supervised the proof tests and range shots, often from horseback. When naval visitors came unexpectedly, they would get "to see the style we sling here," as Dashiell, in cowboy outfit and with several days' growth of beard, rather proudly noted.[23]

Despite the sometimes flamboyant behavior of the fair-haired inventor, Dashiell's drive and energy produced a steady flow of accomplishments at the Proving Ground. He supervised the construction of the valley firing positions, semi-underground shelters called gun-proofs, and target butts, magazines, and instrument houses. Dashiell personally designed and supervised the construction of dwellings and was able to spend the winter of 1891–92 in a small cottage he had built on the hill above the valley. He ordered a telegraph line to Washington, obtained equipment such as chronographs and range finders, cleared the line of fire for the over-river range shots, drained marshes in the valley, and planted crops to feed the oxen, horses, and mules.[24]

Dashiell's driving determination angered neighbors, employees, and contractors, but his work drew favorable notice throughout the Ordnance Bureau and more widely through the Navy. Dashiell not only got the Proving Ground open and operating in short order, but he also went on to set up regular procedures for the routine testing of powder, shell, plate, and guns. In setting up regularized routines, he made further contributions to the emerging systematized structure for innovation. The handwritten and then typed form letters that he composed to report on those tests became the standard, imitated by his successors for two decades, when officers in charge began to use preprinted forms for their reports. Furthermore, the long-range cumulative effect of his routine work, although not explicitly evaluated at the time, was crucial to ordnance progress in the 1890s.[25]

STRUCTURED INNOVATION UNDER WAY: TESTING OF GUNS, ARMOR, SHELLS

The isolated, heavily wooded Indian Head Proving Ground quickly began to fulfill the Bureau's need for a center of naval ordnance development. Dashiell's work in management and in technical supervision called upon him to play new roles that would continue to be crucial as the institution developed: commanding officer with supervisory powers and technical director with responsibility for leading the process of innovation.

Dashiell began with a small staff of an assistant officer, a pay clerk, and an enlisted gunner, supplemented with a civilian labor force drawn from surrounding farms that varied in size with the season. Local farmers and rivermen served as a seasonally available labor force—often as unreliable, in Dashiell's opinion, as the notorious "river pirates" engaged by the Bureau of Yards and Docks to arrange construction, transport, and pilings. While the records surviving from the period are not precise, the total complement of staff, including the "commuters" from the region, ranged up to about thirty men.[26]

The tests Dashiell supervised at Indian Head included types of experiments that in a later era would be conducted at several very different types of facilities. One of the functions was "acceptance testing," in which gun, mount, shell, or armor would be tested to determine whether it met specifications. When armor exceeded specifications, performance figures from Indian Head could lead to a cash premium for the manufacturer. Tests on experimental guns were more in the nature of prototype evaluation, leading to suggested modifications derived from repeated firing of the weapon. Dashiell tested powder lots from Newport not only for acceptance or proof but also to determine explosive force, deterioration in storage, and other characteristics. By modern definitions, the early Proving Ground performed test and evaluation, research and development, prototype design modification, and specification testing. Sometimes several of those various types of experiments would be conducted at the same time. The later vocabulary that would make distinctions among the various types of research and testing had not been created, and all such experiments in the 1890s would be subsumed under the concept of "proving" ordnance materials. A close review of Dashiell's varied routine work gives hints of the emerging differentiation between types of tests and their functions in the innovation process.

As a specialist in gun mechanical design, Dashiell produced many minor modifications and improvements, noted at the time by his superiors at the Naval Gun Factory and at the Bureau of Ordnance and quietly incorporated by the system. Such improvements were not personally patented but simply adopted as Bureau standards. On a nearly daily basis, Dashiell sketched improvements in gas-ejection devices, breech and extractor designs, and loading and gunner seating arrangements. Folger and his successor, Captain W. T. Sampson, referred Dashiell's ideas to Professor Philip R. Alger, the ordnance and mathematics instructor at the academy. Upon Alger's approval, bureau chiefs personally initialled or "endorsed" Dashiell's recommendations with a pencilled "OK," making them, in effect, what the Navy would later call change orders or technical modifications on plans. The bureau chief would then instruct the officer in charge at the gun factory to put Dashiell's alterations into production.

Dashiell, in his three years at Indian Head, made scores of such recommendations; the vast majority were implemented, improving the smoothness and efficiency of naval ordnance. Without setting up the complex system of planning that would characterize such innovation in later decades, the Navy moved slowly toward a system in which the design changes suggested by a testing officer would be authorized and implemented by the organization through a team effort involving experienced officers, an academic specialist, and draftsmen, machinists, and metal workers. In this fashion, hints of the corporate mode of invention began to emerge.[27]

Dashiell recommended modifications in 10-inch gun recoil mechanisms for the *Miantanomah's* gun mount, which used a water-loaded hydraulic system, identifying leaks and tracing them to the bearing surfaces of valves. He made minor changes in the piston system, which Folger approved as standard. The mounts for 10-inch guns on the *Monterey* also required alteration, and he identified water leaks in several of the valves, which he ordered overhauled on the site. With reference to his own rapid-fire breech, Dashiell made a number of recommendations, including the standing positions of the crew, length of trigger lanyard, oiling schedule, lengthening of the recoil, redesign of the shoulder piece used by the gunner, and minor changes to the recoil springs and slides. He designed a tool for removing piston springs, and Commander O'Neil at the Washington Navy Yard ordered it manufactured and supplied to every ship carrying the new Dashiell-breech fitted guns. Dashiell experimented with gun sights and new breech plugs and extractors for 6-inch guns. He suggested redesigned shell casings and a half-dozen modifications for the 10-inch guns to be installed on the *Maine*.[28]

The number and variety of the minor changes suggested by Dashiell give some sense of the beginnings of a new pattern. None of these changes was revolutionary. None went through the patent process; none, except his patented breech, was named after him. Rather, each was a minor, incremental change, approved through the organization and implemented by a process which, although conducted in pen, ink, and pencil, had some streamlined aspects which twentieth century ordnance engineers and developers might envy. Decisions as to implementation were not processed through boards, committees, or design desks. The suggestions were made on the spot by the man running the tests, were drawn up and outlined by him, were reviewed by one or two experts who signed off on them, and then were ordered into effect by the bureau chief who controlled the necessary funds. The fact that Folger was Dashiell's immediate superior and that Folger also was in line authority over the factory that would build the devices made for a simple and effective system of taking an idea to completion in short order. The dates on the documents reveal that the time from suggestion to implementation at the shop level was usually a week or less, an astounding performance by the standards of a century later. Budget was not a problem, since the creative work and the decision process were funded out of regular payroll for "proof of ordnance."

Larger decisions faced a more formalized process. Although the Navy ordered the patented Dashiell breech for 4- and 5-inch guns, it kept searching for a design for the 6-inch breech loading rifle. In February 1892 Dashiell reported on a test of his own breech adapted to fit the 6-inch gun. After noting that in "every instance, the mechanism has operated with entire satisfaction," he suggested, with uncharacteristic modesty, "The mechanism is thoroughly satisfactory to me, but for a full report on its merits I would request that a Board report on a competitive test between it and that [already] in service." Without saying so, he appeared to recognize that his own judgment would hardly be independent in the matter of his own patent and sought organizational endorsement.[29]

A board headed by Captain Howell reported to Secretary of the Navy Tracy on the competitive tests held under their supervision at Indian Head. The standard service breech fired ten shots in five minutes, but the Dashiell design jammed at the sixth round. Instead of making a decision, the board adjourned for about three weeks while the Dashiell breech was refitted at the navy yard. On the next try, the Dashiell breech got off ten rounds in slightly less than three minutes, and the board advised its adoption for new 6-inch guns. The Navy nevertheless adopted very few Dashiells for the 6-inch guns, continuing with the older standard and with several alternate semi-automatic designs developed at the same time.[30]

Dashiell tested an alternate rapid-fire gun, produced by Driggs Ordnance company, in 1892. A number of preliminary problems showed up that he reported in detail. In particular, the breech could only be closed with great difficulty, and the ejected case popped out with such force that it could crush a man's hand if he carelessly held onto the training wheel. The owners of the Driggs company were present for the tests and apparently appreciated Dashiell's description of specific mechanical problems. They implemented his recommendations on the private level in much the same informal fashion as the Bureau put into shape his suggestions at the Naval Gun Factory.[31]

Secretary of the Navy Tracy worked to secure cooperation from Carnegie and Bethlehem in producing nickel-steel plates, developing domestic sources for the new alloys. As American companies produced the plates, they shipped them to Dashiell for testing in 1891 and 1892. Some of the methods, routines, and arrangements that Dashiell established for the testing of the new armor plate also became the standard at the Proving Ground and would be followed as organizational procedures until 1920. The sections of plate, measuring eight by ten feet and weighing up to twenty-five tons, would be off-loaded at the dock, moved by crane and a short rail track, and mounted on butts directly across from the gun batteries. At pointblank range, the plates would be "attacked," as Dashiell put it, by shells ranging from four to ten inches in diameter. Dashiell recorded the weight of the powder charge, the velocity of the shell in feet per second, and the degree of penetration of shell into or through the armor. The heavy butts, constructed of twelve-by-twelve-inch timbers, would be demolished after several rounds had penetrated the armor, requiring suspension of firing while a crew rebuilt the target structure for new plate.[32]

Experimental steel alloyed with a low percentage of nickel from Carnegie and Bethlehem works resisted armor piercing and explosive charges well. In September of 1891 Dashiell was impressed with the "nickel-steel" from both firms: "The results of the last rounds with armor-piercing shell show beyond argument that in cruiser combat the vitals of the ship with a nickel steel deck are fully protected, while those of the ship with a steel deck are at the mercy of an enemy using armor piercing shell."[33]

Dashiell's armor tests at pointblank range were impressive to watch. Folger arranged that an Armor Board test be attended by several dignitaries, and on November 14, 1891, barely a year after opening the Proving Ground, Dashiell hosted a group that included Folger, two senators, Secretary Tracy and the Assistant Secretary of the Navy, Professor Alger from the academy, and two Army generals. Dashiell ran a series of tests of armor-piercing shell against the new ten-inch thick nickel-steel plate, with Secretary of the Navy Tracy giving the firing commands.[34]

In 1892 Dashiell tested a "Harveyed" plate. Augustus Harvey had developed a process of hardening steel with surface carbonization, and the Navy wanted to know whether the process would produce a superior armor. Dashiell, with his usual enthusiasm, called the test the "most important armor trial that has ever taken place." Ten-inch shells disintegrated against the plate. Harveyed plate was lighter than corresponding nickel-steel plate, weighing 20 percent less for the same thickness, and its hard surface broke up the shell rather than absorbing the impact.[35]

These two tests represented a victory for Secretary of the Navy Tracy, who had conducted a two-year campaign for congressional support and to obtain bids from Bethlehem and Carnegie for the new products, including the nickel-steel armor and the Harveyed steel armor. Tests on imported samples in 1890 at the old proving ground at Annapolis and the tests on the American products at Indian Head in 1891 and 1892 marked stages in the long effort to build the American armor industry for the Navy. Benjamin Cooling, in his close study of this struggle, *Gray Steel and Blue Water Navy*, shows how the beginnings of the "military-industrial complex" can be discerned in the Navy arrangements with Bethlehem and Carnegie of this period. Cooling reveals that the steel companies were not particularly eager to engage in the expensive retooling necessary to produce the new alloys, and Tracy offered prices which, in effect, subsidized the required expansion and construction.[36]

When proving shells, rather than armor, the targets were made of less expensive "mild steel" which shattered dangerously under the impact, throwing large fragments several hundred yards and endangering personnel and dwellings on the grounds. Dashiell "earnestly recommended" that nickel-steel plates be adopted for projectile tests as a matter of safety.[37]

Dashiell and the officers in charge who followed him at Indian Head faced a basic technical problem in their empirical tests of armor, powder, shells, and guns, reflected in the problems of bursting target armor. Armor could be tested only by shells; the penetrating power of shells could be tested only by armor. Similarly, the best test of powder as a propellant was to use it in a gun; the best test of a gun came with repeated firings. Setting standards in this context became a continuous process of comparing results from test to test. Dashiell's meticulous reports on early tests, recording all details of powder charge, velocity, gun performance, prior firings of the gun, and armor grade, thickness, and damage, helped set standard base-points as well as establish acceptability in procurement of powder, shell, gun, and steel.

The difficulty of empirically testing shells and guns against each other showed up repeatedly. Once, Dashiell rejected one lot of shell cases with primers provided by Driggs Ordnance company in 1892, because the primers repeatedly misfired.

William Driggs thought that the gun, not the shell, should receive the blame. "I judge," he immediately noted, "there must have been something wrong with the gun. The fact that the failures were all at the beginning of the test leads me to think that the firing pin must have clogged in some way. If lard oil had been used in the mechanism and the weather was cold the results would be just as you found them. . . . There must have been something peculiar about the gun."

Dashiell disagreed: "These theories would seem very plausible did the facts bear out the assumed hypotheses. (1) Nothing was wrong with the gun. (2) The failures were *not* all at the beginning of the test. (3) Lard oil had not been used in the mechanism and the weather was not cold." In short, he blamed Driggs' primer cases.

The next day, however, Dashiell tried more of the cases in another gun and concluded that the primers would fire if the tension on the spring-loaded firing pin was increased. He noted, "I would therefore respectfully recommend the acceptance of these cases should the Bureau intend making the springs heavier, otherwise their rejection." Such problems, in which the error of a shell might be blamed on the gun, or the error of a gun on the charge or the shell, would recur over and over through ordnance testing. Dashiell understood the issue and constantly argued for safe, efficient, and uneroded guns to make the tests standard.[38]

By 1892 the Bureau of Ordnance had decided to pursue a dual research pattern in shell development. On the one hand, large shells loaded with bursting charges would be developed to help in the clearing of the unarmored portion of enemy ships; on the other, the development of high-velocity hardened shells would be pursued to perfect armor-piercing shells. Both types of shell would require testing, all at Indian Head.[39]

Dashiell recommended that armor-piercing shells be constructed of Harveyed steel—"The success of the Harveyed shell over the untreated in getting its bursting charge through the plate was fully shown." He also recommended that tubular Harveyed steel be used for the construction of shells with a bursting charge. Such shells would be ideal for the smaller caliber rapid-fire guns useful in clearing the lightly armored or "unprotected" deck areas of enemy ships.[40]

In 1893 Dashiell moved on to an assignment at the Philadelphia Naval Yard, while he worked on securing the accrued royalties from the patented breech. Lt. Newton E. Mason, Dashiell's successor from 1893 to 1896 at Indian Head, continued the work of Proving Ground improvement and kept up testing of ordnance and armor. While he maintained the routine testing schedule, Mason gave more attention to the human management side of the new facility. Mason offered a few specific suggestions for gun design improvement. His modifications, when offered, were implemented much as Dashiell's had been. But Mason gave far more energy to improving the working conditions, and the recommendations he made in that area set the Proving Ground on a path of staff enlargement and improvement of life at the outpost.

Mason's concentration on management rather than personal innovation reflected a subtle shift in direction. He secured three ensigns to assist him, supplemented by an enlisted gunner and a laboratorian or chemist, and kept on Dashiell's pay clerk, together with the small civilian work force. Mason, like Dashiell, found the isolation difficult and immediately reported that "great trouble has been found in keeping good workmen and leadingmen down here on account of there being no place for them to live, even without their families, and several very desirable men have declined to come down here on that account alone." He recommended the construction of cheap brick dwellings, improvement of the water supply, and the installation of an electric plant. By the end of his tenure, all of his improvements were in place or under construction.[41]

Mason noted the practice of asking the laboratorian to serve as a sort of pharmacist-medic, and he praised this employee's emergency work in treating the frequent small wounds incurred in the course of gun testing. But the nearest doctor lived six miles away, and the delay in getting medical help in cases of serious accident struck him as extremely risky. On Mason's insistence, a series of assistant surgeons and an apothecary reported for duty as medical officers.[42]

At the same time that Mason worked to improve the pioneer conditions which Dashiell had accepted, he began work testing armor-piercing shells against curved nickel-steel plates. An eight-inch curved nickel-steel plate for the *Monterey* resisted angular fire with a wide range of striking velocities and calibers "very well indeed," he noted with somewhat less precision than characterized Dashiell's reports. He recommended curved nickel-steel armor for gun turret emplacements.[43]

Quietly, with no inventions to bear his name, Mason had moved the Proving Ground in the required direction. Although later officers in charge would continue to complain about the isolation and harsh living conditions, the fact that he had concentrated on improving those conditions established another set of precedents which later officers would emulate. One had to assemble a team and make them relatively comfortable if the work was to proceed, and Mason took practical steps toward that end.

1889–1896: ORDNANCE ISSUES AND THE PLACE OF INDIAN HEAD

The late 1880s and early 1890s generated both revolutionary and more gradual, incremental changes in ordnance and armor. Changes in one technology required changes in another. In a technical world in which the measuring yardsticks themselves changed, only a careful system of setting multiple standards could determine the best path of progress. The Proving Ground moved to the forefront as the place for such tests.

Nickel-steel armor and Harveyed surface hardening were among the most revolutionary changes in naval technology in the period. When Tracy set up a system of bonus payments for exceeding specifications to encourage the firms to produce the new alloys, the Navy had to have a controlled, objective, and organized system of standardized tests. Those tests could be conducted only by comparing the penetrating power of similar shells fired from similar guns against the various grades of armor plate. An isolated, yet nearby locale was essential, run by ordnance officers and possessing the intellectual and technological resources for careful, measured evaluation of the product. Indian Head met those requirements.

Improvements in propellants loomed, stimulating inventors like Dashiell and Seabury to develop a rapid-fire breech that could take advantage of them. The anticipated development drove other changes as well. The experimental formulas for smokeless powder all burned more slowly than the older black powder, requiring a longer gun barrel so that all the powder would release its energy before the shell left the muzzle. By lengthening barrels and converting to smokeless powder, more penetrating effectiveness could be achieved with each diameter of gun. By the turn of the century, a 12-inch gun with a smokeless powder charge could fire an 850-pound, uncapped projectile that would penetrate 17.92 inches of Harveyed nickel-steel at three thousand yards; the 1890 model 12-inch gun, with a shorter barrel, could penetrate only 13.79 inches. Similar comparisons of all the standard diameter guns showed improvements of 20 to 30 percent in depth of penetration over the decade. Using projectiles capped with hardened steel, another 15 percent penetration could be achieved. Such data, collected over the decade, reflected the complex interaction of propellant, gun design, and armor metallurgy. The Proving Ground had run the tests and collected the information. The Navy was on the way to realizing a system that simultaneously stimulated progress through a system of rewards, measured the progress, and provided the facts required for purchasing the best product.[44]

Each aspect of ordnance progress would involve a similar weighing of alternatives which could be properly conducted only with a sustained collection of data. The solitary inventor, hunched over a sheet of paper by candlelight, continued to work through the period, but his was not the way of the future. Rather, the conflicting claims of advocates had to be sorted out by an objective, continuing, and standardized system. Dashiell, the lone inventor, had also established the institution and the methods for that new way of doing things.

Even the apparently simple issue of whether or not to load shells with an explosive charge had been debated for years, and it would remain an open question well into the twentieth century. Although explosive-loaded shells had proved themselves effective since their first use against wooden ships in the 1820s and 1830s, the development of more effective steel armor had complicated the issue, since explosive shells would destroy themselves on the outside of the ship. Some gunnery experts argued for solid shot against armored portions of enemy ships. But clearly, explosive loading remained superior for the lightly armored or unarmored sections of a ship. In exploring the pros and cons of explosive charges for shells in 1900, Professor Alger at the Naval Academy utilized facts meticulously gathered at Indian Head Proving Ground by Dashiell and Mason.

In an article for the *Naval Institute Proceedings*, Alger spelled out the problem. On the one hand, an explosive shell, if it penetrated armor, would have more destructive effect inside a ship. That advantage was offset by the danger in handling such shells and by the fact that they might explode on contact with enemy armor, wasting the explosive force on the outside of the ship rather than inside. Delayed action fuzes to postpone detonation until after penetration would be ideal, but they remained unreliable; sensitive fuzes added to the danger of premature explosions in handling and loading and in the gun barrel. In 1900 Alger sided in this perennial debate with the solid shot advocates, precisely because of the improvements in metallurgy "proved" at Indian Head. He argued for an increase in the penetration power of shells by hardening and by increasing the velocity of the shell, so that solid shot would penetrate enemy armor, rather than to pursue the more dangerous course of high explosive-loaded shell.[45]

Combining the new technologies of longer guns, smokeless powder, and hardened metals for both armor and shells would require that all the elements be tested with and against each other. Although Dashiell did not have the ideal personality for cooperative work, he succeeded in establishing the necessary framework for testing and began to accumulate the required data. Mason, with his more humanistic approach to management, improved the working conditions while keeping up the requisite testing schedule.

Below the level of the revolutionary technological changes such as new guns and new metals were the hundreds of small but essential incremental advances in weaponry which were necessary to put new guns into working order. At this less dramatic and more workaday level, another aspect of the new organizational style evolved unnoticed. Routines had to be established. Within the routine testing, day-to-day "fixes" had to be worked out. Dashiell had the ability to focus on small questions of mechanical operation such as the tension of a spring, the characteristics of a bearing surface, the machining of a part, and the ease of operation by a gunner. As the Bureau implemented his suggestions, it used a system of harnessing creative ability, not through the process of independent invention on personal time but through the regular, routine work of the Proving Ground and Bureau operation.

The development of procedures, the design of reporting formats, and the setting up of channels through which recommendations would rapidly flow for approval were all part of this hidden but essential aspect of institution building. Dashiell's minor changes and modifications, as they were implemented through the Bureau, foreshadowed the more nameless and organizational approach that would characterize such work in coming decades.

On the larger issues of armor, powder, shell, and gun design, the Proving Ground began to produce the facts on which to base ordnance purchases as the fleet modernized. The gun that bore Dashiell's name served as a notable memorial to the old style of individual innovation. In a less obvious fashion, however, the facility that he founded and that Mason improved at Indian Head and the careful scientific and administrative procedures that they established there, both for incremental innovation and for data collection, laid the foundations of a more lasting monument. The organization and the methods would grow and alter under their successors into an evolving, more structured approach to research, development, testing, and evaluation of ordnance.

CHAPTER 2

Early Powder Manufacturing and Ordnance Testing

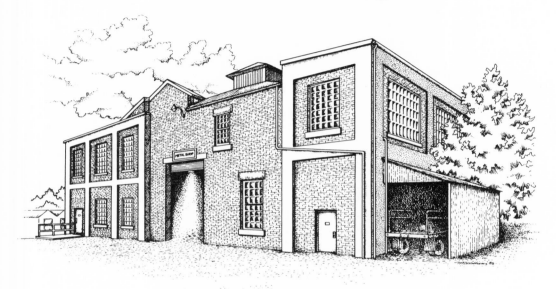

Boiler House
Building 113
1899

O N THE EVENING OF 30 April 1898 Admiral George Dewey ordered the Asiatic Squadron, consisting of cruisers *Olympia*, *Boston*, *Baltimore*, and *Raleigh*, together with two gunboats, to steam quietly into Manila Bay. At 5:40 the next morning, Dewey's broadsides raked the ten ships of the Spanish fleet, anchored off Cavite Point. The battle was over in seven hours. All ten Spanish ships were either destroyed, silenced, or captured. All of Dewey's guns, including twenty 5-inch rapid-fire "Dashiells," as those with the patented breech were called, fired with cocoa powder.[1]

The victories of the U.S. Navy in the Spanish American War at Manila Bay in the Philippines and Santiago Bay in Cuba not only put to the test the weaponry and ships developed over the past decade but also launched the United States as a world power. The guns and armor developed and tested at Indian Head since 1891 by the Bureau of Ordnance would continue to shape American foreign and military policy.

According to some reports, the Dashiell 5-inch rapid-fire guns performed well at Manila Bay, compared with the older models. Lieutenant C. G. Calkins, navigation officer of the *Olympia*, noted "her five-inch battery was much superior to her turret guns in accuracy and endurance, as it was in rapidity of fire." Smoke remained a problem, he noted, both in obscuring the telescopic lenses and in limiting the view from the main 8-inch batteries. "But the five inch rapid fire battery redeemed the situation by firing as many shells per gun, during the hottest action, as the two turrets were able to expend in a much longer period." The *Baltimore* action report detailed many problems with the older 5-inch guns, which had not been upgraded with the Dashiell breeches, including broken firing pins, stuck mechanisms, and problems with shell-case ejection. The guns fitted with Dashiell breeches aboard *Olympia*, by contrast, did not suffer those mechanical problems.[2]

The Spanish-American War brought not only victories and some demonstration of new equipment but also many embarrassments, revealing deficiencies in ships, armament, supply, and organization. The fact that the Spanish navy possessed more smokeless powder than the Americans was just such a revelation. The failure to have an adequate supply of smokeless powder for the new rapid-fire guns may have contributed to the mixed success of American gunnery. As a result of the short war, the Bureau of Ordnance undertook to supplement the production of smokeless powder by du Pont with a full-scale production plant of its own. The new plant, it was hoped, would not only ensure more production but would serve as a yardstick by which to measure the price and quality of the private production.

Three days after the Battle of Manila Bay, Congress passed a bill authorizing the Navy to proceed with the construction of a powder factory at Indian Head, to be devoted to the production of smokeless powder. Commander Albert R. Couden, who succeeded Lieutenant Newton E. Mason in command at Indian Head, continued armor, gun, and shell tests at the Proving Ground and pushed the construction schedule for the new factory. Two years later, on June 16, 1900, Couden's successor, Lieutenant Joseph Strauss, could announce that the factory went into production with its first lot of "I.H.S.P."—Indian Head smokeless powder, S.P. lot #148. The earlier 147 lots had been produced by the Navy at Newport, by du Pont, and by smaller companies.[3]

Strauss launched a new era with the new century. Gun propellant manufacturing soon came to be the dominant mission of the Indian Head facility. As Indian Head's missions evolved to include a wide variety of developmental work in the production of propellants and explosives, those new missions could trace their ancestry back to the Powder Factory and its chemical laboratory and testing programs.

POWDER UNDER DEVELOPMENT

Before the plant opened at Indian Head, the Bureau of Ordnance had spent a full decade in what would now be called research and development, experimenting at Newport to produce a safe and stable powder and to convert a laboratory process into an industrial-scale operation, first for use by du Pont and other private firms, and then into plans for the Navy-owned and Navy-operated factory at Indian Head.

Smokeless powder had been under development for decades. The first smokeless powder had been introduced in Austria in 1852 and used successfully for about ten years. The early smokeless powders used some form of cellulose, such as wood or cotton, which had been processed with nitric acid, or "nitrated." The product was a new compound, nitrocellulose, an early form of plastic which, when burned, performed much like black powder. An early nitrated wood pulp smokeless powder, invented in Prussia, had been widely used for sporting purposes. Alfred Nobel had developed a smokeless powder using nitroglycerin in 1888; Hiram Maxim had patented and produced a smokeless powder using nitrated cotton linters in the same year, which the British called cordite, with beeswax or vaseline as a stabilizer. Both types of smokeless powder, either based on nitrocellulose alone, as a single-base powder, or combined with nitroglycerin, in a double-base powder, differed from the older powders. The new powders included chemical compounds of organic origin and were not simply mechanical mixtures of inorganic substances like the old sulfur-carbon-saltpeter mixes of black powder.

The Nobel formula was regarded as too dangerous for Navy use since it was based on about 40 percent nitroglycerin and 60 percent nitrocellulose; the Maxim formula, which contained about 15 percent nitroglycerin and a soluble form of nitrocellulose, needed refinement to assure long-term stability. The released energy of the new propellants was four or five times that of the old powders, and because the gas was released more gradually than with black or brown powder, a more uniform pressure could be maintained in the bore of the gun and a higher velocity of projectile could be achieved.[4]

Throughout the decade of the 1890s the ordnance community knew that smokeless powder was under development, and each step of progress in that field was eagerly followed by not only gunnery officers but also the Navy as a whole, as frequent "professional notes" and articles in the *Naval Institute Proceedings*, *Scientific American*, and British journals reported powder news. Each year through the 1890s the Secretary of the Navy reported to Congress on the steps of progress at Newport.[5]

At Newport, Charles Munroe, the civilian chemist in charge of the laboratory, worked closely with the Wilmington, Delaware, and Carneys Point, New Jersey, facilities of the E. I. du Pont company in development of smokeless powder, calling on the Indian Head facility for different types of developmental tests. Although Munroe tested many formulas, a problem with most of the experimental lots was deterioration in storage. Since powder for any sustained war would exceed current production, stockpiling in magazines would be required. But the only way to determine the deterioration rate would be to store powder and then to test it over an extended period. No known law would allow for precalculating the rate of deterioration. However, the simple empirical process of storing powder and taking samples out of storage for periodic tests was time consuming and dangerous, as various lots mysteriously sweated, changed color, and gave off pungent fumes. A "surveillance magazine" at Indian Head received small lots of torpedo factory "S.P." or smokeless powder, as early as 1892, for just such long-term evaluation. At Indian Head, Dashiell reported on tests of fresh and stored smokeless powder, along with "acceptance tests" on current deliveries of the older black and cocoa powders. Dashiell, Mason, and Couden directly supervised research and development experiments, more routine "proof tests," and tests that established specifications such as the appropriate charge for various size guns. Other tests evaluated which shape of powder grain performed better and compared perforated with unperforated grains in velocity and power.[6]

Smokeless powders that contained nitroglycerin, following the Nobel or Maxim formulas, had other defects besides long-term instability. When a gun was overheated and a powder charge was loaded in the breech, the charge would absorb heat from the barrel, sometimes causing premature ignition that could destroy the weapon and slaughter the gun crew. Serious accidents with overheated machine guns had convinced several of the European powers to avoid smokeless powder with any nitroglycerin in it and to search for a powder that was both smokeless and as safe as the old black powder. The history of accidents in European plants manufacturing smokeless powder also added to American concerns that any powder containing nitroglycerin should be avoided.[7]

Hiram Maxim, although settled in Britain, was an American and had hoped that one of his formulas, all of which contained some amount of nitroglycerin, would be purchased by the U.S. Navy. In Britain he won a lawsuit which determined that his powder, although containing nitroglycerin, was a separate formula from Nobel's and others, and that he had exclusive rights to a double base smokeless powder rich in nitrocellulose. He worked with du Pont Company, hoping to get them to adopt his patented powder and to sell it to the Navy, but they declined, following the Navy's suggestion to develop a powder without any nitroglycerin at all in it. Shortly after Maxim heard that du Pont had declined to cooperate, he met with Bureau Chief Folger at the Murray Hill Hotel in New York. As Maxim himself recounted the story in his autobiography, Folger approached him, rubbing his hands and chuckling.

"So," he said to Maxim, "The du Ponts would not make your powder, would they?"

"No," replied Maxim, going on to suggest that du Pont might be cooperating with another firm.

"No," said Folger, "I stopped it; it *was* something else—I stopped it. I told them that I had a much better powder." The powder that Folger "had" was the formula under development by Munroe at Newport.

Maxim was disgusted and complained to the Secretary of the Navy, as Seabury had about the Dashiell gun, of Folger's high-handed decision in favor of products whose development he had arranged in-house rather than those by outside inventors. In powder, as in armor and guns, however, Folger and Tracy agreed that the search should proceed until an American product, rather than a European one, could be perfected. As he had with Dashiell, Folger sought American inventions, and ones that the Bureau could control. Soon du Pont manufactured experimental batches, following the Newport formula, and sent their lots for testing to Indian Head along with the Newport lots.[8]

The torpedo factory at Newport went into powder production with some of the earlier formulas, trying for a powder based entirely on nitrocellulose. The torpedo factory reached a capacity of 64,000 pounds a year by 1892. Du Pont also geared up for production, using information supplied by Munroe. The Newport facility tried a variety of forms for the grain of the powder, increasing its power by shaping it in "macaroni" shaped granules with perforations to increase the rate at which gas was generated, although the British believed multiple perforations made the powder too dangerous for handling. Experiments with the perforated powder gave encouragement to those who sought to develop a bursting charge for explosive shell, as well as to achieve a smokeless propellant powder.[9]

The Bureau's early powder research was severely set back by a fire which destroyed the Newport pilot powder factory on July 3, 1893. By 1896 a new factory consisting of six buildings reopened on the torpedo factory grounds for the pilot manufacture of smokeless powder. The Newport facility was crowded, however, and fear of explosion and fire which might spread to other buildings led the Navy to look for a safer location with more acreage for its future work.[10]

Munroe departed his assignment at Newport in 1893, leaving the work in the hands of assistants, including Navy Lieutenant John B. Bernadou, and a young chemist, George W. Patterson, who had graduated with a degree in chemistry from Worcester Polytechnic Institute in 1888. After two years serving as instructor at his alma mater, Patterson moved to Newport, where he worked through the decade as a civilian employee under both the internationally famous Munroe and Bernadou as they explored the powder problem. Bernadou continued Munroe's experiments with the manufacturing process, as well as the various formulas and types of granulation. Patterson would stay with smokeless powder for fifty years, later earning recognition as one of the nation's foremost experts on the subject. From 1900 to 1940, when he retired, he worked at Indian Head, and he was the dominant civilian figure in the history of the facility for four decades.[11]

By 1896 the Newport team had settled on a small-scale manufacturing process involving these steps:

◇ Picking and drying cotton linters, or fibers
◇ Nitrating the cotton linters by dipping in an acid mix
◇ Boiling the resulting nitrocellulose to wash out the excess acid
◇ Pulping to wash out impurities
◇ Poaching or stirring in alternate boiling and cold water baths with sodium carbonate to purify further the nitrocellulose
◇ Testing for stability, using a form of litmus paper
◇ Wringing to get most of the water out
◇ Dehydrating, to remove remaining water and replace it with solvent alcohol
◇ Adding ether
◇ Pressing the resultant nitrocellulose pulp into blocks and forcing them through dies to extrude ribbons and cords for cutting into powder.[12]

Although the pilot process had to be modified and perfected, the steps described in 1896 remained the manufacturing process used in producing smokeless powder for more than ninety years.

Early in 1898 the Bureau ordered complete blueprints and plans of the powder factory at Newport and used them to plan a new facility to be constructed at Indian Head following the same process steps worked out at Newport. At Indian Head Couden carefully went over the plans, suggesting relocations of some of the buildings and potential arrangements for loading coal and other supplies to a separate wharf, in order to avoid interference with the gun-barrel loading and unloading at the dock which Dashiell had built. He specified that no two buildings be built near each other and laid out the plans so small hills would mask one from another in case of fire. He planned an extension of the electric railroad from the valley wharf and arranged water cisterns on the tops of small hills to provide for water pressure. Couden was unclear as to the sequence of powder manufacturing steps, and his original plans, he admitted, could not reflect the needs of a powder factory unless he fully understood the process. In 1900 Patterson transferred from Newport to Indian Head to provide the know-how.[13]

Meanwhile, the search for the ideal smokeless powder or "S.P." formula continued. In 1896 a formula published by Dmitri Ivanovitch Mendeleev, the Russian chemist already well known for developing the Periodic Table of Elements, was translated from the French by Lieutenant Bernadou and later published in the *Naval Institute Proceedings*. The Russian navy had ordered powder using Mendeleev's formula as early as 1892 and had great success in test firings through 1893. Experiments with the Russian formula at Newport continued through the Spanish American war, but sample lots tested both for power and for stability at Indian Head did not precisely conform to Mendeleev's predictions.[14]

The question that Mendeleev addressed and which several researchers approached simultaneously late in the 1890s was to determine which percentage of nitration of cotton would give optimum stability and propellant force or velocity. Mendeleev demonstrated that nitrocellulose with a 12.44 percentage of nitrogen would give a greater volume of gas than any other and was the chemical ideal. But empirical experiments at Indian Head demonstrated that the mean percentage of nitrogen in the nitrocellulose could be increased to as much as 12.80 percent without adversely affecting the final tested velocity of the powder. Strauss reported that work by Patterson showed that minute quantities of ether or alcohol solvent always remained in the powder. The solvents slightly slowed the burning rate and accounted for the good performance of the powder that had been produced with a slightly higher nitration rate than the chemical ideal. By 1899 the Navy had settled on a specification which allowed a nitrogen content in the range between 12.45 and 12.80 percent. This specification change reflected the first of many major individual contributions to the technology of powder production by George Patterson.[15]

POWDER FACTORY: INDUSTRIAL MANAGEMENT

In 1900 powder manufacturing by the Navy in its own facility was a major break with the past. Although the Navy had operated "factories" before 1900, such as the gun factory in Washington and the torpedo factory in Newport, the

Indian Head plant represented a somewhat different kind of operation. This was to be the first major chemical factory operated by the Navy. Like the gun factory, Indian Head's powder factory was to serve as a full-scale plant which could produce part of the Navy's need and, at the same time, serve as a yardstick by which to measure the price and quality of the private production sector. Private plants would continue to produce most of what the Navy purchased. The early powder production at the Newport Torpedo Factory was an attempt to establish a pilot facility, based on a laboratory, where the processes for manufacture could be worked out and tested on a small scale. But Newport could not be expanded to the scale of the du Pont plants and therefore could not serve as a proper yardstick. For these reasons, the Indian Head facility was the first setting in which the Navy had to take on all the considerations involved in large-scale, modern chemical manufacturing processes.

Those management and engineering considerations spanned a wide range. Naval officers dealt with recruitment of personnel, compensation, supervision, and safety; procurement of equipment; and construction of plant and utilities. They moved into new areas such as organizing and ensuring adequate supplies of chemical raw materials and setting up quality assurance systems for both purchased and internally produced ingredients as well as for final products. They had to learn how to set up cost accounting systems to insure efficiency of production and for the evaluation of the final cost of the product. And in a completely new effort for the Navy, they had to design plant facilities so as to ensure the flow of supplies to mixing, blending, and other processing points in compatible quantities and at compatible rates. The growing private chemical industry of the era faced similar problems. In the first decade of the twentieth century, several fields of specialization in industry developed to handle such complex enterprises, including the system of management developed by Frederick Taylor to make efficient use of manpower, the field of chemical engineering which converted laboratory processes into practical industrial-scale production facilities, and the field of cost accounting which provided methods for determining the costs of products by analysis of the time and cost involved in components, raw materials, and process stages.[16]

The Navy opened its Powder Factory under the management of naval officers, traditionally trained as generalists, with backgrounds in engineering and in the command of men. To some extent, Bureau of Ordnance officers were already specialists, but they had developed specialized backgrounds in gunnery, metals, explosives, and armor. The skills they showed in chemical engineering, management efficiency, and cost accounting were largely homegrown, not learned in academic or commercial settings but rather developed out of their Navy careers or on the spot at Indian Head. Many of the ideas and planning tools were learned by the military officers from the civilians they employed at the Powder Factory. The process of moving to industrial-scale production required organized, large-scale laboratories to conduct the needed quality assurance tests, tests of alternate processes, and experiments to solve problems of inconsistency and product variation, which would further accelerate the Bureau of Ordnance effort to change over to a modern research structure.

Some of the procedures and concerns that arose from factory management were already familiar to naval officers who had managed large civilian work forces at navy yards and at other shore establishments. The Navy had a reservoir of experience in dealing with problems of personnel, contractors, housing, safety and health, and structured management of semiskilled and unskilled workers. Yet the production at Indian Head would raise civilian personnel management by naval officers to a new level of complexity. Although naval officers were used to command of men in situations of risk, the chemical plant would require new methods of procedural control in the face of dangerous processes.

Similarly, naval officers had experience in procuring equipment and setting up utilities such as water supply, roads, railroad links, telephone and telegraph systems, and electric power plants and transmission lines. Such facilities for modern industrial operations had been installed in many naval shore establishments. Officers and enlisted men in the Bureaus of Ordnance, Construction and Repair, Engineering, and Yards and Docks could draw on a rich background to plan and build what economists would later call the "infrastructure." Officer in Charge Strauss opened a new coal-fired electric power plant in 1900, followed shortly by a pump house for an artesian water supply capable of 75,000 gallons a day and a local telephone network with twenty-five "stations."[17]

Utilities expanded so rapidly that within a few years the tangle of lines and pipes became a mystery for the officers, and the beginning of a shift to reliance on civilians showed up in various odd ways: "At present," noted Officer in Charge A. C. Dieffenbach, only three years after the powder factory went into operation, "from the absence of the necessary personnel to carry out such a work, no plan exists of the underground piping, sewerage or drainage, or telephone and other lines, and it is only through the knowledge of the whereabouts by old employees that these can be traced."[18]

DOUBLED CAPACITY

Strauss worked under orders to double the plant's capacity from the original planned quoto of one thousand pounds a day. While expanding equipment to meet the new demand, he had to double the shifts to try to reach the quota. The original buildings had to be modified, utilities expanded, and equipment added. He put men on two eight-hour shifts in the factory that made the nitric acid consumed in the production process, at the same time that he installed new equipment in the same factory. He noted that "everything about the factory has had to be duplicated, and this without interfering with the output of the original plant." He doubled the capacity of the drying houses, added recovery tanks to the solvent recovery house, doubled the capacity of the press house, and oversaw a series of other changes.[19]

After the first full year of operation, 1900–01, Strauss reported somewhat apologetically that production had been 250,000 pounds of powder during the fiscal year, with the manufacture of 181,500 pounds of ether and 546,000 pounds of nitric acid. "While this result cannot be called entirely satisfactory, some allowance must be made for that fact that it was the first year in which the plant

TABLE 1
Tenure, Early Officers in Charge

Ensign R. B. Dashiell	1890–1893
Lieutenant N. E. Mason	1893–1896
Commander A. R. Couden	1896–1900
Lieutenant J. Strauss	1900–1902
Lieutenant J. B. Patton	1902–1903
Lieutenant A. C. Dieffenbach	1903–1906
Lieutenant Commander J. Strauss	1906–1908
Lieutenant Commander R. H. Jackson	1908–1910
Lieutenant Commander J. H. Holden	1910–1913

Source: "Inspectors of Ordnance in Charge," Indian Head Technical Library, Naval Ordnance Station, Indian Head, Md. (For full list of commanding officers, see Appendix 1.)

was operated, and that no employee had ever had the slightest experience in any of the branches involved. Added to this, during all of that time extensive changes were being made in nearly all buildings which more or less hampered the steadiness of the output."[20]

Other aspects of powder factory operation could also draw on the reservoir of existing Navy management skills. Purchasing equipment, organizing supplies of raw materials, and testing purchased materials and final products to ensure quality required skills which officers had acquired in ship operation, in navy yard and station management, and in product specification testing.

THE BEGINNINGS OF CIVILIAN MANAGEMENT

Despite the background that officers brought to bear, the management challenge for the Navy soon became apparent. Naval officers rotated in assignments on two- or three-year tours of duty, usually alternating between ship and shore assignments, to insure that officers with seniority would have a wide variety of experience considered valuable for the top flag ranks. Before 1899 promotion had been strictly based on seniority. Other bureaus in the Navy had moved to a system in which the engineering experts drew long-term shore assignments. But unlike the ship-building Bureau of Construction and Repair, with its "Construction Corps" of naval architects, and the Bureau of Steam Engineering, which utilized engineering duty officers, the Bureau of Ordnance was staffed by "line officers." For such men, naval cruises, not shore appointments, were the key to advancement, and the shore assignments would always be temporary. The quickly rotating assignments of officers in charge at Indian Head are shown in Table 1.

Although no explicit record has been discovered stating why Strauss was recalled for the rather unusual second term, it is quite possible that the reason was simply that he had successfully overseen the original expansion in 1900–01, and when in 1906 the addition of a sulfuric acid plant was planned, he seemed to have the requisite construction experience. He showed his versatility when he worked with New Jersey Zinc, the contractors for installing the plant, cadging leftover equipment, brick, pig lead, and other parts and supplies from them. He worked out a ninety-day extension to their contract when machinery deliveries delayed the performance of their contract through no fault of their own.[21]

By 1908, well into his second term, Strauss could once again successfully report expansion, this time with a much higher goal. He had overseen the construction of a 1 million pound magazine and a 1.2 million pound nitrate of soda storehouse, together with a new ether vault, a powder reworking plant, and other minor buildings. This phase of expansion came at the peak of naval growth in ships and shore establishments in Theodore Roosevelt's second term as president. The new "Great White Fleet" displayed its strength in a round-the-world tour in 1907–09. The powder plant expansion was one aspect of the shore side of Roosevelt's major push to develop the Navy.[22]

With the exception of Strauss, all officers in charge served one term, not to return. The principle of rotation in office was even more extreme at the next naval officer level down, where ensigns and lieutenants often put in a few weeks or months of service at Indian Head between other assignments.

From the point of view of the Bureau of Ordnance, one virtue of the rotation system was that it provided training in Proving Ground procedures and, to an extent, in the technology of the new powder. The Bureau not only used the junior ranks at Indian Head to groom future officers in charge but also explicitly regarded it as a good place for training. Even so, under the quasi-voluntary system by which naval officers req uested assignments, it was hard to fill the necessary slots at Indian Head. The Secretary of the Navy noted the problem in 1902: "As a school of instruction the proving grounds should be a most attractive place for officers desiring to improve their knowledge of its special-ties, but the Bureau finds much difficulty in obtaining officers for this duty."[23]

Training in smokeless powder storage and use was essential to the Navy, no matter how reluctant the officers felt to learn the processes of pulping and poaching. Indeed, naval officers at first were highly suspicious of the new powder, and one group aboard the *Olympia* jettisoned overboard a large quantity of powder because its odor had changed in the magazine. Cautiously, they retained a small sample for later analysis. As it turned out, the supply had been perfectly safe. A tour of duty at Indian Head, it was hoped, would expose a whole cohort of recent Naval Academy graduates to the new propellant and perhaps would prevent such embarrassment in the future.[24]

To take advantage of the training opportunity at Indian Head, the Bureau in 1904 established a regularized sixteen-month postgraduate program, in which young officers divided their time between the bureau headquarters, the gun factory, and the proving ground, with short details at "manufacturing estab-lishments." Frequently, ensigns or lieutenants, such as Dieffenbach, Ralph Earle, Andrew C. Pickens, Claude C. Bloch, Henry E. Lackey, Herbert F. Leary, and Garret L. Schuyler, would serve briefly at the Proving Ground under another commanding officer, go off for several alternating sea and shore as-signments, and then, a decade or more later, put in tours at Indian Head as officer in charge with the rank of lieutenant commander or commander. The later distinguished careers of some of the line officers who served as officers in charge at Indian Head revealed the assignment became a stepping-stone to the office of bureau chief, and thence to admiralship and greater, national responsibility.

Some who passed through Indian Head to serve in high-ranking posts included Strauss, who later planned and directed the North Sea Mine Barrage in World War I, and Harold R. Stark, officer in charge, 1925–28, who later served as Chief of Naval Operations, 1939–42. Details on these and other line careers of the officers in charge are provided in Appendix 1. The cohort of ensigns who passed through between 1904 and 1910 and who later served as officers in charge were products of the postgraduate training program set up at Indian Head, that was specifically designed to familiarize younger officers with ordnance and ordnance testing.[25]

The training course emphasized practical experience with gunnery. A few years after Strauss's second tour at Indian Head, when he was chief of the Bureau, he argued against substituting a long postgraduate course at Annapolis for practical experience to be gained at the Proving Ground on the short training tour.

> . . . As long as we have a scientific instructor at the Proving Ground such as we have there now in the person of Schuyler, it would be best to lengthen the period at the Proving Ground. . . . To my mind the greatest need for ordnance men is the grounding obtained by experience, something that must seep in day by day as a result of practical work. In former years, the necessity for this very thing gave rise to the term "ordnance ring" which, after all, meant that those men who had become familiar with ordnance matters in detail were asked again to serve in similar work.[26]

Strauss himself was apparently reassigned to Indian Head precisely because he had the required day-to-day experience. By following the pattern of assigning junior officers to Indian Head and later reappointing them to command the post, the Bureau of Ordnance could patch together a pattern of pertinent experience but not continuity of command. Although many would achieve notable results in their two- or three-year stays, none centered their careers on improving or transforming the Powder Factory as did engineering officers at the Engineering Experiment Station or construction corps naval architect officers at the Model Basin. None stayed long enough to define an "era" in the growth of the station, although clearly Dashiell, Mason, Couden, and Strauss, more than some of the others, left their imprints in facilities and practices. The Navy regarded ship-shore rotation as a means of insuring that the shore establishment remained sensitive to the needs of the fleet.

The line officers, many of them talented and efficient managers, on their way up in their careers, took on the assignment as they would other ship and shore tours: come in, master the problems, take good advice, implement the required changes, carry out the orders from the Bureau, make recommendations, and move on, without devoting a major proportion of their lives to the single post as did the Engineering and Construction Corps officers at those Bureaus' experiment stations.

For those in the eighteen-month postgraduate ordnance course, the proving ground experience was part of the training, but even for them powder factory work was not a central assignment. Others, like Schuyler, on longer tours rather than on the short course, specifically noted their preference for proving ground duty over the powder factory operation. Testing ordnance, armor, and guns provided firing experience and technical knowledge which would pay off in line duty as gunnery officers, especially if one were later to be called upon to fire the weapons in combat. Such duty was far preferable to the demanding and unrewarding tasks of managing a chemical plant, with its personnel, safety, and financial headaches. Whether involved in proving ground or powder factory work, the social isolation at Indian Head made the post unattractive for many, a problem that persisted over the following decades.

For such reasons Strauss, as officer in charge, and the Bureau as a whole soon made the distinction between powder factory work which ought to be civilian-run and the military proving ground work appropriate to ordnance officers. In his first tour at the Proving Ground, Strauss requested "a good practical powder maker" who would enable "the few officers stationed at Indian Head to give more time and attention to the regular duties of the proving ground." Clearly, as early as 1902 the industrial side held less attraction than the "regular" duty of gun and armor testing.[27]

Despite the attraction of short tours for training purposes, the disadvantages of rotation in service from the standpoint of managing an ongoing institution were readily apparent to the officers themselves. Only continuity of management in sufficient numbers could build familiarity with the diverse and complex problems in operating the Powder Factory. Strauss and the Bureau understood very well the problems of discontinuity of control. Strauss particularly noted that without continuous service, various investigations in the armor area could not be brought to resolution: "It is to be regretted that [officers'] period of service cannot be made longer. The frequent changes render impossible any continuous effort at the solution of the many problems that present themselves; the history and spirit of previous investigations are lost and the time taken to arrive at conclusions necessarily lengthened."[28]

In the context of officer turnover, it was inevitable that control of the industrial operation would have to shift into civilian employee hands, and by 1901–02 the chief of the Bureau of Ordnance began to lay the groundwork for that shift of management, as suggested in his section of the *Report of the Secretary of the Navy*.

This work has heretofore been and is now being carried on by the officers and chemists attached to the proving ground, who give as much attention to it as their other duties, which are very exacting, permit; but it is not possible for them to give it that constant attention which it requires. Moreover, officers are continually changing, and but few at a time are ever attached to the proving ground. The manufacture of smokeless powder is a study of itself and requires familiarity with a special branch of knowledge, which is by no means general.[29]

Even in the naval officer generalists' area of expert knowledge—the command of men—the military was quite willing to relinquish control. Supervision of the civilian work force became more complex as the plant expanded and spread out. Unlike a ship, where all supervision could be done by eye and voice, the plant

TABLE 2
Continuity of Civilian Personnel 1901–1906

Personnel Category	1901	1902	1903	1904	1905	1906
Chemist	Patterson ---
Asst. Chem.	Kniffen ---/Storm ----------
Electrician	Staughton -- *
Leadingman	Lloyd --
Draftsman		[job opened 1903–] Rainsford --------------------------------------
Leadingman		[job opened 1903–] Dement --
Leadingman		[job opened 1903–] Olmstead ------------------------ *

Source: Strauss to Chief of the Bureau of Ordnance, 2 April 1907, RG 181, Acc. 9959, 2605-0-1, Philadelphia National Archives Branch.
* = resignation; Patterson, Storm, and Lloyd all stayed on through the 1930s.

was uncomfortably laid out for the naval system of command, as noted by the bureau chief in his comment on the problem.

> The Bureau believes that a foreman who is a practical powder maker should be allowed for the smokeless powder factory. The plant consists of 28 buildings scattered over some 300 acres of land, many of the buildings being widely separated from each other. The necessity of constant supervision and of competent superintendence is very great not only as a means of safety and economy, but in order that the various processes of manufacture, upon which the integrity of the powder depends, may be closely watched and faithfully performed in the manner prescribed by the specifications.[30]

Under such circumstances it was very difficult to keep things shipshape. In 1903 Officer in Charge Dieffenbach noted: "The lack of officers to assign sufficiently subdivided duties has necessarily resulted in considerable deterioration."[31]

In the first few years after the Powder Factory opened, the transition to civilian management and control was already evident. Although the officer complement rapidly revolved in the first six years of powder factory operation, civilian employees already put in greater continuity in service. Simply by staying on the scene, the civilians became the permanent staff. The relative longevity of some of the civilians is revealed in Table 2.

By 1907, the supervisory civilian work force increased to 12, including those shown on Table 2, together with Powder Factory Foreman Alex Cruickshanks and 4 magazine attendants. Including all the supervisory people, the total civilian work force numbered over 120, well up from the floating force of 30 or so under Dashiell and Mason when Indian Head had no factory. Not only chemist Patterson, who took the leading role in chemistry, chemical engineering, and plant cost accounting, but also Cruickshanks and their 10 civilian supervising employees all showed far more continuity of employment than was possible for the officers in charge.[32]

The only uniformed member of the staff with any extended continuity in office in the period was "Ordnance Writer" or Pay Clerk J. J. Gering, who served under Dashiell with a petty officer rating and was still on duty at the Powder

Factory in 1922, having moved up to the rank of lieutenant in that year. Gering, who managed the clerical work for about thirty years, received quiet recognition—pay raises and promotions; but his extraordinary longevity at Indian Head was clearly unique among those in naval uniform. After 1908 he was regularly regarded as one of the "old-timers" and was the only one on staff who could remember the Dashiell days. His unique status at Indian Head as an enlisted man eventually promoted to officer status only gives emphasis, by contrast, to the norm of short tenure and advancement for the line officers.[33]

Under such a regime of military rotation and civilian tenure, it was inevitable that those with more experience, longevity, and roots at the Powder Factory would come to take over its administration. But the emergence of civilian control did not accidentally evolve out of the relative longevity of the civilian crew and the contrasting temporary tenure of most of the officers. Rather, it was the consequence of a conscious and concerted effort by the officers to have the factory side run by civilians and the proving ground side run by officers. Like Dashiell, his successors knew that their careers would benefit from sea duty and from on-shore gun work. But none would see a future in factory management, especially in the unique chemical plant. For these reasons, the Powder Factory at Indian Head pioneered a pattern of civilian management which other naval shore establishments would not adopt until decades later, if ever.

MANAGEMENT PROBLEMS: INFORMATION, STAFF, COSTS

On occasion, naval officers and civilian employees sought to discover exactly what process or procedure worked in the private sector. Sometimes they succeeded readily; at other times they had to resort to what amounted to industrial espionage. Sometimes they ran into a stone wall despite their best efforts at finding out how the private sector solved a particular problem, simply because private firms regarded such information as proprietary and did not care to share it with the Navy for fear it might leak out to private competitors.

In the period immediately after the factory opened, George Patterson was the only college-trained civilian on the force, and as an experienced chemist he took an ever larger role in the complex decisions required to improve the factory and its product. The military officers worked closely with Patterson in attempting to get information for use at the new facility. Officer in Charge Dieffenbach travelled with Patterson to the Army's Sandy Hook Proving Ground in New Jersey, using the occasion to examine the sulfuric acid works of New Jersey Zinc Company in Hazard, Pennsylvania, Harrison Brothers and Company in Philadelphia, E. I. du Pont in Wilmington, and Repanno Chemical Company in Gibbstown, New Jersey, in order to determine which company's system should be emulated at Indian Head.[34]

The exchange of information between the Navy and the private sector was not a one-way proposition, and what a later generation called "technology transfer" was more of a series of two-way exchanges of details regarding process and machinery. After Patterson adopted a procedure to cut down on sodium carbonate in the nitrocellulose poachers and recommended a shorter boiling period, du Pont adopted the improved procedure less than a month later.[35]

There were certain limits in these exchanges, however. While du Pont seemed to be willing to share its processes with the government plant at Indian Head, the managers of du Pont did not care to have the material disseminated to other firms. Thus, when Officer in Charge R. H. Jackson sought blueprints of the International Smokeless and Dynamite Company (a du Pont subsidiary) in order to plan for once again doubling output, the bureau chief had to assure du Pont that the plans would be "considered strictly confidential."[36]

Appropriations of $300,000 through 1909 helped in this phase of expansion, which included a new sulfuric acid house and a new and larger electrical generation system. Jackson's successor, J. H. Holden, could finally report the completion of the new setup, including new boiling and poaching tubs, new pulpers, a new nitrating house, and an expanded electrical power system with boilers, turbo alternators, cooling towers, and an electrical switchboard and metering system to determine how much power was consumed in each phase of manufacture. The metering would be essential for cost analysis.[37]

With the hectic pace of expansion in several waves through the first decade of the twentieth century, Indian Head faced a severe manpower recruitment problem for its civilians. Southern Charles County did not have a skilled labor pool. Many of the civilians recruited to work in the difficult chemical technology were local rural residents, and each process had to be reduced to a set of procedures that could be followed "by the book" at a time when the "book" was constantly under innovation and when the equipment changed from year to year.

Strauss and his successors were very conscious of the problem of skill levels. "Nearly every laborer at the powder works," he noted, "is unskilled in any trade. They are mostly Charles County farm hands. Very few of them reside in the near vicinity, and when bad weather cuts off communication from the surrounding districts, it is sometimes necessary to shut down one or more departments of the works."[38]

More qualified experts or experienced supervisory people were reluctant to relocate in what was still perceived as a remote location. Patterson's assistants Kniffen and Storm were lab technicians or laboratorians. Not until 1912 could Patterson secure the assistance of another college-trained chemist, and even then the recruitment was difficult. In that year the Navy arranged to transfer Walter Farnum, a 1904 Harvard graduate, with eight previous years of experience at the torpedo factory in Newport, where he had run the experimental powder plant there. Farnum was at first reluctant to take the job at Indian Head, which he regarded as so remote as to deserve hardship pay "like the Philippines." Furthermore, Farnum was none too happy with what he saw as a demotion to "assistant" status, but accepted the post when he was allowed to occupy one of the houses constructed originally for the naval officers on the heights above the valley.[39]

Supplies and heavy equipment continued to come in by boat or barge rather than over land. Officers in charge attempted to acquire a workable launch and more regular runs of river boats from the navy yard, only gradually getting the transport on a scheduled basis. Even so, ice and storms interrupted even the food supplies to the local grocers, and complaints about the high prices charged by merchants who were able to ship goods aboard the government transport led to local employee protests, in which Patterson and the leadingmen participated.[40]

Like Dashiell, several of the later officers in charge did not enjoy working with a civilian work force. Used to the full-time employment of sailors aboard ship, they found the concept of holidays off with pay abhorrent. One noted: "The cost of administering this station is enhanced by the charge debited to it of labor which is not performed. Fifteen days' leave with pay, in addition to legal holidays allowed by law, increase the expense annually nearly $7,000, and make quite a sum, which ought to be separately appropriated and reduce the showing on maintenance."[41]

Housing on base rapidly deteriorated. In 1903 the officer in charge noted that Cottage No. 1, which had been built by Dashiell, "is becoming less and less habitable, due to the shock of discharge of heavy guns, and this building will eventually have to be abandoned."[42]

COST ACCOUNTING AND CHEMICAL ENGINEERING

The cost accounting ideas worked out by the officers in charge reflected the work of Patterson. Although the private sector supplied smokeless powder at seventy cents a pound, Patterson calculated the cost of Indian Head smokeless powder in 1903 as about forty-five cents a pound. He included in this charge all maintenance and repairs plus a monthly write-off for depreciation and equipment of 5 percent per year, not including the clerical and supervisory force of naval personnel. His calculations did not include interest, insurance, or depreciation of buildings. In these regards he was following standard accounting practice and making explicit the basis for his method so that true comparisons could be made with the commercial sector.[43]

The Bureau of Ordnance used elementary cost analysis in planning expansion and in justifying expansion to Congress:

> As mixed acid is one of the most costly ingredients used in the manufacture of smokeless powder, the Bureau feels called upon, in the interest of economy, to recommend the erection at Indian Head of a sulfuric acid plant having a capacity for producing 10 tons per day. . . . The average cost of mixed acid, by purchase is about $22 per ton, whereas it can be manufactured at a cost of about $12.40 per ton; hence the importance of installing a plant. Moreover, the questions of transportation, delay of tankers, demurrage on cars, etc., which now frequently occur will be obviated and the Bureau will control all the necessary important operations in connection with the manufacture of smokeless powder, which in itself will be a great advantage.[44]

Patterson continued to use cost accounting methods through the decade to spot bottlenecks, to identify cost items, and to work at reducing the cost of powder. By 1911 the price of newly manufactured Indian Head powder was down to 33.619 cents a pound (over one million pounds at $350,000), reflecting decreased costs in the manufacture of ether, sulfuric acid, and nitrocellulose itself, and in the fuel cost per kilowatt hour of electricity generated and consumed in the production of powder. In sum, Patterson could report that "improvement in systems of accounts, machinery, appliances and methods ... has made it possible to further decrease the cost materially although the cost of labor has steadily increased during this period."[45]

Like cost accounting, chemical engineering for large-scale industrial manufacture was in its infancy. Few managers, in or out of uniform, were familiar with this fledgling field. Nevertheless, Patterson soon worked out some of the essentials, combining a knowledge of the chemical process, the machinery involved, and the nature of the work for the laborers. In an early analysis of whether to adopt a centrifugal system to work alongside an existing alternate "pot system" of nitration of the cotton fibers, Strauss reflected Patterson's consideration of multiple factors noting, "A careful study of the two methods in advance to my mind, show many advantages for the [pot system]. First it requires 50 per cent more labor by the centrifugal method; second, when a 'fire' takes place, 24 pounds of pyro are lost, as well as about 900 pounds of acid (in the pot system these figures are reduced to 4 pounds of pyro and 50 pounds of acid)." The wasted "pyro" he referred to was pyrocellulose, or cellulose which had been nitrated below the minimum of 12.45 percent. He noted the centrifugal system wasted more acid. However, the centrifugal system "is neater and less hard on the men, and the wearing out of the extractor is not as great as the breakage of pots." In any case, the factory operators looked forward to the arrival of the centrifugal equipment so they could compare the two systems in operation as well as increase production. Each change of equipment and method involved similar discussion of interacting factors in terms that were lucid and understandable to the generalist-trained line officer but that also showed a sophisticated grasp of the engineering problems in a chemical plant.[46]

PROVING GROUND: GUNS, ARMOR, SHELLS

Proof of guns, armor, and shells proceeded at the Proving Ground through the period. The piece of land immediately to the south of the station, Stump Neck, had several residents, particularly the elderly Eli Gaffield, who complained of the shells passing overhead. In 1900 and again in 1901, Strauss recommended its purchase or lease "to obviate the annoyance to people now residing in the vicinity when shell pass close to this point." Over a thousand acres at Stump Neck were acquired from the Gaffields in 1901, and a marine barracks for station guards was installed there. Soon, however, the passing shells endangered the barracks, and troops had to be marched out of the line of fire.[47]

Officer turnover and understaffing presented the officers in charge with difficulties in carrying out the required tests of guns, armor, and shells. Strauss and his successors constantly argued for longer terms and for more officers, with only gradual growth in response to their pleas. Nevertheless, Strauss could report on semiautomatic rapid-fire guns, on the question of explosive shells, on attempts

35

to vary the black-powder ignition charge for smokeless powder, and on the ability to shorten the barrel lengths of guns while using the more powerful and slower burning smokeless powder.[48]

Through the first years of the century, the tests kept up on ordnance improvement, with various attempts to improve breech-loading speed, to adjust guns to smokeless powder pressures, and to select an effective machine gun. The Proving Ground ran experiments with electric firing of guns, rather than percussion; officers continued work with armor-piercing projectiles, either loaded or unloaded with explosive. Proof of some of the explosive armor-piercing shells in the valley, with fragmentation of both shell and armor, caused concern to the growing on-base village of housing on the hills above, as shell fragments fell near the residences.[49]

Work continued in the testing of "Harveyed" steel and "Kruppized" steel, as well as experimental Carnegie nickel-steel for decks. Other processes for hardening steel began to come thick and fast, including the Ely process and a system of compound plates developed by the "Cosgrove" process.[50]

By the end of Strauss's first term, the Navy started to argue for moving the proving ground work away from the growing settled area of the Powder Factory. The proving ground, noted the chief of the Bureau, is "by no means as satisfactory for its purpose as it formerly was. The great increase in the power of guns in recent years, and their greatly extended range, renders a more isolated location necessary for proving and ranging them." The chief noted in his report to the Secretary of the Navy as early as 1902 that firing down the channel endangered river vessels. The proving ground would hold off until river traffic passed, but the situation was beginning to echo the problems of Annapolis. Nevertheless, the chief reported, "The Bureau is not yet prepared to recommend a change of site, but is convinced that the time is not far distant when the matter will have to be seriously considered." In fact, the issue continued to be discussed for the next two decades.[51]

Meanwhile, funding for land purchase and a complete facility was difficult to obtain, and the Proving Ground continued side by side with the growing civilian community of workers at the Powder Factory, firing over the marine barracks and down the sometimes crowded Potomac River. It was an uneasy and somewhat nerve-wracking time for the residents. With civilians at work in the powder factory, where a fire could be fatal to a whole crew, the sound of an explosion would be frightening. With the constant firing of heavy guns along the river, with concussions wracking the flimsy frames of the first wooden cottages, and with occasional armor or shell fragments landing in the neighborhood, nerves were on edge.

When Dieffenbach came aboard he was perturbed and very explicitly spelled out the problem for his superiors:

> Aside from the argument advanced in favor of a change of site for the proving ground—namely the necessity of a less obstructed channel and longer range—the change could be made to include a different arrangement of relative locations of battery and working ground. At present the battery is located on the side of the ground which brings the line of fire over the heads, or nearly so, of everyone working on the ground except the immediate gun

servers. As there is always danger from flying mouth cups, stripped rotating bands, etc., it is required at every shot for all to take shelter.[52]

EXPERIMENTAL WORK

Beyond Patterson's contributions to the beginnings of civilian management, he also introduced a set of laboratory attitudes in which we can see further evolution of the later research emphasis of the facility. Chemists and other researchers at the time clearly distinguished between what they called "routine" work on the one hand and "experimental" work on the other. The first decade of the twentieth century saw quite an output of "experimental" chemical work at Patterson's laboratory.

In 1907 Patterson noted that "the usual work has been conducted in the laboratory consisting of a very large number of analyses, more or less routine," which he did not attempt to quantify or describe. However, he went on to specify "special experimental work," all of which related to the production, storage, and inspection of powder. Experimental work included:

◇ Effect of electric light on powder
◇ Hydrolysis of nitrocellulose impurities by acids
◇ Effects of salt in water in powder manufacture
◇ Establishing a method for the detection of mercuric chloride in powder
◇ Work with surveillance tests

Patterson gave special notice to the problem of alcohol and water drying of powder, which he noted had been "long and patiently investigated." The attempt to scale the process up from the laboratory to the commercial or factory level had failed "for reasons that so far have remained obscure." He promised continued work on the problem. In 1908 special experimental work included further tests on stability and, more significantly, establishing a standard litmus test paper for powder testing and supplying litmus paper for both Army and Navy powder inspectors.

In 1909 Patterson again reported on a wide range of "experimental work," including

◇ improvements in tests and methods of analysis
◇ effects of stabilizing agents on powder
◇ effects of different methods and temperatures on drying powder
◇ methods of solvent recovery[53]

Almost all of the experimental work that Patterson noted in the period bore a direct relationship to the powder production and was designed to produce a modified industrial process, to set methods of quality assurance of the product, to assist in "scaling up" a process from laboratory to production, or to refine the methods of evaluating stored powder. Whether such a factory operated in the private sector or under government ownership, it was essential that it develop a research arm, if only to ensure efficiency in choice of methods. The close relationship of such work to the powder factory mission was apparent, and it was clear that a very full schedule of original experimental work could be maintained as part of the production mission. Patterson set up his laboratory to do the kind of work taken on by the larger laboratories of private firms like du Pont.[54]

Very few of the full reports of Patterson's experimental work survived from the early period. However, two of Patterson's technical papers from 1909, which he sent to the Bureau for publication approval with the International Congress of Applied Chemistry, show the areas and techniques of his work. One, "Detection of Mercury in Explosives," described his experimental work, using an electrolysis method for accumulating a sample which could then be tested by spectroscopy, in detecting the presence of unwanted metallic mercury in nitrocellulose powder. The method was superior to other methods in that it could work with a very small sample and was reliable and simple where a considerable number of samples had to be tested. The method had been in use at Indian Head for two years before he submitted a description of it for publication.[55]

Patterson prepared a second report which had far more significant long-run impact—a study of methods of testing stability of smokeless powder. He began his review of the literature with the remark that "the old saw of 'keep your powder dry' has been changed to 'keep your powder cool and test it frequently.' " Patterson reviewed several existing stability tests of powder, which involved heating the material at various temperatures to simulate and accelerate the "conditions of decomposition of the powder in service." The standard tests included one developed by the British explosives inventor Abel, another by the French expert Vieille, and the "German Test," which required heating the powder to 135 degrees centigrade.[56]

Patterson had been working on evaluating the various testing methods as early as February 1906, when he requested information through the Office of Naval Intelligence regarding the actual methods employed in the German tests. Later that year, two complete testing outfits were purchased by the U.S. Naval Attache in Berlin and shipped to Indian Head aboard the North German Lloyd ship *Graf Waldersee*.[57]

Patterson introduced and described a new surveillance test that required heating a sample of powder to 65.5 degrees centigrade and holding it in storage for a month. By 1907 laboratorian Storm was dutifully conducting hundreds of the tests, including the German and the Vieille test, as well as the "heat test" developed by Patterson, regularly filling out a form for each lot of powder tested, with results from all three tests. This procedure allowed comparison of European results with American ones, cross-tabulation of the different measures of stability, and comparison of the methods for consistency of results.[58]

All of the tests proceeded quite simply: a sample of the powder would be heated at each of the standardized temperatures. The time at which each sample began to give off fumes would be noted and compared with the standard. Powder that fumed too soon would be rejected; that which fell in the standard range was acceptable. Under the Patterson test, acceptable powders would take about thirty days before they emitted fumes; under the higher temperature German test, stable powder would emit fumes after only about five hours. While requiring a longer test period, the Patterson test gave consistent results. The 65.5 degree test worked out by Patterson became the standard "Navy test," employed to check smokeless powder stability over the next decades.[59]

One of the problems with the litmus paper required in stability testing was that the color could be judged subjectively by different observers. Patterson directed the work on developing a standardized color-tile against which the litmus paper colors could be judged, and in 1906 one hundred of these tiles were supplied in lots of five, to allow naval "inspectors of powder" at various magazines and industrial sites to conduct the same test in the same fashion.[60]

Chemists were by no means the only researchers at the turn of the century to distinguish between "experimental" and "routine" work. As early as 1903 and 1904 Dieffenbach reported on series of experiments with ordnance, heading a whole section of his annual report with the term "Experimental Work" under which he included testing new instruments, looking for means of stiffening cartridge bags, and testing a variety of new gas check devices.[61]

The use of the term "experimental" to identify research was not simply restricted to a coterie of like-minded, innovative officers. Rather, officers throughout the Navy used the term "experimental" over and over in the era to report a variety of types of research including developmental work, product evaluation, testing methods and standards refinement, establishing of basic tables of performance, and a variety of projects adjusting one technology to another, such as powder to guns.

When other naval bureaus opened facilities at the time to conduct such work, the very names reflected the concept of "experimentation"—David Taylor's model basin at the Washington Navy Yard was called the "Experimental Model Basin"; George Melville's new facility for the engineers at the old proving ground site at Annapolis was the "Engineering Experiment Station." Ordnance, too, wanted to institutionalize the experimental approach. In 1907 Newton Mason, by then a rear admiral and serving as chief of the Bureau, requested "an Annual Fund for Experimental Work in the Development of Ordnance Material," and spelled out the results of recent experiments as well as a program for new ones. Work already accomplished included doubling the striking power of 1905 12-inch guns compared with those of 1895, improving the resisting power of armor by 50 percent, and improving the penetrating power of projectiles by a similar ratio. Such achievements, Mason pointed out, "resulted from careful, painstaking experimental research along a great variety of lines. Such research demands a body of observers who can give their undivided attention to the work." Army Ordnance, Mason complained, had an experimental fund. But the Navy had funded its ordnance experimental work out of regular appropriations for purchase.[62]

Further work on gun erosion, testing of smokeless powder properties, developing specifications for guns, and testing the penetrating power of 12-inch guns, together with a range of other developmental work, would be taken up if the funds became available. After unsuccessfully requesting $200,000 in both 1907 and 1908 under the Theodore Roosevelt administration, finally, in 1909, under President William Howard Taft, Mason secured congressional approval for $100,000 for ordnance experimentation, together with the $250,000 for powder factory expansion. This expansion was the less spectacular side, the follow-on shore phase, of the expansion that had produced the Great White Fleet. The Bureau put the money to work on a variety of projects, including

◊ study of the behavior of powder in guns"—interior ballistics"
◊ investigation of the effect of shell shape on flight of projectiles
◊ study of gun erosion
◊ study of projectile rotation, as affected by bands and erosion
◊ perfecting of bursting charges for shells, together with fuzes
◊ studies of the angle of shell impact[63]

Of the $100,000 appropriated, at least $78,000 went to Indian Head projects. In very explicit language, Officer in Charge Holden identified the new program of experimental work as distinct from the routine testing of the proving ground. In the period 1909–11, a dollar went a long way. Over 75 percent of the budget was spent on material, less than 25 percent on labor; no overhead was charged to the appropriation. With both labor and material costs held down, over one hundred separate projects were studied in the thirty months following July 1909. To implement the broad program outlined by the Bureau, Holden directed projects with Maxim fuzes for explosive charges in major and medium caliber shells, tests with guns against armor mounted on the decommissioned ram *Katahdin*, tests with experimental shells, cartridge cases, armor, and new types of explosives. He also directed work with fireproof powder bags, saluting charges, and "augmented rotating bands" for shells in eroded guns, and work with the Davis "torpedo gun." Holden particularly drew attention to tests for developing tracers and tracer fuzes, tests of cartridge cases, tests for developing major and medium caliber detonating fuzes, and tests for determining the loss in velocity due to erosion, relative to number of rounds fired and weight of charge. Holden admitted to occasionally slipping in a bit of research on the routine budget: "In all cases where such action is possible routine proof work is combined with experimental work in order to reduce cost and to render possible experimental work and investigations for which there would otherwise be no opportunity."[64]

In directing a wide range of experimental projects, Holden was carrying forward the sorts of innovative work that Dashiell and others had conducted. But under the regular funding obtained for Proving Ground experiments which Mason, as bureau chief, secured after 1909, experimental projects would be separately identified, and a more regular system to produce technical developments would be put in place. When Dashiell had recommended a modification in a gun he was routinely testing, the work was the product of an energetic and creative talent, implemented by the Bureau. Dashiell's many suggestions and the procedures for putting them into production at the gun factory had reflected the beginnings of a system of technical modification. By the time that Holden and Patterson regularly reported on scheduled and programmed studies, a further change had set in. They had planted the seeds of research which would flower in later years.

ZEALOTS AT WORK

The success story of the Powder Factory outlined by Patterson in his 1911 cost analysis was all the more remarkable when it is recognized that it was achieved against a backdrop of continuing naval officer turnover, a rapid shift to civilian management, personnel recruitment difficulties, innovation of process, and essentially pioneer living conditions. Experimental work in chemistry and in gunnery was beginning to mature from Dashiell's style of ad hoc modifications to longer range, regularly budgeted developmental work, which in several areas produced new standards, tests, and procedures for the handling of new propellants. When the added harassment of living under a constant barrage of gunfire is kept in mind, the growth and productivity of the factory was a tribute to the diligence and "zeal" of both civilians and naval personnel, to utilize a word which nearly every officer in charge during the era chose to describe the spirit of the staff.

The zeal, which so many officers noted, was a real ingredient. It was fostered by the fortuitous circumstances that had created a research-oriented mission. The powder factory brought with it as a necessary arm the chemical laboratory. The factory and its laboratory simply could not be operated except by a chemist, and under the Bureau of Ordnance with its line officer control it was inevitable that the chemist would be a civilian professional. Thus the factory and its laboratory necessarily brought into ordnance research the orientation of a trained scientist with his interest in conducting experimentation, analyzing problems, and operating of the factory not simply as manpower management but as complex interaction of process and people.

The choice of Patterson was fortunate. He began his career at Newport in the "research and development" phase of smokeless powder, then took on factory design and layout because Couden realized the need to understand the process to plan properly. Then, because of the Bureau of Ordnance need for a manager, Patterson was thrust into a role which merged features of "director of operations" and "technical director." Patterson grew in his career with the development of smokeless powder. His choice to stay with the Navy facilities, rather than to move to the private sector, was personal, but that choice was to the Navy's great benefit. By this unplanned series of events, Patterson became an early version of the civilian scientist turned administrator. Decades later, that model would characterize many defense laboratories. The pattern that, after much study in the 1960s, would be selected as the ideal method of operating military research structures had developed without planning, out of the necessities and personal choices of the period.

Patterson soon emerged as the leading specialist in smokeless powder and gave the rest of his professional life to the Indian Head facility. When Patterson's selection was coupled with the interest in "experimentation" which characterized the whole generation of naval officers around the turn of the century, the effect was to accelerate and shape the development away from the individual innovative style. Because he chose to make his career at Indian Head, Patterson would leave a lasting imprint on the method of operation. He brought the ability and the desire to expand on the "experimental" side as much as possible, while maintaining the "routine" side as an efficient operation with assistants whom he trained himself and, eventually, with other college-trained chemists. The scale of operation of the factory and the large volume of both experimental and routine work required a team, not an individual approach.

The career path of the line officers through Indian Head also served to favor the institution, despite the problems of discontinuity. When former Officer in Charge Mason moved up to bureau chief, he would remember his former command and advocate a regular budget to fund a program of experimental work there. Later, Strauss would follow the same path, serving as chief of the Bureau of Ordnance from 1913 to 1916. With such support from the top, the Proving Ground could implement the beginnings of a system of experimentation.

When an early version of a "research, development, test and evaluation" budget became available in 1909, Indian Head was in a position to take advantage of it and move forward, converting the zeal of officers and civilians into a flow of accomplishments.

CHAPTER 3

Ordnance Technology in the World War I Era

Standpipe
Building 130
1899

PRESIDENT WOODROW WILSON sat on the aft deck of the presidential yacht *Mayflower*, chatting with his personal physician, friend, and golfing partner Dr. Cary Grayson. It was Thursday, 3 July 1913, and after their light summer lunch, the two men expected a quiet trip up the Potomac to Washington, where Wilson would put the finishing touches on his scheduled Fourth of July speech at Gettysburg.[1]

Meanwhile, at the Proving Ground, Lieutenant Garret L. Schuyler readied a 14-inch gun for its second firing of the day. Knowing the gun's six-mile down-river range, the lookout at Chicamuxen Point phoned in the all-clear to Schuyler and reported the position of *Mayflower* in the channel, about a thousand yards west of the line of fire. Schuyler assumed those aboard "would think the proving of one of our largest guns a somewhat unusual and interesting proceeding to witness."[2]

At 2:00 P.M., Schuyler ordered the gun fired. The shell passed safely down range as expected, but its rotating band broke loose. As the shell whistled overhead, both the president and Dr. Grayson noticed that the band splashed down about three hundred feet away. Alarmed by the screech of the shell, the crew rushed on deck to ascertain the trouble.[3]

According to newspaper reports, Wilson turned to Grayson.

"The Proving Ground at Indian Head," he noted calmly, "presents a very dangerous menace to navigation."

Grayson agreed. He had served as Acting Assistant Surgeon at Indian Head for fourteen weeks in 1903 and remembered the place well.[4]

"It would be a good plan," Wilson added, "to do away with the testing of guns there, selecting some spot where there would be no danger to passing ships."[5]

But Commander Newton Alexander McCulley, commanding officer of *Mayflower*, reacted more angrily.

"In my opinion," stated McCulley in a confidential report to Josephus Daniels, Secretary of the Navy, "there was little or no danger, but the impropriety was manifest."[6]

Franklin D. Roosevelt, the young Assistant Secretary of the Navy, quietly investigated the incident a week later, asking for reports from both McCulley and Lieutenant Commander Jonas H. Holden, officer in charge at Indian Head.[7]

Commander McCulley noted that altogether the Proving Ground fired three of the 14-inch rounds. McCulley denied newspaper reports that a part of any shell fell within the vicinity of the ship.[8]

At first, Holden was indignant. "It is greatly regretted," Holden said, "that this incident in which [Lieutenant Schuyler] was apparently prompted by zeal in the prosecution of his work should be considered an impropriety. . . ."[9]

Roosevelt finally extracted a sort of apology from Holden: "If any inconvenience was suffered by those on board the *Mayflower* it is greatly regretted, as it is appreciated that a feeling of supposed danger productive of alarming rumors is highly undesirable." However, Holden could not resist a parting shot: "A projectile passing through the air several miles away," he noted with disdain, "would alarm some while others accustomed to firings would understand conditions and suffer no inconvenience or unease."[10]

Roosevelt wisely ordered the case closed, and the papers were dutifully filed.[11]

The shots that caused a news flurry in the doldrum summer of 1913 were neither the first nor the last to arouse concerns about the safety of the Proving Ground. Despite continued efforts of officers in charge, bureau chiefs, and the Navy Department to expand or relocate the range, final separation of powder factory and weapons experimental work would not come about until 1921, when a "Lower Station," supervised from Indian Head, was opened at Dahlgren, downriver on the Virginia side.

In the meantime, the original mission of the Indian Head facility further evolved. The work moved gradually away from simple proving of guns and armor to include standardization of shells and powder. George Patterson attempted to expand the chemical research program, both routine and experimental, that he had set up in his first decade there. That effort was impeded by orders to mount a vast expansion of powder production to meet a changed perception of the relationship between government and private roles in armaments. Following policies set by Wilson and Josephus Daniels, the Navy worked to decrease dependence on private firms, especially those that dominated steel and powder production.

That major domestic political struggle, reflecting disagreements over the nature of corporate power, was soon overshadowed by events in Europe. Although World War I broke out in Europe in August 1914, the United States remained officially neutral until April 1917. By the time the United States entered the war, Indian Head was a major producer of smokeless powder. The war, with its new weapons and its introduction of the submarine and the airplane into naval warfare, placed new demands upon both the chemical laboratory and the growing ordnance testing abilities of the station. "Captured enemy equipment," including armor-piercing shells, bombs, and specialized ordnance such as flares and smoke-barrage shells, required examination and analysis. The Navy's attempts to modernize further the powder plant and to expand work became severely frustrated by labor shortages and massive bottlenecks of transport and supply. The near-chaos throughout the American economy brought on by the lack of industrial preparedness would be severely felt at all naval shore facilities. The situation at Indian Head was made worse by its isolation, wage ceilings, and the continuing shortage of trained labor.

For Indian Head, the years of the two Wilson administrations were bracketed by two shots, the one which missed the president and the last shot fired before all gun testing moved to Dahlgren. Between those two punctuating events, the period was one of crises, readjustment, and evolving redefinition of mission.

RANGE SAFETY

Officers in charge from Lieutenant Newton E. Mason in 1896 through Lieutenant Commander Jonas J. Holden in 1913 had viewed the down-river range as risky. Considering the nature of the work, the safety record was good. One worker died from injuries on the short rail line, and three men tending a gun which blew apart had been killed, but fortunately no one down range had been struck. The list of complaints about near misses had grown more serious, however, as the volume and range of testing increased.[12]

Ever since Mrs. Gaffield had complained about the effect of overhead shots on her elderly husband at their quiet country residence on Stump Neck in 1899, the Proving Ground had attempted to steer a difficult course. Such protests might help secure funding for more land or a better range, but they could also support arguments for closing the facility altogether.

As the power of smokeless powder and the range and diameter of shells increased during the years which followed, the number of incidents had mounted.

In 1908 a shell struck the water forty feet from a Standard Oil tug pushing a petroleum barge. The company's Baltimore office politely requested the War Department to "see if something cannot be done to prevent shells being fired while boats are passing up the river." War referred the problem to Navy for a reply.[13]

The Secretary of the Navy had to mollify Congressman John Hull who spoke up in May 1909 about danger to nets and fishermen in the vicinity of High Point. As chairman of the House Committee on Military Affairs, Hull passed on the report that one fisherman had his net damaged by a "fragment" of a shell, again, probably a rotating band. The Secretary ordered a shift in the range, but the problem was not so easily solved.[14]

Through 1910 and 1911, testing of the 14-inch guns with improved smokeless powder kept extending the range. Holden explained to headquarters that, when the 10-, 12- and 14-inch guns fired down the long range, the projectiles passed within one thousand yards of Chopawamsic Island on the right, or Virginia, side and about five hundred yards from Budds Ferry on the left. Sometimes projectiles impacted closer to the shores, and thus the angle of shot could not be changed either to the left or right. Without a major increase in territory or a relocation, the same sector of the river had to be used.[15]

On Stump Neck, a peninsula across the Mattawoman Creek from the Station, private land holdings, together with a barracks for a small troop of marine guards which had been built on the Navy-owned section of the property, were particularly sore points. "Ordinarily," Holden noted, "when the flight of projectiles is regular and smooth, the personnel living on Stump Neck . . . are considered to be in positions of comparative safety." Inefficient shell bands or worn rifling caused erratic flight of the projectiles. In such cases, he noted, the shells "either wobble badly or tumble, and in either case fall very short and ricochet widely in either direction."[16]

Holden reported that in one three-month period two 12-inch projectiles had landed on Stump Neck, and that several hit in the water, one within one hundred feet of the Navy-owned dock on Stump Neck. Gunners ranged and angled the shots for a twelve thousand yard flight down river, but they sometimes fell on Proving Ground property and in the surrounding inlets. Holden pointed out that if the Navy's Bureau of Ordnance used the official rules employed at the Army's Sandy Hook station in New Jersey, all firing would have to be terminated at Indian Head. Thus, Schuyler's shot over the presidential yacht, while creating a short flurry of news interest in the summer of 1913, was only one of a whole series of worrisome incidents.[17]

The warnings from Holden and his predecessors had been heard at Bureau headquarters but with little result. Commander Volney Chase headed a special board to investigate the Stump Neck problem. He noted that "until a better range can be established," the officer in charge should be given clear authority to order the marines out of the line of fire and that regulations should be promulgated to that effect. The marines did not report directly to Holden, handicapping the operation. Adopting a somewhat philosophical tone uncharacteristic of official Navy reports in the era, Chase remarked, "It is not in human nature to render continued and implicit obedience to any kind of orders," and therefore, to reiterate existing regulations would do little good. Chase recommended that the marine barracks be closed and the detachment evacuated. The marines stayed, however.[18]

The next year, another board recommended the purchase of thirty-eight hundred acres of privately held land, in a long strip south of Stump Neck along the Maryland side of the Potomac, to allow for overland shots and for a greater margin for error in eastward angle variation of down-river shots. The Navy assumed that the land could be purchased for under $30 an acre, and asked for $200,000 to purchase the land and to begin improvements, shifting the batteries to take advantage of the proposed addition. In support of the acquisition, Admiral Twining, chief of the Bureau, explained the "extent of the embarrassment" suffered by the constrained range. He noted that a spot for loading railroad ties aboard barges on the river gave particular trouble, as the barge operators sometimes refused to be towed out of the line of fire; when they were towed, they would put in claims for the time lost. On another occasion, Twining reported, shells had ricocheted across the river to Quantico, causing "some consternation over there and real danger to the country." Twining accepted the fact that the Navy had no right to limit commercial traffic on the river. The best option, he concluded, was the proposed nine-mile extension along the shore.[19]

Despite Twining's arguments that extension would be cheaper than relocation, Congress did not appropriate the funds. The purchase of land and building a new facility, despite the apparent support from the casual remark reported from President Wilson, simply would run against the effort to maintain a low profile for the military during the Wilson administration.

The Navy could only round out some disputed boundaries on Stump Neck itself. Whiskey Point, a promontory on Stump Neck, had become a hangout for the off-duty marines. A local couple operated a bootleg juke joint there in a shanty on an unsurveyed corner. The lack of jurisdiction prevented eviction, and other small private holdings on the island added to the risk that someone might be hit by stray shots.[20]

The Navy acquired a small 3.3 acre parcel on Mattawoman Creek in two stages, authorizing it in 1912 and finally buying the land in 1918. The rounding-out process would be accomplished during the war years. In 1918 the Navy acquired 427 acres of Hopewell Farm and another 845 acres released by a variety of landholders and confiscated. The land added by the end of 1918 brought the total acreage, including marshlands, to some 3,208 acres. Even with these additions, which provided lots of space for the powder factory, officers' housing, and the valley proving ground, the problem of the long range over the marine barracks at Stump Neck and the risk to river traffic continued. The Navy abandoned its plans for an extension nine miles along the river front, opting instead for a complete relocation of the proving ground during the flurry of expansion brought on by World War I, when the Dahlgren site on the Virginia side of the Potomac was acquired. Construction delays and concerns with interrupting the program would prevent the full transfer of gun testing to the new location until well after World War I.[21]

STANDARDIZATION OF SHELLS AND POWDER

While Holden and the other officers in charge struggled with the issue of proving ground safety, they continued their full schedule of routine and experimental work. In ordnance, they sought to equip new ships with larger guns, putting into the fleet long-barrelled weapons designed for smokeless powder. Meanwhile, they worked to develop a wide range of explosives, signal flares, and smaller weapons.

Some of the standardization required experiments that smacked of pure, rather than applied, research, in that the results would set general principles and patterns which could only be determined empirically. In January 1913, for example, Holden requested and received permission to conduct a series of experiments to determine the effect of changes of center of gravity within projectiles upon the ballistic performance, particularly the range of the shot.[22]

By March, after a series of 10,000-yard-range shots with modified projectiles, Holden could report that relocation of the interior cavity and of the center of gravity of a projectile had no "material effect on flight." He suggested that the projectile specifications permit manufacturers to arrange the interior cavity in any fashion they desired, as long as they met strict standards on weight, charge, and shape.[23]

Ordnance experimentation covered literally dozens of topics in the last years before World War I. Indian Head worked on a variety of shells, fuzes, gun mechanisms, gun mounts, and experimental armor plates, as well as on such new developments as tracer shells and fuzes, guns to be mounted in naval balloons, and new types of electrical firing circuits. The "Physical" laboratory which tested powder lots for safety, ballistics, and pressure worked completely independently from Patterson's Powder Factory chemical laboratory, which did chemical tests on the powder composition of Indian Head-produced powder as well as contractor lots. Both laboratories studied hundreds of separate lots each year.[24]

When experimenters measured powder in a gun, they still faced the classic "experimental yardstick" problem which had plagued Dashiell and Mason. Holden tried to transform the problem from a difficulty into a virtue. He claimed that up to twelve different types of tests might be combined in one firing of a round from a gun. He looked at the issue as a financial and management question, rather than as a question of research logic. The procedure of running simultaneous tests saved money but created an accounting nightmare, as the cost of the operation had to be attributed in some pro rata fashion to each separate set of experiments. Nevertheless, he liked the obvious efficiency of combining several tests in one event. The same test, Holden claimed, could simultaneously check several of the following: projectile flight, proof of gun, powder lot ballistics, primer lot proof, tracer or tracer fuze, projectile seating, gun mount, sight or lock, fuze, cartridge, or armor plate. While it is doubtful that all such tests could be effectively run on the same fired round, it is clear that more than one item could be measured, as long as the experimenters held several elements standard or they adjusted the data to an established standard, using a carefully calculated "fudge factor."[25]

Holden's motive was clearly to squeeze more research out of the budget by combining several types of data collection on each particular shot. A lack of sensitivity to the multiple interacting factors could lead to serious mistakes, however. In 1916 some of the data assigned to various smokeless powder lots proved wrong when gunners used the powder in practice fleet firing. Following the guidelines established at the Proving Ground, fleet gunnery officers assigned a certain size charge of a specific lot of powder to a major caliber weapon. When they fired salvos with the calculated charge assignments, the shells did not disperse properly, and in some cases the "resulting pressures in new guns were uncomfortably high." Fortunately, no guns blew up from the incorrect assignment of pressure readings, but fleet officers complained of the erroneous Indian Head figures on proved powder lots. There was something wrong with the evaluation of the powders, as demonstrated in target practice. On investigation, Officer in Charge Lieutenant Commander Julius Hellweg traced the problem to the use of worn and eroded guns at the Proving Ground. The guns, as yardsticks, simply could not be trusted to be consistent, especially from one gun to another. Hellweg ordered a series of "powder reproofs" and reassigned new pressures for the doubtful lots, working in an adjustment factor that took into account the eroded guns. Hellweg then assured the Bureau that "the proof of all powders is now being conducted with sufficient appreciation of the effects of erosion so that no material errors in powder assignments should ever be made in the future."[26]

The increasing worldwide naval arms race stimulated more experimental work through the pre-war years. Work for foreign governments picked up, as Proving Ground staff modified and ranged guns for Argentine ships and tested experimental armor for the Italian navy. However, incremental improvements in weapons and shells for the U.S. Navy generated the most experimental work at the Proving Ground. Some of the projects focused on development of special-purpose shells, such as tests on the "audibility" of saluting and signal guns, work on smoke-producing shells, evaluations of shrapnel anti-personnel shells "as used in Nicaragua," and studies of a variety of armor-piercing shells.[27]

Other experimental work ranged over the whole field of ordnance, including tests against conning towers and turrets, studies of angling shots to improve oblique penetration, studies of shell acceleration, and testing of electric primers. In 1914 alone, Hellweg reported on more than fifty categories of experimental ordnance work, including antiaircraft shells, rockets for passing tow lines to wrecked vessels, recoilless guns for mounting aboard aircraft, and work with CO_2 fire extinguishers, and ammonium picrate or Explosive D for explosive shell filler. Explosive D had high chemical stability, was insensitive to shock, and did not combine readily with metals to form more sensitive compounds. Thus it was safe in shells, and such shells could be fitted with delayed action fuzes with another, more sensitive explosive to set off the charge. During the war, the demand, both by the Army and the Navy for this new explosive, would increase vastly. By the time World War I broke out in Europe, Indian Head had already experimented with some of the technologies that would be decisive in that war, including not only Explosive D but also some new bomb, mine, and torpedo designs brought in by aircraft and submarines. In addition, the laboratory at Indian Head would examine a number of minor advances in armor, shells, fuzing, detonators, explosives, and specialized ordnance such as signal shots, rockets, tracers, and smoke screen shells.[28]

Routine ordnance work continued in the period 1911–14, as Holden and

Hellweg ran hundreds of acceptance tests for armor-piercing projectiles each year and "proof" tests for hundreds of separate lots of smokeless powder. After Hellweg's 1916 adjustment for gun erosion, the fleet could begin to trust the charges to fire consistently.[29]

POWDER EXPERT AT WORK

George Patterson continued as the key civilian researcher during the second decade of the twentieth century, serving at once as a general manager of the powder factory plant, director of research, and leading chemical engineer for the expanding and changing powder production facility. Although not operating on a designated experimental budget, Patterson conducted a wide-ranging experimental program and built a team out of the powder factory's funding.

In 1912 Patterson ran the chemical laboratory with a staff which consisted of Assistant Chemist Walter Farnum and eight laboratorians or lab assistants. As the Indian Head Powder Factory expanded, Patterson divided the organization. Farnum moved to run powder production, and Patterson remained as his supervisor while personally running the chemical laboratory. By 1916 Patterson upgraded the laboratorians to assistant chemists and added six laboratory helpers and two clerks. The volume of routine and experimental work they produced through the years of the first Wilson administration increased with the size of the staff and included literally thousands of routine tests and dozens of experimental projects.[30]

Through the pre-World War I years, Patterson worked at the problem of stability of smokeless powder. The French battleship *Liberté* suffered a magazine explosion in 1911, causing concern in American ordnance circles and in Congress, since the French smokeless powder formula was similar to that of the United States. Patterson had worked on two different approaches to the stability problem. In 1905 he had experimented with the addition of rosaniline dye to the powder; eighty-three indexes of powder produced with the dye, designated "S.P.R.," had been manufactured through 1908. The purpose of the dye was to give an indication of the deterioration of the powder by color change resulting from combination with the evolving acid.[31]

In 1908 Patterson recommended dropping the rosaniline dye process and adding diphenylamine to the powder as a stabilizer, following a German innovation. Indexes of powder were designated "S.P.D." after 1908 to indicate the addition of diphenylamine. By 1912 the specifications for American smokeless powder included the diphenylamine stabilizer. None of the stabilized powders that were properly stored had to be retired, some staying in storage for as long as twelve years without loss of stability. Thus, when reports of the *Liberté* disaster alerted line officers to the danger of unstable smokeless powder stored in shipboard magazines, the Bureau of Ordnance could report that, thanks to Indian Head research, American stabilized powder was far safer in magazine storage. The Navy also required ships' officers to run regular shipboard testing programs on a monthly basis, using violet paper supplied from Indian Head and a simplified version of the "German" test which Patterson perfected to verify the stability of powder in magazines.[32]

Patterson had defined the specifications and the tests, had developed and manufactured the methyl papers, and had designed the apparatus for shipboard evaluation. The manufacture of the test papers would remain a minor product of the factory through 1989. Furthermore, he had overseen the production of the powder itself. By the year 1912 Patterson was known as the Navy's "Powder Expert," with some justification. Later Indian Head publicity materials referred to him as "Dr. Patterson," reflecting the fact that in 1932, his alma mater, Worcester Polytechnic Institute, granted him an honorary doctor of engineering degree in recognition of his work over the years as the Navy's foremost "Powder Expert."[33]

Patterson's experimental work in the years before the United States' entry into World War I, like that of the physical laboratory that concentrated on ballistics and gun design, showed a range of interests. Much of the research related to the powder manufacturing process, such as the study of new stability tests, further development of litmus paper tests, and experiments with new pieces of production equipment including settling tanks. A few other tests dealt with new weapons, such as Maxim fuzes, tracer shells, and primers, and product tests of such items as powder bags and shell lacquers. Even before the United States entered the war Patterson scheduled a few studies of British-captured German equipment, such as high-explosive shell and torpedoes, as well as work on British fuzes and powders.[34]

Although Patterson reported on a varied schedule of experimental work through the period, the routine work increased at a far greater rate to keep up with the analyses and tests needed in regular powder production and in testing sample powder from private firms. In 1915 Patterson reported, as routine, thirty-seven hundred complete analyses, including separate tests on smokeless powder samples, cotton, alcohol, spent and mixed acid, ether, and a miscellaneous collection of explosives. By contrast, his experimental projects numbered less than eighty. Although some of the experimental projects would involve a series of separate tests, the sheer numbers suggest that routine work, as defined by Patterson, represented the great bulk of the work of his laboratory through the period.[35]

As Patterson and Farnum worked to increase powder production during these years, the time for experimentation declined. But the increase in production during the decade brought Indian Head to equal, surpass, and eventually replace the private firms as suppliers of smokeless powder to the Navy in the last year before the United States entered the war.

RELATIONSHIP WITH PRIVATE INDUSTRY—THE ROLES REDEFINED

Under Woodrow Wilson and his Secretary of the Navy, Daniels, relations with Navy suppliers changed from the cordial ones of the Roosevelt and Taft years. Theodore Roosevelt and Taft and their Secretaries of the Navy, in the years 1901–12, had viewed the Powder Factory as a supplement to, and a gentle price-guide for, the private sector. In 1908 Roosevelt's last Secretary of the Navy, Thomas Newberry, summed up the relationship. Because of Indian Head, the government, he noted, "may be considered nearly independent of the Du Pont companies." The companies, he noted, "have never taken advantage of the situation to charge exorbitant prices." Consequently, the Navy planned to run Indian Head at less than full capacity, "so that enough work will be left for the private plants—to encourage their owners to maintain them in condition and to retain their valuable forces of chemists and workmen."[36]

At the end of Taft's term, Secretary of the Navy George von Meyer argued for continuing the production of powder by private firms, on the grounds that they "are always making experiments" and that they always reported what they found out. "I am not in favor of the government doing it entirely," he told Congress. The cooperation between large private firms and the Navy had flourished since the days of Tracy and Folger.[37]

But Wilson's secretary, Daniels, was less willing than Newberry and von Meyer to accept the prices set by private firms in armor, shells, and powder. Daniels was particularly incensed when Midvale, Carnegie, and Bethlehem submitted identical bids of $520 per ton for armor. He believed the bids should be truly competitive, and the submission of identical bids struck him as collusive. With straight faces, the officers of the firms asserted that the identical bids were coincidence. When Daniels asked them to resubmit and to avoid "coincidence," they again gave the original bids. When Daniels confronted them with a competing price of $460 per ton offered by a British firm, the three major American firms again all met the new price with an identical bid. Over the next years the firms more or less equally divided the sale of "Class A-1" armor at a set price of $454 a ton. The three firms had practiced the identical-bid system since Tracy's administration in 1891. In the Roosevelt and Taft years, the Navy Department had continued to divide the purchase of armor about equally between the three companies. The virtue of the system was that it kept all three firms in production; if one had submitted a lower bid, it would have eliminated the other two, eventually creating a single-firm monopoly. In a choice between collusive bidding and a one-company dominated field, the Navy had always accepted the identical bids, explicitly recognizing the need to maintain a large-scale armor capacity to be prepared for wartime expansion. But Daniels was outraged and worked to promote a third path: constructing a government-owned armor plate factory.[38]

Daniels threatened to turn all shipbuilding, weapons, powder, and armor manufacturing over to Navy-owned facilities. Daniels "earnestly recommended" that the Navy open its own armor factory and increase the appropriations for the torpedo factory, the gun factory, and the powder factory. In each case he developed plans to transform the institution from a price yardstick and developmental center into a sole-source, government-owned and government-operated supplier. Daniels was as indignant about private control of smokeless powder as he was about the steel companies' pricing. "The department," he reported, "is forced to buy too large a quantity from the powder trust at exorbitant price." By "powder trust," Daniels referred to the E. I. du

Pont holding companies, which controlled all private smokeless powder production. Under Daniels, the Indian Head powder plant would become a major symbol in the debates over private ownership of arms production, profit making from weapons, and the proper role of government-owned and government-operated manufacturing establishments.[39]

In 1912 and 1913, at the height of the antitrust activity of the progressive era, reformers viewed the rise of national corporate enterprises in moralistic terms. Some later historians would come to analyze the formation of United States Steel, du Pont, and other major corporate entities in structural terms, focusing on the long-range efficiencies which derived from economies of scale. Alfred Chandler and others pointed out that such firms could rationalize markets, standardize products, "institutionalize innovation," and efficiently organize supplies and transportation. It was for just such reasons that Benjamin Tracy had worked with Carnegie and Bethlehem in the 1890s, funding the establishment of nickel-steel and Harveyized steel capacities. Close cooperation with du Pont in the 1890s, too, had been based on the need to develop a relationship with a reliable, large-scale manufacturer. In 1912 early defenders of the "powder trust" or the du Pont holding company used very similar arguments. They pointed to the value of a large, diversified, and strong enterprise to the national defense. Critics like Daniels, however, saw the suppression of competition as evil. He believed that, without competition, the du Pont company had begun to charge "extortionate" prices. Thus, the question of what represented a realistic and fair price for smokeless powder lay at the center of the disputes over the "powder trust" and at the heart of the debate over how to organize the twentieth century American industrial system.[40]

In the 1912 election, in which Roosevelt ran on an independent Progressive Party ticket against Republican Taft and Democrat Wilson, campaign rhetoric focused on the question of the "trusts." Wilson generally held in the New Freedom doctrine that the danger of large corporate enterprises was monopolization and unfair price gouging and that the large "trusts" should be dissolved to keep opportunity open for smaller firms; Roosevelt argued for government regulation which would accept consolidation but guarantee fair practices by the new corporate giants. In either view, the question of what constituted a fair price was essential.

The key to determining a fair price for smokeless powder was the cost accounting method, whether developed at Indian Head or in the private firms. In the decade from 1900 to 1910, Patterson had worked out a standard system for determining the cost of the powder and constantly sought means of reducing that cost through process efficiency, manpower management, recruitment and retention of the needed skilled labor, and continued study of American and international manufacturing processes. The cost reductions at Indian Head were real, but the private sector continued to argue that governmental formulas for determining Indian Head powder cost were not properly calculated, particularly with regard to overhead.[41]

Patterson and the Navy did not ignore overhead. In 1911 Patterson included in his cost per pound such items as leave and holiday pay, superintendence and clerical pay, labor and materials expended for repairs, the cost of the experimental laboratory, and a cost figure for electric power consumed. The only major element of "overhead" which Patterson did not include in his costs was depreciation on the capital investment. Had he done so, however, the cost of powder would have been raised only by two or three cents per pound, keeping the government cost at about thirty cents per pound compared to commercial prices over sixty cents per pound. Of course, an imponderable calculation in the overhead question was how to attribute the cost of the "management" staff at the Bureau of Ordnance and at the office of the Secretary. Had a reasonable proportion of their salaries been calculated as part of the national management of the Indian Head facility in order to compare it more fairly to large corporations, those costs might also have raised the price by a cent or two per pound. Daniels and the Bureau of Ordnance went through just such calculations to account for every conceivable overhead charge, including pay, allowances, leave and pensions for officers, and interest on plant investment, raising the adjusted price to just over thirty-four cents a pound, still far below the price charged by private industry.[42]

The careful analysis of the cost of Indian Head powder supported arguments that the du Pont monopoly overcharged. However, the response of the government could take any of several alternate paths: increased reliance on government manufacture, dissolution of the trust, or the use of the Indian Head price as a guideline for bidding. Since the Army and Navy were the only domestic purchasers of smokeless powder, the company needed the market. The actual policy choice represented a blending of the various alternatives.

The year 1912 saw the wrap-up of a five-year Sherman Antitrust case against the "powder trust," leading to the alternative of partial dissolution of the monopoly. Investigators had learned that, over the decade between 1902 and 1912, du Pont had merged with or otherwise eliminated sixty-five rival explosives firms, gaining control of about 70 percent of U.S. explosives production and 100 percent control of all military explosive purchases, through the du Pont holding company. Through an international cartel agreement, foreign firms would not undersell du Pont in the United States, and du Pont reciprocated in their markets. The district court of Delaware ordered the company divided into three separate entities, with two of the former du Pont concerns to engage in smokeless powder production.[43]

A second approach reflected price control through regulation, using the Indian Head figures as a yardstick for calculating a fair price. After a separate congressional investigation in 1912 into the "powder trust," Congress inserted a clause in the naval appropriation act of 4 March 1913 requiring that the Army and Navy accept no bids from private firms for smokeless powder above fifty-three cents a pound; furthermore, the act required that Indian Head produce at full capacity and that the Navy purchase from private firms only the remaining unfulfilled need. Considering that the Indian Head powder had been produced for less than thirty cents a pound, compared with the du Pont prices that had slowly dropped from eighty to sixty cents over the decade, the Navy expected to save millions of dollars.[44]

While retaining the concept that Indian Head would provide the price guide, the practical effect of Daniels' policy was to convert the Powder Factory

TABLE 3

Navy Smokeless Powder Deliveries, July 1913–June 1914

Supplier	Pounds Supplied		Percentage
Indian Head			
New Powder	2,338,448		
Reworked Powder	1,013,940*		
Total Indian Head		3,352,388	50.5
du Pont			
Carneys Point Plant	1,559,358		
Haskell Plant	605,535		
Total du Pont		2,164,893	32.7
International Smokeless Powder & Chem Co.**			
Parlin Plant		1,111,737	16.8
Total Powder Supplied		6,629,018	

Source: U.S. Department of the Navy, *Report of the Secretary of the Navy* (Washington, D.C.: U.S. Government Printing Office, 1914), p. 234.
*Indian Head supplied all of the reworked powder.
**International separated from du Pont by court order, June 1912.

into the Navy's sole provider of smokeless powder for a brief period. The rapid move from private monopoly to government monopoly, followed by a return to joint private and naval production, is revealed in a review of the Indian Head powder production through the Wilson-Daniels years.

POWDER PRODUCTION

Admiral Twining, chief of the Bureau of Ordnance, inquired what facilities would be needed at Indian Head to supplant private production entirely. Patterson and Holden investigated the issue, and Holden assured Twining that with sufficient appropriations, in about two years the Navy could make all its own powder.[45]

During 1912 and 1913 Patterson and Farnum rebuilt the powder factory for the required expansion. By 1914 the Navy could report that Indian Head was manufacturing more than half the Navy's purchases, as shown in Table 3.[46]

As Indian Head stepped up production, the major challenge facing Patterson and Farnum was similar to that which had faced Joseph Strauss more than a decade before—maintaining full production while retooling. By 1914 Indian Head production of new powder climbed to over 2.3 million pounds and reduced the cost to less than twenty-eight cents a pound. The next year the figures were over 3 million pounds at less than twenty-six cents a pound.[47]

Soon, Daniels could report that private contracts for powder had been further reduced. During 1914–15 the Navy purchased 3,112,868 pounds from the Carneys Point, Haskell, and Parlin private plants, but Indian Head supplied 3,984,978 pounds or 56.2 percent of the need. The Navy left only one contract with du Pont outstanding, for 790,000 pounds, while du Pont began to fill European orders.

TABLE 4
Powder Production 1910–1919, Indian Head

Fiscal Year	Reworked Powder	New Powder	Total
(July–June)			
1910–1911	760,486	1,041,648	1,802,134
1911–1912	968,632	1,467,281	2,435,913
1912–1913[a]	964,267	1,823,369	2,787,636
1913–1914	1,013,940	2,338,448	3,352,388
1914–1915[b]	887,403	3,097,575	3,984,978
1915–1916	893,076	3,327,329	4,220,405
1916–1917[c]	459,829	6,020,989	6,480,818
1917–1918	334,933	5,977,681	6,312,614
1918–1919[d]		799,243	799,243

Source: Typescript annual reports, 1911-19, in RG 181, Acc. 9959, 2103-1, Philadelphia National Archives Branch (PNAB).
[a]Inauguration of Wilson, March 1913
[b]WWI began, Europe, August 1914
[c]U.S. Declaration of War, April 1917
[d]Armistice, November 1918

The government's order with du Pont remained unfilled over the next year, and the Navy's *total supply* for fiscal year 1915-16, just over 4.2 million pounds, all came from Indian Head. Patterson's prediction in 1912 that the Indian Head Powder Factory could make all the Navy's powder had come true, in one sense. But that accomplishment was achieved only because the total Navy purchase was reduced from over 6 million pounds to about 4 million pounds. By 1916–17 Indian Head production almost equalled the total of both public and private production in 1913–14. By that time, however, the nation was at war, and the old consumption predictions were inadequate. The statistics of production at Indian Head through the years are presented in Table 4.[48]

As can be seen from the raw figures, Patterson and Farnum doubled production of new powder between 1910 and 1914, and then more than doubled it again by 1917. Reworked powder was older powder which had become destabilized; the guncotton was soaked and washed, then re-mixed with new stabilizer and released in separate lots as reworked powder with a special designation. When reworked powder is included in total production, the output of 1917 was over three and a half times that of 1910. With the rapid consumption of powder during the war period, April 1917 through November 1918, there was less old powder available for reworking, and Indian Head, like du Pont, ran around the clock to meet the demand through new powder production.

Regardless of the difficulty of developing a true comparison between the cost of powder produced in the government-owned facility and in a corporate-owned plant, Indian Head served as a keystone in the Wilson-Daniels progressive critique of the corporate structure. In the case of smokeless powder, the companies refused to be guided by the government standard, continuing to produce powder at sixty cents a pound and then finally selling some to the Navy at the legally mandated limit of fifty-three cents a pound. The Navy under Daniels had quickly moved to replace the private production, using the Indian Head factory. For one year, 1915–16, that goal had been achieved. After that, the demands of war put Indian Head back in the role of supplementing, not substituting for, the private sector.

The Navy's plan to establish a government-owned armor plant to replace similarly the Carnegie, Midvale, and Bethlehem suppliers was based upon a close study of the manufacturing methods of those firms. Again, working out a careful cost accounting system, the Navy calculated that it could produce "A-1" nickel-steel armor plate for about $100 per ton less than the private firms. As the Navy worked to put the factory into production during the war years, the three firms continued to submit identical or nearly identical bids. However, the Navy did not complete the armor plant in Charlestown, West Virginia, until after the war, and it never got into competitive production.[49]

INDIAN HEAD GOES TO WAR

The first doubling of powder plant production came not as a wartime project but as a consequence of the dispute over the cost of privately produced powder in the years 1912–14. The second burst of production came in response to the demand in 1917, as the war used up ship supplies and reserves, and Indian Head, as well as the private firms, hastened to fill the Navy's needs. Most of the other aspects of expansion at Indian Head during the decade can be traced to the effect of the war. However, much of that expansion came too late to be of immediate value.

Maintaining an official stand of strict neutrality, Wilson and Daniels hesitated to mobilize private companies during the period of official American neutrality, 1914–17. While adding ships to the fleet, the administration did not address the larger question of organizing the private industrial sector. Assistant Secretary Roosevelt preferred a more activist approach and a more organized mobilization, but official policy did not sponsor any major upgrading and increase in industrial capacity for war production through 1915 and 1916. The nationwide "preparedness" movement went forward without official blessing and in an uncoordinated way as American firms filled European war orders. Thus, when war came to the United States, overnight expansion and attempts to adjust the lack of coordination brought the economy to a near standstill.

At Indian Head, as in other sectors of the military and industrial establishments, rapid attempts to provide manpower, new transport and utilities, new factories, raw materials, and weapons fell short of expectations. The creation of a host of boards and committees through 1917 and 1918 slowly brought order to the effort. On the whole, the wartime-induced increase in capacity did not come into full readiness until the war was nearly over.

Captain Henry E. Lackey, officer in charge from 1917 through 1920, transformed the Indian Head facility, at once improving the Powder Plant and finalizing the removal of gun work. In his efforts to attract and hold the needed skilled labor force, Lackey would improve material conditions at Indian Head, finally addressing a range of emerging social problems about which his predecessors

TABLE 5
Testing Program, Indian Head, 1915–1919

Type of Test	Number in Fiscal Year			
	1915–16	1916–17	1917–18	1918–19
Guns:				
Major Caliber	43	50	33	44
Intermediate Caliber	243	299	851	3,063
Minor Caliber	85	150	235	344
Total Guns	371	499	1,119	3,451
Smokeless Powder:				
Pounds	297,094	338,779	558,168	831,033
Lots	150	212	263	522
Primers, in Lots	427	1,500	1,789	1,970
Cartridge Cases, in Lots	292	346	*	1,271

Source: *Naval Ordnance Activities, World War, 1917–1918* (Washington, D.C.: U.S. Government Printing Office, 1920), p. 250; derived from Report, August 8, 1919, RG 181, Acc. 9959, 2125-1-3, Philadelphia National Archives Branch (PNAB). Data for 1915–16 added and 1916–17 corrected from published version by reference to typescript "Annual Report Fiscal Year 1917," dated August 29, 1917, in RG 181, Acc. 9959, 2103-1-2, PNAB.
*Information not available.

since Mason had complained but which they had failed to resolve directly, including safety, housing, health, education, and communications concerns.

The war itself brought on a rush of wartime tests of guns, powder, primers, and cartridge cases, that would exacerbate the river-safety issue. The increase in testing is shown in Table 5.

The increase of work created physical overcrowding as well as raised hazards. Within a few weeks after the declaration of war on 4 April 1917, Lackey ordered a new battery and a new magazine constructed. The testing crews worked on construction, adding new areas for parking guns, new platforms for shells, and a new target butt which could withstand repeated armor tests. Lackey regarded the new target butts for 14-inch armor-piercing shell as the greatest single improvement in the first months of the war, pointing to the fact that they could speed up the whole testing process. The older butts, as in Dashiell's day, had to be rebuilt between tests, holding up firing while the work crews struggled with the heavy timbers.[50]

Lackey's approach to the river safety issue was forthright and solved that problem, at least for the duration of the war. After April, he obtained an order by the War Department closing the river for certain hours of the day. Lackey noted that otherwise it would have been impossible to keep pace with the proof work brought on by the war.[51]

The increase in routine proof testing, together with a continued program of experimental work, required a reorganization of the testing force. Lackey named Lieutenant Commander C. L. Lothrop as proof officer and Lieutenant Commander A. G. Kirk to head the experimental program, administratively recognizing the distinction between experimentation and routine work. Experimental work continued on some of the pre-war lines and branched into new areas. Kirk conducted experiments on machine guns, howitzers, anti-aircraft guns, and Davis torpedo guns. New shell work included smoke, line-carrying, 16-inch armor-piercing, and "asphyxiating gas" shells. Kirk ran tests on new fuzes and explosives and a wide but miscellaneous collection of aircraft bombs, incendiary bombs, tracers, gas masks, depth charges, and experimental armor. By the simple expedient of delegating administration of routine and experimental work to two different officers, Lackey was able to increase the capacity of the Proving Ground in both areas.[52]

Despite the fact that the Daniels plan to supplant private powder production with the Indian Head factory had largely succeeded, the adjustment of the factory to the war was, like the adjustment of the Proving Ground, a story of too little too late. In 1917 the plant faced a series of difficulties, according to Lackey: the factory had just emerged from the disruptive effect of the expansion of facilities; transportation was barely adequate to the needs of the station; and the peacetime complement of marines was too small to provide an adequate guard against "marauders." The fear of marauders may have been a war-scare induced phobia, but the other problems were genuine enough. War contracts with foreign countries had drained skilled labor away from Indian Head to private firms of all kinds.[53]

When the Navy attempted to place new contracts with the private firms for powder, it discovered that the private companies had already stretched their capacity to meet foreign sales. The Bureau planned to expand Indian Head powder production once again, from the range of twenty thousand pounds per day to eighty thousand per day. While attempting to expand production, the chemical laboratory force had to mount an increased inspection program for purchased powder. Raw material supply also became a problem. Nitrate of soda ordered from Chile was diverted to allied countries, and the Navy planned a nitrate plant at Indian Head to fill the need. No sooner had construction started on the plant and at the far larger, electric-powered air-reduction facilities at Muscle Shoals, Tennessee, than the armistice loomed. The Navy ordered the Indian Head nitrate plant construction stopped on 8 November 1918, after the armistice with Austria and before the armistice with Germany.[54]

Even with the labor shortages which held back construction, Lackey made a few notable additions to the facilities. He ordered a Cyclone fence built around the powder factory and had the marine guard increased. In 1917 he opened an experimental plant to produce ammonium picrate, or Explosive D, the stable explosive useful as projectile filler. By June he could report production of D up to about 500 pounds a day, despite the usual problems of dealing with unskilled labor and breakdowns in machinery. The next year Farnum reported production of more than 1,000 pounds of Explosive D a day, with a yearly rate of about 240,000 pounds. Aside from the new D works, however, no additional factory facilities were completed after April 1917 and before the armistice on 11 November 1918. Lackey attributed the continuous delays in construction to the difficulty of getting construction materials delivered, as well as to the labor shortage. He had planned for a railroad link, but it too was not completed in time to assist in bringing in materials for the war effort.[55]

Valued employees left, not simply because of the general opportunities but because the "rival powder and acid companies," as Hellweg called the private firms, actively recruited the oldest and most experienced men. These in turn would return as recruiters for their new employers. In the period 1915–16, 333 employees out of a total average force of 805 resigned for better jobs. Some of the groups walked off *en masse*. "One morning," Hellweg reported, "at about 8 o'clock, the watch at sulphuric acid plant walked out. We were compelled to keep the leadingman and others from the watch on till noon next day till other men could be brought in to relieve the watch." Another time, the nitrators walked off at midnight, leaving the plant "crippled."[56]

The Powder Factory suffered from the labor shortage, particularly as Patterson and Lackey searched for men to fill the "common laborer" rating. The problem had several causes, Lackey believed: the generally great labor demand nationwide, the higher wages elsewhere, the "unsatisfactory living conditions" at Indian Head, and the fact that contractors engaged in construction work at the Station absorbed much of the local labor supply, leaving few for the Powder Factory operation.[57]

While Lackey could do nothing about the nationwide factors, and very little about the pay scales, he could deal with the question of living conditions. Lackey noted in 1917: "First and foremost the Station suffers from not having a village near large enough to accommodate the force necessary to carry on our work."[58]

Lackey was embarrassed to note that "men from laborers to mechanics, clerks and assistant chemists have reported for work but upon looking over the living conditions as to quality, sufficiency and cost of same have declined appointment and left on the next boat." The turnover of personnel, even when not the next day, meant that the facility had become a "training station," where staff left once they had been "broken in."[59]

To help solve the housing problem, Lackey ordered an ambitious construction program of a new hotel, several barracks for civilians, and some new homes for both officers and civilians. He sought Bureau approval for establishment of "messes" or eating halls run by clerical and chemical staffs for their own use. All of these measures slightly helped to relieve the situation. In the winter of 1917–18, some of the enlisted men stayed in tents; by mid-1918, temporary barracks near the marine facility on Stump Neck provided a solution of sorts. But Lackey's major solution came to bear the next year. He developed a plan for one hundred new homes, laid out in a village pattern, to be built on the "Irving farm" area, that had been acquired in 1890 and which lay along the boundary line between the station lands and the small community of Indian Head, that developed near the gate. He planned a post office, a schoolhouse, and a set of dormitories. The housing project was put up and operated at first by the U.S. Housing Corporation, a civilian wartime agency established to relieve the housing shortage near shipyards and other mushrooming industries. The school was completed, and most of the housing was ready in the fall of 1919. Again, a major project came to fruition after the war was over, but in this case the project was finished and used immediately. The construction of housing for civilians on the military base itself broke precedent; the on-base housing built up since Dashiell's day had been intended for naval officers, and the few civilians resident there, like Farnum, were regarded as special exceptions. To construct a complete civilian community, with planned school and library, would soon raise questions about police jurisdiction, taxation, and self-government, none of which fit into the Navy's usual way of managing shore establishments, where residence facilities were usually restricted to men in uniform and their families.[60]

However, Lackey hoped to "build up a real community spirit and to make the Proving Ground one of the most attractive industrial communities under the Bureau of Ordnance." To that end he set up a community council which would elect representatives to discuss problems, particularly those growing out of the government-owned housing project. Like the manufacturing plans of the war years, the social planning did not come to fruition until after the war was over.[61]

Even with the housing under construction, the personnel question became desperate. By June 1918 Lackey attempted to calculate the shortage that had been experienced over the year since June 1917. He calculated a total "shortage" of 268 men. His method was to indicate the number of positions advertised for and the number of men who reported, with the difference representing a shortage. The greatest disparity between demand and supply was in the common laborer category, where only 281 men reported for 430 advertised positions; but shortages came up in a whole variety of skilled trades, such as leadburners, machinists, firemen, and sheet metal workers.[62]

Despite the headaches with personnel turnover and construction delays, Patterson's workload in the chemical laboratory increased vastly with the war, raising the routine work to over ten thousand separate analyses a year. Like the naval officers under Lackey, Patterson continued to mount a hectic experimental program. In the first six months of the war, he reported on experiments with Maxim fuzes and with ammonium perchlorate, which was an experimental explosive from Britain. He studied captured German guncotton, Russian boosters, and zeppelin cartridges. He worked with aircraft bombs, torpedo charges, antiaircraft shell, improved daytime tracer shell used in aircraft firing, and a wide range of other explosives, detonators, and powders. By 1917 Patterson's laboratory staff climbed to an average of ten assistant chemists, ten laboratory helpers, and two clerks, although some of the personnel kept moving in and out through the job slots.[63]

In retrospect, the increase in powder production achieved by Patterson and Farnum during the decade seems a decent accomplishment. In the light of the quadrupled output between 1911 and 1917, the pronounced disappointment of both Lackey and Patterson at not reaching a ten million pound goal in 1918 is at first difficult to understand. But the planned additions in capacity, which were contracted for and partially built in 1917–18, would have boosted production to the planned goal had they been finished on time. Rather than resting on their laurels with the achievement of a six million pound capacity, they built towards a much higher goal. The disappointment sprang from seeing the construction almost completed in time.[64]

The planned railroad—a twelve-mile link to White Plains—like many other improvements brought on by the war-induced expansion, came too late to aid in the war effort. Once established, the Navy-operated spur railroad could bring in heavy equipment, but it never operated as a passenger link. Lacking a decent highway, personnel continued to travel to and from Washington by river boat, as they had since Dashiell's time. The contract for constructing the railroad was issued 15 July 1918. It opened six months after the Armistice, 29 May 1919.[65]

EVOLUTION OF MISSION

Although officers in charge had advocated a change in the proving ground range and its separation from the Powder Factory as early as 1901, the funding for a new proving ground did not come until the war years. The Bureau purchased land for the Lower Station, another twenty miles down the river, on the Virginia side. Dahlgren included 1,463 acres and allowed a "clear water range" of nearly ninety thousand yards, or over forty miles, down the Potomac nearly to the Chesapeake Bay. Construction began on 28 May 1918, and proof of some guns began there in 1920. Construction at Dahlgren began well before the Armistice, but complete transfer of gun work from Indian Head to Dahlgren could not be scheduled until August 1921. The Dahlgren facility then operated for over a decade as the Lower Station of the Indian Head facility.[66]

In nearly every respect, the attempt to improve hastily the Indian Head facility to meet the expanded needs of the war did not succeed in time. Powder production, already increased because of the antitrust policies of Daniels, could not be expanded to the scale demanded by war needs. The housing project, the railroad link, and the move of the proving ground to a new location, while they could be initiated during the war, could not be completed because of labor and material shortages. The only wartime expansion of facility which went into full production during the war was the Explosive D plant. Behind that major success story were hundreds of day-to-day accomplishments in the study of guns and powder. However, the chaos of the nineteen months of American participation in the Great War prevented effective massive upgrading of powder producing capacity.

Yet the legacy was there. When the Lower Station opened for gun work, it marked a stage in the evolution of mission which had begun in 1900. Indian Head had been left with a fenced-in, expanded powder plant, a separate Explosive D factory, a developed research capacity for both experimental and routine work in chemistry and ordnance, a rail link to Washington, an on-base village of homes, and a school. The shift of the proving ground, together with the laboratory expansion, had redefined Indian Head; no longer the place to test guns, it had been converted into a factory and a laboratory, with its own civilian housing. Lackey's solid administrative accomplishments during the war, even though fraught with disappointments, had addressed all of the fundamental problems of safety, security, transport, social conditions, and research organization. The years which followed would bring naval disarmament, the Great Depression, and gathering war clouds in Europe. Survival, adaptation to new research priorities, and preparedness would be the challenges of the coming decades.

CHAPTER 4

Indian Head in Ordnance Research, 1921–1941

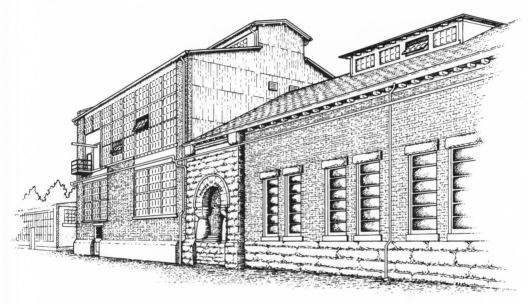

Powder House
Building 111
1899

F OR UNITED STATES naval ordnance research, the decades between World War I and II were bracketed by two dates, punctuated by the sounds of guns and bombs. With the move of the proving ground to Virginia, the last proving round was fired at Indian Head on 21 July 1921. Coincidentally, on the same day, less than one hundred miles to the east, in the Chesapeake Bay, Billy Mitchell bombed the captured German battleship *Ostfreisland*, testing the effect of aircraft bombs against modern warships. Twenty years later, war had broken out in Europe again, but this time the United States rearmed while remaining officially neutral. American anti-aircraft weapons would be put to the test of combat on 7 December 1941, as American seamen fired at carrier-launched Japanese planes from the ill-fated ships anchored in Pearl Harbor.

Between those two dates in 1921 and in 1941, the 1920s and 1930s at Indian

Head outwardly seemed a period of quiet doldrums. The effort to expand facilities in World War I had only partly succeeded. Despite the opening of the railroad link to White Plains in 1919, the Station remained isolated, with a two-hour trip by launch upriver being the common means of transport to Washington. Easy and regular auto and bus travel to Washington did not begin until the opening of a new highway in 1942. Low budgets, naval disarmament, a restricted demand for powder, and international and national policy all boded ill for the facility. Furthermore, the move of the proving ground to Virginia threatened not only jobs on the Station, but the newly incorporated village of Indian Head, with its retail establishments and scattered housing. The town that had grown up outside the gates clearly depended upon the Station for its existence. Townspeople and employees petitioned Congress in 1921 for the return of proof testing from Dahlgren. Local politicians, such as Congressman Sydney Mudd from Charles County, supported the petition effort, but to no avail. When the last echoes of the guns died, the flow of government payroll threatened to fade away as well.[1]

The community of Indian Head survived, however. A tightly knit society emerged, with the long-term civilian Powder Factory employees at its heart. The story of the town itself is the focus of the next chapter.

The survival and success of the naval station at Indian Head during the interwar years was not the result of political activities by the Station employees and their advocates, however. The Station's survival reflected the fact that the institution had begun to find a role for itself, as one part of a loosely coordinated research establishment in naval ordnance. By later standards that coordination was primitive. Even so, there was a pattern in the apparently random collection of ordnance research items taken up by the chemists at Indian Head and proof officers at Dahlgren.

Ordnance innovation in the interwar years lacked some of the drama of the preceding decades and of the decades that would follow. In the 1890s the developments of smokeless powder, improved armor, and rapid-fire guns had made that period one of revolutionary ordnance innovation. The two decades from 1900 to 1920 saw the creation and expansion at Indian Head of a Navy-owned and Navy-managed factory to produce propellant, another major step forward. Later, the period during and after World War II would see the beginnings of the revolutionary shift to the rocket to replace powder and gun as the major surface-ship system for delivering warheads.

The 1920s and 1930s, however, were a period of far less spectacular progress. Rather than "leaps forward," a set of smaller and less highly publicized advances characterized ordnance in the period. As in all technological fields, periods of revolutionary ordnance innovation alternated with relatively slow periods of incremental change, of regular and gradual improvements within the limits set by the last revolutionary advance. Thus, from a technological as well as a human perspective, the interwar period at first would appear to be one of little drama, of quiescence and survival of a sleepy backwater. Some of the memoirs and internal histories of the Indian Head facility convey just such an impression—a closed proving ground, silent guns, boarded-up housing, and an underutilized and increasingly outmoded plant.[2]

However, a close look at Indian Head during this period of incremental advance shows quite a different picture. The success story of the decades was not simply survival but the beginning of adaptation to a more complex world of technology and significant progress on a variety of weapons, as the Navy sought to arm a balanced fleet that would include the relatively new submarines and aircraft as well as the traditional surface ship. Weapons change took place through modifications and models—"mods" and "marks" in naval nomenclature. The Navy adapted to a world in which assigned staff made policy decisions, allocating scarce money to fund specific, sometimes very minor improvements. During the same period, in the private sector, major corporations attempted to establish laboratories to produce scheduled, even annual, model changes and improvements in product. The Bureau of Ordnance moved in that direction, and Indian Head reflected the change.[3]

After 1920, individual inventors rarely attached their names to ordnance developments. There were two reasons for this shift from the individual to the organization in weapons progress. One was that incremental advances were by their nature less striking and less identifiable. A modification from one mark or model of a weapon to another, and then to a third and fourth might be quite minor, and the evolution of a device through successive changes, rather than through a sudden invention, would commonly require input from many individuals and teams over an extended period.

A second reason for the growth of an organizational rather than an individual approach was the interdisciplinary nature of many of the ordnance improvements in the era. A successful innovation required people with knowledge of propellant chemistry, ballistic physics, machining, shop techniques, practical gunnery, metallurgy, and sometimes other disciplines such as optics, photography, or electrical engineering. In the 1890s Robert Dashiell had a working knowledge of such varied areas. But by the 1920s the advances in all ordnance technical fields and the required degree of specialization required to master each, generally precluded the "renaissance man" approach of the individual after World War I. In these decades the Bureau of Ordnance rarely associated an individual with a particular improvement, but more often regarded the improvements as the product of one or more teams. Consequently, the history of the period is less one of particular individuals but of groups, laboratories, facilities, and projects. Rather than inventions, the Bureau sought modifications; rather than inventors, it sought team players. Those teams were scattered in several locales in a loose network. That network would survive and play a continuing significant role when the cycle turned from incremental change to revolution and back to incremental improvement in later years.

The network had no clear structure. The Bureau of Ordnance in the 1920s assigned projects to Indian Head which dealt with propellants, explosives, and special purpose devices such as signal pyrotechnics. Indian Head studied weapons without regard to which "platform" carried the weapon, whether surface ship, aircraft, or submarine. Nor were the various parts of the innovative process clearly separated, as under a later generation's "systems approach," into pure research, exploratory development, prototype development, production engineering, and test and evaluation. Sometimes one, sometimes another laboratory would take up various aspects of the ordnance research process.

Coordination, such as it was, came at first from the "Powder Desk" at the Bureau—Desk F, and then, after 1920, from the "Research Desk"—Desk Q. Veterans of a later naval organization might not recognize the ancestors of program offices of the Naval Sea Systems Command in these ordnance "desks," but Desks F and Q made decisions as to which projects in ordnance improvement should be studied and ordered the work at the various facilities.

The Bureau of Ordnance decided which weapons should be worked on, and where. An examination of the forces that drove the Bureau and its choices of priority will show how a combination of experience, technology transfer, and applied science worked together. Solutions often took years to find, but the cumbersome system of ordnance research produced progress. The choices and the forces behind them shaped three related research facilities which would later grow into essential parts of the Navy's ordnance research and development capacity: Indian Head, Dahlgren, and the Experimental Ammunition Unit (EAU) which became the Naval Ordnance Laboratory at White Oak. Indian Head was the direct parent of the other two institutions. The three facilities took the first steps towards different identities as they responded to the research agenda.

FACTORS SHAPING THE RESEARCH AGENDA

The twenty-year period was one of intense planning and uneasy weapons development by all the major powers, despite international efforts to stave off another world war through such devices as the League of Nations, the Kellogg-Briand Pact, and naval disarmament conferences. Planners in Britain, the United States, and Japan nevertheless recognized that if war came, the potential enemies would have made vast improvements in weapons. But which weapons? In an era of scarce funding, all major powers had to make difficult choices as to exactly which lines of improvement to fund, which goals to pursue. The American choice, particularly in thinking of a potential war with Japan, was to improve on traditional weapons and to achieve a balanced fleet which would incorporate long-range "fleet" submarines and aircraft carriers. Development of aircraft, bombs, anti-aircraft weapons, and air doctrine went forward vigorously in the years between the wars. Such strategic thinking would trickle down to the decisions shaping the work agendas at Indian Head, Dahlgren, and the EAU.

Perhaps the most notorious debates over new weaponry in the United States during the interwar period were those centering on the colorful advocate of a unified air command, General Billy Mitchell. It was clear that the Navy sought to develop its own strong air arm, building on the experience gained in World War I, and to research the various ways in which faster and better planes would change the nature of sea warfare, as well as to develop gun defenses against aircraft attack. Within the Navy, Admiral William A. Moffett in the Aeronautics Bureau, Admirals William S. Sims and Bradley A. Fiske in the Line, and chiefs of the Bureau of Ordnance, including Admirals Ralph Earle, Claude Bloch, and Harold Stark, all wanted to develop doctrine on the future employment of such weapons. Earle, Bloch, and Stark had each put in time as officers in charge at Indian Head before their service as chief of the Bureau.

While developing the new air arm, such ordnance officers who ran the Bureau continued to order research on guns and armor. There were solid and particular lessons to be derived from the war, and especially from the battle of Jutland, 31 May to 1 June 1916. Jutland had represented the kind of battle central to the U.S. Navy's strategic thinking since Mahan. The battle had been a clash between the main battle fleets of two major enemies, Britain and Germany. Although much of the German fleet returned somewhat unscathed to harbor, the battle had established British surface domination and had shown the wisdom of the strategists' concern with the ability of a main battle fleet to engage the enemy's main fleet. In particular, specific aspects of the engagement yielded lessons about gun and armor performance, not only for the British but for the American Navy.

In the period between the wars, gunnery officers re-fought, over and over, the battle of Jutland, attempting to draw lessons from this first battle of modern dreadnought against dreadnought. The lessons they learned would later be applied in World War II. In that later war there were three great battle fleet engagements between the United States and Japan. Although at Midway and the Phillipine Sea engagements, aircraft would prove most crucial, significant big-gun duels at Leyte Gulf put to the test the work of gun, shell, and fuze designers in the interwar years. Similarly, guns played crucial roles in night actions in the Solomon Islands, and the 5-inch, 38-caliber, multipurpose gun proved effective as an anti-aircraft weapon in hundreds of engagements. In the island-hopping campaigns of the Pacific theater, as well as in landings in Europe, the gun, whether major or minor caliber, proved essential in preparing the way for amphibious invasions.

During the interwar years, Navy planners sought not only to integrate the submarine and the aircraft into the fleet to achieve a balance but also continued to work on the traditional weapon that could prove decisive in main battle fleet engagements—the gun. Recognition that ship-borne guns would also play a vital role in defense against air attack led to developmental work on the 5-inch and smaller caliber weapons. Quietly, without the publicity that attended the bombing tests conducted by Mitchell or the newsworthy launching of aircraft carriers, incremental work on improvement of guns, shell, fuze, and propellant would continue, eventually to prove vital in the next war.

Knowing the major role of both submarines and aircraft in World War II, it is difficult to imagine the significance attached to Jutland by ordnance specialists in the interwar period. A review of a few of the elements of that battle shows how it could reveal needed improvement in ordnance. Jutland had engaged capital ships against each other and represented the test of nearly fifty years of development of steel ships, major armament, and armor. Modernization of the gun clearly could benefit from a practical, blow-by-blow study of that great naval engagement.[4]

Some of the specific ordnance problems revealed by Jutland concerned magazine and loading arrangements. Several of the British ships had been destroyed as a chain reaction of exploding charges down through the ammunition elevators ignited the magazines. Both the British and the Germans made improvements in fire doors and elevators during the war, following the spectacular explosion of magazines during Jutland.[5]

Although the British had practiced in peacetime gunnery exercises at ranges under twelve thousand yards, both navies at Jutland fired at ranges from sixteen thousand to nineteen thousand yards. Practice ranges would have to be increased; accuracy of fire at the longer ranges required new training and new techniques.[6]

Jutland revealed, as no other battle could, that German fire control was superior to British, and that the Germans could score a better proportion of hits. German gunners got off more shots in shorter periods than did the British. From the British point of view, the battle proved the need for better speed of firing, better aim, and reduced dispersion of shot. Furthermore, after Jutland, the British were disappointed to learn that most of their armor-piercing shells exploded on contact rather than penetrating before explosion. The British concluded that the angle of impact was the problem, since hits against armor at great range usually impacted at an angle to the armor. British fuzes, like American fuzes, had been designed for right-angle impacts. A study of battle damage showed that 15-inch shells did not penetrate even 6-inch armor when their impact was as much as twenty degrees from a right angle. Thus, improvement of penetration and prevention of explosion on contact became other high priorities in the British fleet, and here, too, American ordnance designers pursued the same goals during and after World War I. The very establishment of Dahlgren reflected the U.S. Navy's recognition that post-Jutland guns would require ranges much greater than available at the old Indian Head Proving Ground.[7]

More central to the problem of their poor record, the British sought means of reducing dispersion of shot and improving fire control. Practice rounds would have to be loaded with identifying dyes and would have to explode on impact with water in order to improve drill, so that "shorts" and "longs" could be properly identified, especially when several batteries fired on the same target. The causes for dispersion of shot would need to be identified and technical means found to produce closer grouping of shots.

Still other "lessons" of the battle of Jutland that could apply in future surface actions had impact on ordnance objectives in the interwar period. During its withdrawal, the German navy conducted a successful night action using a combination of fire-controlled searchlights and star shells. The British worked to perfect illuminating shells. Again, the American Bureau of Ordnance followed suit. The American author and professor of Naval Architecture at the Massachusetts Institute of Technology, William Hovgaard, explicitly emphasized in his major text on naval ships, published in 1920, that the Germans had been the first to use star shells. Another student of naval warfare pointed to successful German use of smoke screens and stressed the need to improve that old technology.[8]

At Jutland and in land engagements, German smokeless powder fired at night did so with only a "dull red glow" rather than the blinding flash associated with British and American grades of smokeless powder. Such a consideration was more crucial to the Army, which often engaged in night artillery barrages, than to the Navy. Naval gun exchanges at night had been rare, since moving enemy ships could elude targeting. However, the Germans' successful combination of star shells, searchlights, and relatively "flashless" powder, showing that capital ships could score hits on enemy ships at night, stimulated both British and American pursuit of a propellant which could be fired at night without temporarily blinding the gunners. American naval interest in flashless powder would be revived later, in World War II, when the development of radar gun control made night firing even more reliable. However, in the 1920s the Navy was less interested in flashless powder than the Army, and by 1928 the Bureau decided not to devote limited resources to this product any further. The reason for this decision was that in order to reduce flash, there had to be some increase in smoke; smoke remained "objectionable" until the advent of radar.[9]

SPECIFIC PROBLEMS: LESSONS APPLIED

The degree to which the concerns raised by Jutland shaped the research objectives of the American Bureau of Ordnance in the early 1920s is revealed by the specific agenda established at Desk F, the Powder Desk, of the Bureau of Ordnance, and then transferred in the postwar years to Desk Q, the Research Desk. Desk F put together a list of problems calling for ordnance study by the EAU. Examination of the specific problems reveals not only the day-to-day research concerns of the specialists at that unit but the priorities established by the Bureau for the researchers at both Indian Head and Dahlgren.

Immediately after the war, the Bureau of Ordnance established the EAU with the idea of continuing research projects which had begun during the war, some of them dispersed through a variety of contract and university laboratories. The Bureau created the EAU to continue chemical and mechanical weapons research conducted during World War I at the American University Station of the Chemical Warfare Service, located on the university's campus in Washington, D.C. The Navy had taken the laboratory from the Bureau of Mines for the duration of the war, intending to transfer personnel from the American University group to make up the EAU; but the organization was taken by the Army and transferred wholesale to Edgewood Arsenal, Maryland. The Edgewood move delayed the Navy's plans to replicate the wartime research group until late in 1919.[10]

The Bureau first established the EAU at Indian Head, but during 1919 and 1920 the mechanical side of the research was transferred from Indian Head to the newly completed Mine Building in the Washington Navy Yard, with its better access to shop equipment. For five years the EAU was split between its chemical branch at Indian Head and its mechanical branch at the navy yard. In 1924 F. F. Dick, the EAU chemist, transferred from Indian Head to the navy yard. However, many aspects of the chemical research begun under the EAU continued in the Chemical Laboratory at Indian Head after 1924 as part of the "experimental" work conducted by George Patterson. Similarly, the proving ground after its shift to the Lower Station at Dahlgren continued to test experimental products developed by the EAU. Later, the EAU at the Washington Navy Yard would form the core of personnel transferred to White Oak, Maryland, where the Naval Ordnance Laboratory was established during World War II.[11]

Most of the first eighteen specific problems assigned to the EAU in October

TABLE 6
Experimental Ammunition Unit Problem List, 1919

1. Safety Adapter for Mark 7 Fuze
2. Safety Chuck for Mark 7 Fuze
3. Burning Time of Short Powder Trains
4. Energy Decrement Device
5. Variable Delay Detonating Fuze
6. Akimoff Delay Detonating Fuze
7. Aircraft Float Light
8. Light and Smoke Pyrotechnic Shell
9. Marker Shell
10. Water Impact Fuze
11. Spotting Projectile
12. Aircraft Smoke Bomb
13. Smoke Shell
14. Day and Night Tracers for Anti-aircraft Shells
15. Marker Float for Use Aboard Ship
16. Combined Colored Smoke and Light Pyrotechnic Shell
17. Star Shell Mixture
18. Super Quick Detonating Fuze for Spotting Shell

Source: Naval Ordnance Laboratory, *History of the Naval Ordnance Laboratory, 1918–1945: Scientific History*, vol. 3, (Washington, D.C.: United States Navy Yard, 1946), report no. 1000, narrative history no. 131c.

1919 on its founding, and which continued on the ordnance agenda through the 1920s, closely reflected the "lessons," or at least the questions raised, by the battle of Jutland. Without Jutland in mind, the list would look like a random collection of odd ordnance projects; in the light of Jutland, the list was a timely attempt to resolve some of the leading ordnance problems of the time. The original October 1919 agenda for the EAU is presented in Table 6.

Four of the problems represented continued work from the American University projects: the pyrotechnic mixture for marker shell and aircraft float light, the development of marker shells and the float lights themselves, and the combined colored smoke and light pyrotechnic shell. When establishing the EAU and providing the list, the Bureau at first explicitly placed the highest priority on the long delay detonating fuze and the spotting projectile. At Indian Head during 1919–20, Patterson regularly included reports on the EAU work along with his other "Experimental" projects more directly related to powder production, while the ordnance proof officers regularly reported on the related tests of developmental star shells, spotting shells, time fuzes, and delayed action detonators.[12]

The relationship between the EAU's postwar research problem list and the reassessment of Jutland is even clearer when the tactical function of each of the specific projects is considered. Projects dealing with fuzes and time delay would assist in the question of deflagration or partial premature explosion of shells; projects dealing with pyrotechnics and marker shell would be of assistance in reducing the pattern and improving the accuracy of fire, which the British regarded as their greatest failure at Jutland. The marker shell, colored pyrotechnic shell, water impact fuze, and "super quick fuze" in particular would be required for improved practice drills. Three of the projects spoke to specific needs for new or improved ordnance raised in the standard works and journal articles on the battle: star shells and two smoke devices. On the original 1919 list, only project seven, that dealt with aircraft float lights, and project fourteen, that reflected the need for improved anti-aircraft tracer shells, appear to have been unrelated to the ordnance reassessments discussed in connection with the great inconclusive battle. These two projects, listed by 1919, grew out of ordnance officers' early concern with aircraft.

Other projects under study at Indian Head through the 1920s focused on other aspects of the Jutland lessons, including a variety of efforts to reduce shell dispersion and increase accuracy, to improve deck-penetrating shells, and to deal with the issue of flash from smokeless powder. Each of these technical goals associated with ordnance hardware grew out of deficiencies revealed by the battle on the afternoon and night of 31 May 1916. Officers at Indian Head were well aware of the hot topics in the service. In 1920 Captain J. W. Greenslade, officer in charge, appealed for a special ordnance research officer to review old records in order to shed light on the current problems. "Such questions," he said, "now very much alive in the service, as dispersion, erosion and high angle fire could each occupy the full attention of an officer, if past records and present progress are to be fully and thoroughly understood and kept in mind during current work along those lines." The issues were "alive" from Greenslade's point of view, because the war had brought them to the fore and because the Bureau had explicitly ordered research into them.[13]

The EAU, during both its stay at Indian Head and its difficult years at the Washington Navy Yard, struggled with problems of inadequate funding and poor personnel morale. The separation of the chemical and mechanical sides of the unit in its first five years hampered coordination, and a shop force had to be trained to develop the new fuzes and shells. During the 1920s Bureau-ordered changes in priorities further delayed completion of the original list, with gradual progress being reported on most.[14]

Through the 1920s the EAU had responsibility for work on all types of fuzes, special projectiles, pyrotechnic and signal devices such as rockets and Very signals, and all sorts of smoke apparatus. The unit was specifically charged with studying foreign and commercial products in these categories and with developing working designs, specifications, and instructions for new weapons, with the concept that quantity production could be based on their preliminary work.[15]

Through the decade, new projects were added to the original list. By 1927 many of the original projects were listed as 90 or 95 percent complete. Over its first eight years the EAU had begun work on an additional group of problems, with a few added every year. The expanded list, reflecting problems added between 1920 and 1926, is presented in Table 7.

By 1926 the ambitious list reflected growing concerns with submarines, torpedoes, and aircraft. Few of the new projects were completed by early 1927. The raw listing of projects presents only part of the picture. In the first eight years of the existence of the unit, a total of forty-seven projects were under-

TABLE 7
Experimental Ammunition Unit, Problems Added, 1920–1926

Project	Date Added
19. Special Projectile for 3" Mk 16 Gun	1920
20. Chase Delay Action Fuze	1921
21. Recovery of Projectiles	1921
22. Submarine Recognition Signal	1921
23. Torpedo Superheater Fuze	1921
24. Aircraft Bomb Fuze	1921
25. Anti-submarine Fuze	1922
26. Smoke Candles for Investigation Air Currents	1922
27. Barker Mechanical Fuze	1923
28. Plastic Shell Filler	1924
29. Chilowsky Process for Projectiles	1924
[30, 31, 40 new aspects of day and night tracers, continuation of project 14]	
32. Torpedo Marker Buoy	1924
33. Mk 2 Flare	1924
34. Aircraft Recognition Signal	1925
35. Miniature Practice Bomb	1925
36. 37 mm. Supersensitive Fuze	1925
37. Torpedo Recovery Buoy	1925
38. Launching Tube for Signals	1927
39. Aircraft Distress Signal	1925
41. Submarine Recognition Signal [merged with problem 22]	1926
42. Anti-aircraft Spotting Projectile	1926
43. Medium Caliber Base Detonating Fuze	1926
44. Time Fuze, Fragmentation Bomb	1926
45. Projector for Aircraft (dirigible) float light	cancelled
46. Flexible Base Plug for Armor-piercing Projectiles	1926
47. 4" Nonricochet Projectiles	1926

Source: Naval Ordnance Laboratory, *History of the Naval Ordnance Laboratory, 1918–1945: Scientific History*, vol. 3 (Washington, D.C.: United States Navy Yard 1946), report no. 1000, narrative history no. 131c.

taken. Of these, six dealt with the torpedo and submarine menace; thirteen dealt with aircraft offense and defense. Two had to do with rescue signals. The remaining twenty-six specific experimental problems focused on armor penetration, improved accuracy of fire in drill, smoke screens, and star shells, all of which addressed deficiencies that had been revealed at Jutland.

The experimental projects developed at the EAU provided a steady stream of devices, substances, and materials to be tested at the "Lower Station" proving ground at Dahlgren, which through the 1920s was a division of the Indian Head operation. Mechanically timed fuzes, anti-submarine fuzes, aircraft tracers, illuminating projectiles, and marker shells all required special test firings at the Dahlgren facility, interspersed between regular tests of production shells and guns.[16]

THE VARIABLE DELAY FUZE—A CASE OF
INCREMENTAL INNOVATION

The "prime development problem" for the EAU and for ordnance experimenters at Indian Head and Dahlgren remained the variable delay fuze for major armor-piercing projectiles, according to the now-declassified history of the origins of the Naval Ordnance Laboratory. For fifteen years, all three labs worked on the question of how to develop a delay fuze which would allow penetration of a shell into the interior of a ship before explosion, and the laborious, step-by-step progress is a classic example of incremental innovation. By 1931 the Mark 7 fuze, one of several alternate designs considered, emerged as the standard.[17]

As the EAU and the Indian Head teams struggled with the variable delay fuze, a frustrating factor was "difficulties encountered in the manufacture of relatively small components that required special high-tensile material." The objective of a reliable fuze had to wait for advances in materials and machining.[18]

Improvement of deck penetration and armor penetration required finding a solution to the deflagration problem. If an explosive shell could penetrate armor and then explode inside the enemy ship, it would have a far more devastating effect than a shell which detonated on contact. The secret to solving this problem was to design a fuze which would begin its action on contact, but would delay before detonating the charge long enough to penetrate the armor and explode forty to fifty feet inside the target ship. Two approaches to this problem were possible, one relying on clockwork and the other on a combination explosive-mechanical device which would delay for a fraction of a second the explosion of the charge. A further consideration was that lightly armored targets would require that the shell explode quickly, while heavily armored targets would require a longer delay since it took slightly longer to penetrate thicker armor. For safety's sake, the fuze would be inert during handling, being "armed" just before loading to a setting dependent upon the armor thickness of the target.

The history of the fuze problem for the Bureau extended back before the creation of the EAU. In 1911 a Semple patented fuze, designed for time delay, had been tested against the retired ram *Katahdin*. Marks I, II, and III of time delay fuzes had all been modifications of the Semple patent, and through the following years all showed unreliability, either exploding on contact or acting as "duds" in tests at the proving ground. In 1914 the Proving Ground at Indian Head had tested the Firth fuze against the Semple fuze, finding both unreliable. A six-inch Semple fuze, designated Mark III, was tested in 1915, and it was downgraded from a delay fuze to an instantaneous fuze. By 1916 the proving ground had concluded that the 1912 Mark II Semple design was superior to others but "still not good enough," with low-order detonations, deflagrations, and only occasional runs of successful, high-order detonations after penetration.[19]

In 1919, when Desk F in the Bureau of Ordnance turned over experimental work to Desk Q, they passed on the fuze problem, with a series of observations. Desk F regarded further modification of the Semple Mark II as useless; they regarded use of tetryl as a fuze booster as a good line of research to pursue; and they noted that the British were aiming for a forty-to-fifty-foot penetration before detonation, requiring a long delay. Following the war, the problem was complicated by requiring fuze delays for oblique impacts; but through 1921–22, as the testing was shifted from Indian Head to Dahlgren, a frustrating run of duds and deflagrations led to the conclusion that it was useless to try to improve on the Mark II fuze.[20]

By 1926, after several attempts, Harry J. Nichols at the EAU had developed the "VD7F" or Variable Delay, Mark 7 fuze which showed increasing success even in oblique impacts. By 1928 the Bureau concluded that the fuze was superior to the Mark II, that it met the forty-to-fifty-foot objective, and that it worked reasonably well on long-range oblique shots impacting on decks and sides, producing the appropriate high-order detonations. The Bureau rated it as "satisfactory for service." With improvements, the complex, finely machined VD7F became the basis for the naval standard.[21]

The testing of delay fuzes had consumed over fifteen years, and the original objective had been modified by the British experience. Not only delay and penetration, but penetration at long range, variable delay for varied thicknesses of armor, and operation in oblique impact—all objectives which grew out of the "Jutland lessons"—had guided fuze research at the EAU, at Indian Head, and at Dahlgren through the 1920s.[22]

A further problem involved in assuring that explosive shells expended their force inside, rather than outside, the targeted ship arose out of the nature of the shell explosive itself. Ammonium picrate, or Explosive D (so called because one brand of it, adopted by the United States Army, had been named "Dunnite"), had been invented in 1888 by Alfred Nobel. The Army Ordnance Board had recommended the use of Explosive D as shell filler in 1901 because of its insensitivity to shock and because it was capable of being compressed into shells with minimum danger. Furthermore, it was less likely than the related perchlorate explosives to combine chemically with metals. In 1907 the Navy had accepted Explosive D as shell filler, and during the First World War it had been used in a mixture with black powder for shell loading. In 1917 the Navy had settled on a charge loaded in front with Explosive D and in the rear with black powder for its larger caliber explosive shells, while using a TNT-black powder charge for smaller caliber shells.[23]

Explosive D had the virtue of relative imperviousness to shock and could regularly be ignited only by heat in the form of a fuze of another explosive material. Thus, the Bureau of Ordnance regarded Explosive D as close to ideal for shell loading purposes. A shell loaded with Explosive D would not detonate when accidentally dropped or exposed to violent shaking or "racking" of a ship under the impact of enemy shells. Theoretically, a shell loaded with Explosive D would not explode on contact with enemy armor until the fuze ignited the charge. However, experience during and after the war with Explosive D-loaded shells revealed a problem—many of them deflagrated or gave low-order detonations on contact with armor, even without fuzes. When a wooden plug was substituted for the fuze, Explosive D-loaded shells still gave low-order detonations on contact with armor. In the immediate postwar period, the puzzling question of such premature detonations absorbed researchers at Indian Head, at the EAU, and at Dahlgren.

In 1919 and 1920, both the Army and the Navy took up the issue of premature detonation of Explosive D-loaded shells. In conferences between Army and Navy ordnance specialists, and as a result of experiments at Dahlgren, the Navy concluded that the cause of the premature detonations was heat generated from internal friction caused by the sudden forward shift of the explosive in the shell on impact. Tighter packing, or increased density, might prevent the problem.

A series of tests at Dahlgren with Explosive D, packed at different densities through 1923, settled on a density of 1.48 as ideal (based on specific gravity water = 1.0). In 1923 a Special Board on Naval Ordnance recommended that Explosive D be packed to that higher density. The Bureau of Ordnance decided that older naval shells, packed at a density of 1.34, should be repacked under the higher density as soon as the redesigned variable delay fuzes became available. However, repacking presented special problems. The du Pont company believed that unloading the old Explosive D and repacking it to the higher pressure would be too dangerous. Several of the presses which had compressed Explosive D to the older 1.34 density had exploded, killing or injuring the press operators. The delicate job of removing the explosive, further compressing it, and reloading shells under higher pressure would involve several steps which could expose the charge to heat from friction and pressure. As a consequence, the Army investigated the production of guanidine picrate and other explosives, while the Navy experimented with a variety of other explosive mixtures through 1923 and 1924. H. J. Nichols, at the EAU, recommended a mixture of 5 percent zinc stearate with Explosive D to provide lubrication and plasticity to allow better packing of the explosive at the new density, but that method proved unworkable because the resulting densities varied too much. The Navy decided to continue with the use of the Explosive D, looking into the question of how it could be safely repacked at the required higher density.

Indian Head opened its own repacking facility, and chemical research there indicated that the required density could best be achieved if Explosive D was crystallized to different specifications. As guns were modified, old explosive had to be recrystallized and repacked for the new guns, and Indian Head had continuous work in repacking through the late twenties and into the thirties.[24]

POLICIES, BUDGETS, TREATY, FOREIGN PROGRESS

While the reassessment of Jutland was one driving force in the selection of projects and items, a host of other factors would limit the rate, degree, and organization of the research at Indian Head during this period. In the 1920s and 1930s, the Navy's procurement was dominated by several national policy considerations.

In 1922 the Army Navy Munitions Board was established to coordinate industrial mobilization planning. In many cases both the Navy and the Army had attempted to produce or purchase almost identical products, and they had often competed in the marketplace for the same raw materials. The Army Navy Munitions Board was to divide the purchase of items so that one service would have the responsibility for producing wartime supplies of munitions and other products required by both services. With the setting of this agreement, the procurement of explosives and propellant for both services was put under Army jurisdiction; exceptions to this rule were allowed to keep the Navy in touch with developments in explosives production or when the Navy had a particular interest in a specialized explosive or propellant. Specialized work in torpedoes, mines, and armor penetration clearly fell within the Navy's purview under the agreement. However, in the case of smokeless powder, TNT, Explosive D, and various RDX compositions and rocket propulsion powders, 80 percent of the materials supplied to the Navy came from either War Department facilities or from facilities owned by the government and operated under War Department contracts. The remaining 20 percent came from Navy-owned and Navy-operated facilities or from private contractors. For such reasons, the Bureau of Ordnance F desk became concerned that Indian Head production be kept up so that it not lose control entirely over production to the Army.[25]

Still another constraining factor on naval ordnance development was the set of international agreements coming out of the 1922 Washington Naval Conference. That conference resulted in an international arms limitation treaty setting limits not only on tonnage but also on gun size for various classes of ships. No guns above 16-inch shell diameter could be deployed; in any case, a "holiday" in the construction of battleships, which lasted until 1936, required that navies focus their ordnance improvements on smaller calibers. Preliminary plans to continue the course of expansion which had led from 10-inch guns through 12- to 14-inch guns had to be scrapped, and speculations about the problems of 16-inch and 18-inch guns, begun in the period 1919–21, came to an end.

In 1920 the Bureau of Ordnance briefly considered the possible impact of an 18-inch gun design. In reviewing the factors involved in the 18-inch gun, the Bureau revived the classic argument between the advocates of greater volume of fire against those who argued for greater weight of fire. This debate had come up when the Navy moved from the 12- to the 14-inch gun a decade before. The new gun would be able to throw a greater charge, over a longer range, but such shells would require that fewer be fired. Only increased accuracy or reduced dispersion could make up for the reduced number of shots. Another consequence of the abortive 18-inch discussion was a recognition in ordnance circles that aircraft spotting would be necessary because of the greatly increased range and the reduced volume of fire. Aircraft spotting of artillery would require that the fleet be accompanied by aircraft. Much of the concern with aircraft in the Ordnance Bureau in 1920 and 1921 focused on the ability of aircraft to assist in gun fire control.[26]

Advocates of air power believed that the advent of flight should itself be a limiting factor in ship construction; some predicted that the capital ship was doomed. Yet a review of the experiences of the war did not support that conclusion. In several battles during the First World War, German dirigibles had helped locate British ships; unreliable radio communication had hampered the use of the intelligence. Nevertheless, the role of aircraft in locating ships and assisting in artillery spotting had been demonstrated. Although some followers of Mitchell predicted that aircraft would mean the end of the heavy capital ship, Japan and the United States proceeded with construction of capital ships, within the constraints of the treaty. The conclusion was reasonable. The war had demonstrated that air support was important, not as a means of destroying ships but locating them for the guns of other ships. New capital ships were frequently designed with a catapult-launched spotting aircraft built into the design. Even before the Chesapeake bombing tests against warships conducted in July 1921, the Bureau took up questions of aircraft bombing as well as artillery spotting.[27]

With the Washington Treaty, all work immediately stopped on production of 16-inch guns. While the treaty had not banned such weapons, it limited existing ships to no changes in armor or in caliber, number, or general type of mounting of main armament. With no further battleships going into production, even the sixteens would not be needed. The older battleships had relied on 14-inch guns, and the 16-inch guns under construction had been intended for ships not yet commissioned. The twenty 16-inch guns already produced for the Navy at Bethlehem and Midvale Steel were sold to the Army for its use. Despite the temporary elimination of the 16-inch gun and the speculative 18-inch gun, much of the research which had been spent on the larger calibers could be applied with good results to the smaller calibers, particularly with regard to controlling dispersion of shell fire. While the Washington Treaty imposed limitations on gun size and ship tonnage, the focus of improvement would be on perfecting guns, shells, and armor within the new constraints, rather than moving on to heavier guns and thicker armor.[28]

Still another constraining factor on the research of the Bureau of Ordnance and its facilities such as Indian Head, Dahlgren, and the EAU derived from the generally low military procurement budgets of the period. During the 1920s and early 1930s, the revulsion with war and the tendency to blame the cause of the Great War on "munitions makers" and their international cartels resulted in congressional limitations on arms budgets that severely held back research and development as well as procurement.

Despite the cutback of personnel in the early 1920s, and despite the national policy which reflected a mood of disarmament and isolation, the research projects at Indian Head were diverse and full of dozens of significant success stories. Peacetime demand for smokeless powder production declined, and Indian Head needed only to supply powder for target practice and occasional new ships until the mid- to late 1930s, when the construction of new vessels began to require an increase in supply. Although production was in a period of decline, research work was sustained, vigorous, and pertinent to naval needs.

On orders from the Bureau, researchers at the EAU, Indian Head, and Dahlgren kept in touch with the generally advancing front of progress in a wide variety of fields. There were several sources for such information. One was a close awareness of foreign arms work. Other sources were publications, advertisements, and correspondence with fellow researchers in the private explosives and chemical industries. At all three of the related Navy facilities, researchers made minor improvements in method, applying developments in instrumentation, metals, and chemical research.

The gradual, incremental improvement which characterized all technological work sometimes stimulated and sometimes retarded development in the areas of concern to the Bureau of Ordnance: guns, mounts, shells, explosives, armor, and auxiliary items such as illuminating shells and fuzes and improvements in range and direction-finding instruments. On the one hand, purchase of necessary equipment and new materials in the early 1930s, from mundane items such as lathes, grinders, and motion picture cameras through more sensitive chronographs and strobe lights, advanced the work at the EAU. On the other hand, the effort to perfect the Mark 7 variable delay fuze, the effort to repack Explosive D to a higher density, and the detection of the cause of deflagration of Explosive D shells had all been retarded because of lagging development in instruments, tools, or materials.

American researchers assessed foreign developments, both those of former allies and former enemies. During and after World War I, the Chemical Laboratory at Indian Head studied the wide variety of explosives in use by various foreign armies and navies. In 1920 alone, Patterson reported surveillance tests and studies of Austrian powders and boosters, Brazilian cordite, and French *balestite*, or ballistite, and black powder. Other tests in the early 1920s focused on ignition caps of German percussion fuzes, tests of British machine gun and tracer cartridges, French, British, and German powders, and investigations of British signal cartridges. More directly, British specifications arising from their reassessment of the effectiveness of gunnery influenced American objectives and methods. Whether the work was conducted by the EAU chemist assigned at Indian Head in Patterson's laboratory, or by the rest of Patterson's staff, or by ordnance proof officers at the Dahlgren range, the projects involving foreign materials were often dictated by a review of World War I experience as well as by the looming difficulties presented by aircraft and submarines.[29]

THE AGENDA JUGGLED: SHIPS CHALLENGED FROM ABOVE AND BELOW

The particular projects such as illuminating shells, reduced shell dispersion, the variable delay fuze, the deflagration of Explosive D, and the development of flashless powder, that occupied researchers at Indian Head and at the EAU and Dahlgren, derived from the Bureau's concerns with specific technical issues and long-range strategic thinking. Assessment of Jutland, and new materials and products stimulated progress, but the extraordinary time involved on such items reflected the constraints of budget and the tedious, repetitive tests of slight modifications towards the identification of flaws in design and their correction.

Through the early 1920s the bulk of the work at both Indian Head and Dahlgren showed the evolving orientation of the Bureau. Literally dozens of specific experimental projects related to illuminating and spotting shells, marker shells and time fuze mechanisms, and experimental armor-piercing shells, all of which grew out of classic gun problems. Even more traditionally, the primary focus at Dahlgren remained proof work of guns and mounts, and proof of standard powders. Long-standing problems such as safety of powder in storage, accuracy of fire, penetration of shells, and strength of armor saw incremental improvements, derived either from slow developmental work or from the importation of foreign innovations. That traditional focus came under increasing challenge from the threats to the supremacy of the gun mounted from above, in the air, and from below, in the submarine and torpedo, a challenge reflected in new items added to the ordnance research and development agenda.

Although the Bureau was in fact dominated by the "gun club" of gunnery officers, the receptiveness and curiosity they displayed about air warfare can best be understood by going beyond the limits to the discussion suggested by Billy Mitchell and his supporters, who put the issue as a tension between "progressives" and "conservatives." A review of the actual work by ordnance researchers in aircraft and aircraft weapons, rather than an abstract discussion of whether the battleship was doomed, as presented in the popular press, shows that the gunnery officers were much more interested in the airplane and its possible naval role than their critics would admit. The nature of the actual developmental work conducted reveals that ordnance men viewed the airplane at first as an adjunct to the gun, serving a spotting and target-scouting role. However, at the same time, the research into aircraft at both Dahlgren and at Indian Head during the 1920s showed a lively concern about aircraft bombs and about anti-aircraft weapons.

The record of day-to-day research at the two ordnance facilities managed from Indian Head enforces the view held by scholars of the subject that the naval recognition of the importance of the air arm did not start with Mitchell's famous tests. Rather, the evolution of interest in air, both from an ordnance offensive and an ordnance defensive perspective, could be dated back to the first days of heavier-than-air-flight, and gradually grew through the 1920s along with the focus on more traditional problems.[30]

In 1921 a hangar was completed at Dahlgren, and by 1922 the "Aviation Detachment" there consisted of four officers and a support crew of enlisted men ranging from a low of twenty-four to a high of thirty-two. The detachment conducted test flights, torpedo observation flights, and testing of tow targets. At this point, enlisted men served as pilots, and they observed bomb trajectories, tested gyroscopic bomb sights, and conducted photographic and "pigeon training" flights. These efforts were well under way in the ordnance-controlled facility, the citadel of the "gun club," even before the bombing of the *Ostfreisland*. The Navy continued its experiments with homing pigeons through these years and, appropriately, naval aviators were in charge of the pigeons. Radio communication, as revealed by the German experience, remained an unreliable means of relaying intelligence, and the message-by-pigeon method still competed as late as the early 1920s. The detachment operated six planes in its first year, including two Hispano Suiza 2-Ls, an Aeromarine model 41, and one N-9 flying boat, similar to those which successfully crossed the Atlantic six years before Lindbergh. Most of the detachment's flying time was logged on the Hispano Suizas.[31]

It was the Bureau of Ordnance which organized the bombing tests against the *Ostfreisland* and other ships in July 1921. Despite Mitchell's spectacular disregard for the rules that had been imposed to study the effects of the bombings between each run, the Navy used the tests to focus attention on the need for more developmental work on bombs and anti-aircraft defense. Much bitterness resulted from Mitchell's attempt to portray the Navy as opposed to all aircraft development. The Bureau of Ordnance as well as the rest of the naval establishment moved steadily to the development of aircraft carriers and to a recognition, by the early 1930s, that command of the air would be as important to naval warfare as it would be to land warfare. In seeking to balance the fleet with submarines and aircraft, the Navy recognized the strategic implications of the new technologies that had been partially demonstrated during the Great War, as well as the more narrow, tactical, and specific microlessons of Jutland.

As the Navy developed aircraft carriers, bombs, and anti-aircraft guns in the interwar period, the research at Indian Head, Dahlgren, and the EAU reflected the Navy's shift in emphasis from aircraft as spotters for gunners to aircraft as bombers and fighters. Naval exercises in 1929 off Panama dramatically demonstrated the potential role of carrier-borne aircraft. In mock war exercises, they showed that such planes could effectively bomb and strafe enemy ships and shore installations. Even before that date, however, the Bureau of Ordnance had focused research and development on bombs, anti-aircraft weapons, and tracer shells for aircraft guns and for anti-aircraft use. Deck-piercing bombs had been tested in 1923, along with the possibility of converting drifting mines into aircraft bombs. The same year saw installation of special Bamberger photo theodolites for the investigation of bomb trajectories at Dahlgren. In 1923 several experiments with the 5-inch, 25-caliber anti-aircraft gun tested fragmentation shells against airplane bodies, ranged the fragmentation shells, and tested an experimental incendiary shell for the same weapon. The next year saw further tests of the 5-inch, 25-caliber anti-aircraft gun. Again, the aircraft work, both offensive and defensive, was buried in a large schedule of more conventional gun work, among dozens of projects growing out of the search for improved accuracy of fire and improved spotting and illuminating shells.[32]

Well before 1927, the official schedule of research problems reflected the growing concern with air developments. As the original list of projects assigned to the EAU and to Indian Head for developmental research expanded from 1919 to 1927, the list of items related to aircraft increased from two to thirteen. The researchers worked on separate devices in the categories of bomb fuzes, tracer shells, practice bombs, aircraft distress signals, and anti-aircraft weapons. From the record, the Bureau of Ordnance by no means ignored air developments, and early in the 1920s the experimental workers at the three related facilities—EAU, Indian Head, and Dahlgren—studied this central weapon of the next war, as well as the older weapons put to the test in the prior war.

Newport continued to focus on torpedoes. The submarine-related work in acoustics and radio communication went forward at the new Naval Research Laboratory set up on the outskirts of Washington. Even with such specialized facilities for submarine work, the three ordnance facilities at the EAU, Indian Head, and Dahlgren played a part in several submarine developments as well. By 1927 the problem schedule involved work on torpedo fuzes and anti-submarine fuzes as well as signals and buoys used in practice torpedo and submarine work. Torpedoes and mines were the concern of other units, but Indian Head and Dahlgren worked on the auxiliary items through the 1920s.

INDIAN HEAD—A PLACE IN THE NETWORK

Despite work on aircraft, it is fair to say that the Bureau of Ordnance also "stuck by its guns" and concentrated proof and development work on the traditional areas of ordnance: guns, mounts, shells, explosives, propellants, and armor.

Although the EAU, Indian Head, and Dahlgren all worked on overlapping aspects of the same projects during the 1920s, a degree of specialization within these three facilities emerged by the mid-1920s. With the shift of EAU chemist F. F. Dick from under Patterson to work directly in the Washington Navy Yard with the other EAU ordnance workers during 1924, the specialization became slightly more clear. The EAU worked on developmental designs in the mechanical or hardware side of specialized shells, tracers, fuzes, and bombs growing out of the war-induced studies and the adjustment to aircraft and submarines. Dahlgren, while involved in providing tests on those experimental designs, also pursued an agenda that Dashiell would have recognized, doing proofs of guns, shells, and armor in increasingly separated tests. The addition of a trained physicist to the civilian staff there brought a closer regard for the need to run specialized tests for each kind of research. Indian Head focused on propellants and explosives in the manufacturing of better products, in conducting research into the older problems of stability and burning rates, and in dealing with the possibility of flashless powder.

While no order explicitly dividing the tasks in such a clear fashion has emerged in the files, the separate schedules of work at the three facilities shows some sharpening of focus by 1930. The EAU was the laboratory for ordnance design development. Indian Head, in addition to being a major powder supplier for the Navy, was the chemical developmental and testing laboratory. Dahlgren was the test and evaluation facility for prototypes and for weapons, shells, and propellant lots before they entered the fleet.

After 1923 Indian Head was no longer designated the "Naval Proving Ground" but was called the "Naval Powder Factory." Indian Head focused increasingly on producing good quality, stable powder, studying the stabilizing effect of diphenylamine on nitrocellulose, working on improving Explosive D for shell fillers, and studying new explosive and propellant mixtures, including keeping up with the work done on flashless powder done by Army laboratories.

The Chemical Laboratory at Indian Head, established by Patterson early in the century as an adjunct to the Powder Factory, could proceed as a research institution out of the various budget lines which supported powder production. By the 1920s, it, like the offspring EAU and Dahlgren, represented a survival of World War I expansion into research. The evolution of a research orientation, and a continuing output of projects, reports, and contributions from that facility were testimony to the ability of Patterson and Walter Farnum to adapt to the constantly changing needs of ordnance through the period despite the generally constrained budgets.

At Indian Head, Patterson's experimental work included a range of studies of the stability problem and studies of old surveillance indexes and lots to determine new approaches to the standard question of powder deterioration. Patterson continued surveillance of stored powders, often involving a test project which ran over a ten- or twelve-year period, to determine the degree of deterioration in powder lots put in storage before World War I. One of Patterson's longest experiments began in 1908 (when he was forty years old) and continued until 1929 (when he was sixty-one). Four separate batches of smokeless powder were made, each with a slightly different method of processing, in order to test whether boiling the nitrocellulose after pulping made any difference in long-range stability, and whether or not adding salt or sodium carbonate to the pulp would extend the life. None of the four samples were stabilized with diphenylamine. Patterson concluded that as long as the nitrocellulose was boiled a minimum of twenty-one hours, further boiling did not further purify the product; he also concluded that the addition of salt and bicarbonate of soda had no effect in improving stability. The powder was packed in two different configurations: one for 6-pounder guns and the other for 6-inch, 50-caliber guns. The 6-inch powder showed good stability after twenty years; the 6-pounder powder remained stable for ten years. He warned that stability would be affected by stowage conditions, and any factor which would allow absorption of moisture would reduce stability.[33]

The proving ground had moved to Dahlgren but continued under direct military and administrative control of the Indian Head command through June 1932. By the time it was established as a separate facility, it was well under way with personnel, equipment, experience, and a set of procedures inherited directly from a chain of ancestry back through Indian Head to Dashiell and to the earlier Annapolis Proving Ground. In essence, Dahlgren represented the location for an institution which already had, by 1932, a fifty-year history.

With the decline in new ship construction in the postwar era, fewer guns came to the proving ground at Dahlgren to be "proved" than had come to Indian Head during the World War I years, although target practice continued to require that a steady supply of powder be checked. Experimental and new weapons required continued testing of a small number of guns in the postwar years, ranging from under a hundred to about two hundred per year. Thus, the years of transition to the Dahlgren facility were also years of transition in ordnance testing itself, when the emphasis changed from proving new major caliber guns for new warships to more experimental work with new projectiles and propellants.

While still a proof facility, the Dahlgren group focused increasingly on experimentation, drawing attention to its research accomplishments, rather than to the quantity of guns proved, as a measure of success. Commander Herbert F. Leary, who served as officer in charge at Indian Head during 1928–30, described the Dahlgren research which had led to reduction in the dispersion of shell pattern in the early 1920s. Earlier projectile research, he noted, had focused on designing shells that were strong enough to stand the pressures in the gun and which would be satisfactory in penetration. In the period 1919–22, projectile designers concentrated on the physics of the shell itself, examining the relationship between rotation and moment of inertia about the longitudinal and transverse axis, and examined band design in order to reduce dispersion.[34]

Among the Proving Ground experimental projects while Dahlgren remained under Indian Head jurisdiction were studies of high capacity (HC) shells. HC shells had been utilized on the naval guns mounted on railway cars in the First World War. The concept of a high capacity shell was that the shell would gain in lethal blast by sacrificing wall thickness for a larger explosive cavity. However, HC shells generated several mechanical and chemical problems that found their way onto the Indian Head agenda in the 1920s, requiring redesign of the powder grain itself to achieve the correct pressure in the gun barrel. After many false starts in which one modification would generate further new problems, a more effective grain for the HC shells was produced. The new grain reduced gun erosion and improved the uniformity of velocity.[35]

During the war years, both the Army and Navy had sought to expand powder production, and the increases seen at Indian Head were only part of the national picture. Two vast plants had been built with government funding: the Old Hickory Plant at Nashville, Tennessee, and a second at Nitro, near Charleston, West Virginia. When completed, the Nashville facility was theoretically capable of producing 1 million pounds of powder a day; the Nitro plant could produce 625,000 pounds a day. The valiant effort to upgrade production at Indian Head during the war had finally resulted in a ten million pound per year capacity. The Navy's powder plant at Indian Head would never be able to compete, in quantity alone, with the new War Department factories which had been built from the ground up. While continuing as the Naval Powder Factory, where methods and costs could be monitored under direct Navy supervision, the Chemical Laboratory had firmly entrenched Indian Head in the slowly emerging research and development network of the Navy.[36]

THE 1930S—EXPERIMENTS, PERSONNEL, TECHNOLOGY TRANSFER

Through the 1930s, Farnum took increasing responsibility for the research side of the Chemical Laboratory at Indian Head. He led the investigation into several areas which reflected the effort to keep in touch with developments overseas and in the private sector. In particular, he conducted experiments designed to find a substitute for mercury fulminate as a detonation material in fuzes. He worked with lead azide, and later cuprous azide, testing whether the materials reacted with other metals and would be stable in storage. Other experiments focused on the use of other chemicals for primers by the Army at Frankford Arsenal and experimental chemicals tried out by du Pont and by the Naval Ordnance Laboratory. Although du Pont remained cautious about release of proprietary material to potential competitors, the company was generous in sharing its research with Farnum for naval purposes.[37]

Working with a junior chemist, F. C. Thames, Farnum produced a massive report on the use of magnesium and aluminum in pyrotechnic shells and provided the information for work at Newport and at the Naval Ordnance Laboratory. This work related to the attempt through the 1930s to develop a more effective tracer shell, particularly important in air-to-air and surface-to-air engagements. In 1933 the Bureau ordered the Naval Powder Factory laboratory to review the specifi-

TABLE 8

Indian Head Personnel, December 1939

Department	Commissioned	Enlisted	Civilian
Administrative	2		17
Production			
Laboratory			27
Powder, Acid, & D plants			287
Engineering and Construction	2		279
Transportation	1		53
Grounds	1		104
Supply and Accounting	1	5	
Naval Dispensary	1	5	
Disbursing	1		1
Marine Detachment	1	46	
Proof, Safety, Recreation	1		
Totals	11	51	794

Source: "Organization Personnel Pamphlet as of December 30, 1939," RG 74, Entry 25, NP8/A3, Washington National Records Center, Suitland, Md., p. 12.

cations for aluminum and magnesium powder used in tracers and pyrotechnics. On the basis of this order, Farnum gathered samples and documentation from a wide variety of private and governmental sources, bringing together the information for new specifications through April 1933.[38]

Powder production at Indian Head remained the major task, with the "Powder Line" employing more personnel than any other unit at the Station. By the mid- and late 1930s, the plant produced on the order of three million pounds a year, both to supply the practice needs of the fleet and to outfit new ships as they came into the fleet. Appropriations to cover the cost of the new powder came from both the "Increase of the Navy" budget and from the "Ordnance and Ordnance Stores" budget, with occasional additions from another budget line— money allocated for "Alterations to Naval Vessels." With the coming of the New Deal, funds for "Replacement of Naval Vessels," appropriated under the National Industrial Recovery Act, also provided some support for powder manufacture.[39]

By 1938 and 1939 powder production ran on the scale of three million pounds a year, with increases planned in the event of war. Du Pont, by contrast, planned production on the order of ten million pounds a year. Plans included production to equip battleships 55–61, the new ships being constructed after 1936, as the naval holiday ended and construction picked up.[40]

The steady flow of powder orders, powder reworking, repacking shells with recrystallized explosive, and studies of stability brought the work force up to the range of 800 employees in the late 1930s. At the end of 1939, after war had broken out in Europe but before American entry into the war, 62 uniformed personnel and 794 civilian employees made up the complement at Indian Head.

A close look at the organization of the Station at the end of the 1930s and the division of labor between uniformed and civilian personnel indicates how strongly the early patterns established under Patterson and Joseph Strauss in the first decade had taken ahold. The only department not under the direct control of a naval officer was the Production Department and its laboratory. The "Grounds Department" reported to the executive officer, who was in charge of both Transportation and Grounds. Each of the other departments, even when engaged in such essentially routine tasks as conducting maintenance, operating the railroad, or running the power plant, all reported to a naval officer.

The civilian management structure in the Production Department had evolved into a somewhat complex arrangement, but it still resembled the pattern from 1901. Patterson, now past his regular retirement age and reappointed by special legislative exemption, directed the Production Department as "Head Chemist." Farnum, as "Senior Chemist," ran the Chemical Laboratory. The Explosive D plant, the powder line, and the acid plant were directed by chemical engineers T. C. Jenkins, R. H. Dement, and H. M. Coster. Dement, in operating the powder line, had the relatively large complement of 232 civilians working for him, including 2 "quartermen," 4 "leadingmen," and 2 "leaders." The intentional abdication of management by officers to civilians in this one area had persisted and become a regularized part of the institution.[41]

What a later generation would call "technology transfer"—the movement of improvements across national lines and from the public to the private sector, or from the private to the public sector—operated through the 1930s by a variety of procedures. At the most elementary level, Patterson and researchers at the EAU examined shipments of foreign propellants and explosives, doing chemical analyses to determine the ingredients and proportions. From time to time, the Q desk would ask for a review of information culled through Office of Naval Intelligence gathering of published documentation in foreign capitals or of materials published in the open literature. When foreign and domestic inventors submitted ideas or patented explosives for naval use, Patterson and the EAU would sometimes put together the materials described and conduct tests on them to determine whether the suggested improvement was worthwhile.

Only rarely did Indian Head personnel have the opportunity to go overseas to examine foreign developments. But when such a chance came in the 1930s to attend an international conference in Britain, the specialists kept their eyes and ears open. Farnum was able to learn directly of European advances by attending meetings of the Chemical Engineering Congress in England early in 1936, tacking onto the trip visits to both British and German ordnance and chemical plants, and by attending other conferences on the same trip.

Farnum and chemical engineer Coster also attended meetings of the British Institute of Chemical Engineers and the meeting of the Society of Chemical Industry in Liverpool on the fifteen-day trip. Farnum visited the Imperial Paper Company at Gravesend, Lever Brothers, near Liverpool, and the I.C.I. Ardeer Munitions Factory in Scotland. Farnum went on to the continent, where he visited I. G. Farben at Leverkusen and Frankfurt and other factories at Frankfurt and Darmstadt. Farnum also learned a great deal, he claimed, by associating with the 104 members of the American delegation from private industry. Farnum felt he gained specific, useful information about several topics: water conditioning, methods of determining the viscosity of cellulose developed in Japan, work on lead azide as a substitute for fulminate of mercury in Britain and Germany, fire-proof-ing of wood as used on the *Queen Mary*, and the Bofors explosive "Bonit." In general he found the Scottish I.C.I. manufacturing methods for smokeless powder quite outdated when compared to American methods. By contrast, he was very impressed with the modern research conditions he found at the German works at Darmstadt. From chemists working for private American firms he learned of the use of hexanitromannitol as a substitute for fulminate of mercury and of the use of saltwater-resistant paint manufactured by American Cyanamid. From discussions with American academics he learned of efforts to revive early Austrian methods of making nitrocellulose from wood pulp, which would address the problem of possible cotton shortage during war.[42]

Both Farnum and Coster made excellent use of their time. Their extensive reports suggested that each of them recognized that the Navy would expect the trip to be worthwhile, and they conscientiously noted every contact, including informal ones aboard the return voyage. As supervisor of the Acid Plant, Coster gathered ideas regarding nitric acid refining processes, obtaining copies of patents and following up with the contacts after his return. In an excess of conscience, perhaps, Coster even reported social meetings with private manufacturers in fields unrelated to chemical engineering, such as Buxton, the "inventor of the key-chain."[43]

THE INTERWAR YEARS—QUIET PROGRESS AND CROSS-CURRENTS

In the clutter of projects, many taking years to complete, a few patterns emerged at Indian Head in the interwar years. Research priorities usually focused on specific, sometimes minor improvements; the "big picture," which can be detected from the perspective of several decades later, comprised several different, overlapping, and sometimes contradictory patterns. The Bureau ordered the researchers at Indian Head and its two offspring laboratories to apply the lessons of Jutland and other aspects of the First World War; to take advantage of the aircraft; and to apply scientific method to the solution of some long-standing ordnance problems such as penetration of armor, dispersion of shot, and deterioration of powder in storage. Although the coordination was halting and the planning oriented around specific "problems," the Bureau of Ordnance managed to keep alive within the Navy the research establishments which had been born out of the immense but short-lived expansion in the First World War. The war had created all three: the move to Dahlgren, the expansion of the Chemical Laboratory at Indian Head, and the EAU. The Navy showed foresight in keeping those resources alive and in providing a schedule of innovation for them. Without a concerted plan, the Bureau began to move toward a degree of specialization at the three related ordnance centers that could be adapted in later decades when a more organized approach to research and development emerged in the Navy. Other policy developments led to attempts to establish a workable relationship between Army, Navy, and private manufacturers to be able to upgrade to full-scale production in case of war. The Navy had to patch all this together out of a very limited budget predicated on limitation of armaments and upon providing employment, not on funding research or improving productivity.

The differentiation of function between Indian Head, the EAU, and Dahlgren grew, and the growing specialization of each became slightly clearer. The EAU severed its formal connection in 1923 and Dahlgren became an independent command on 1 July 1932. Despite the organizational separation, the Bureau used all three to work on different aspects of the same issues throughout the period, revealed in the work on fuzes, illuminating shells, and bombs.

The developments at Indian Head as it moved away from proof testing to a role more squarely devoted to powder production and to the chemical side of ordnance research reflected these national, international, technological, and administrative trends. In the day-to-day work of the men and women at Indian Head, however, the concern was usually the problem at hand, not the larger world scene. The problem would be set at Bureau headquarters. Because Ordnance was operated by line officers, the central focus remained guns and shells. For the men at headquarters, the lessons of the war, the potentials of the aircraft and submarine, and the technical implications of the arms limitations agreement would shape the choices of priorities. But for the researchers at Indian Head and at the Washington Navy Yard, the problems came defined by Bureau headquarters; within the limits of staff and budget, the listed problems defined the objectives.

The weapons with which the U.S. Navy entered World War II were modified and improved versions of those used in World War I, greatly augmented by substantial progress in the aircraft carrier, airplanes, and the long-range submarine. Naval planners, who had sought to construct a balanced fleet capable of a distant war in the Pacific, had the foresight to keep research and development advancing on an appropriate range of fronts in the field of ordnance. The World War II years saw spectacular new developments in ordnance, with rockets, the proximity fuze, and improved torpedoes. In other areas of naval progress, the war years also saw rapid developments in electronics and aircraft design.

The accomplishments of the interwar years would be forgotten by contrast with the spectacular developments of World War II. The mundane devices such as armor-piercing fuzes, pyrotechnic shells, Explosive D shells that were operative rather than duds, improved sulfuric and nitric acid plants, propellants adjusted to new guns, tracer shells for aircraft, improved magazine safety and storage, and signal buoys were not the sort of technology that made headlines. In fact, such steps of modernization were hardly perceived at the time outside of the circles of those who actually used the devices and weapons. But these successes through the interwar decades, and the development of skilled groups in institutional settings who worked on these problems, laid the groundwork for the great leaps forward in the war years. By keeping alive a structure for research and holding together an assortment of groups conducting research and development in ordnance, the Navy had made an investment. That investment in facilities and manpower flowered rapidly with the infusion of funds and the crowded agenda of new and more revolutionary research and development problems that came with the war.

CHAPTER 5

Indian Head: The Navy and the Community, 1890–1940

Ether Vault No. 1
Building 165
1899

T HE NAVY CHOSE Indian Head for the Proving Ground in 1890 largely for its easy access to Washington by barge and for its isolation and long down-river range. Before Robert Dashiell could fire a shot, he had to drain the marshes, build a dock, and prepare the land for heavy equipment. The construction work required a good-sized crew, numbering more than one hundred laborers in the first years. Dashiell provided a temporary work camp at the Station to house the Charles County farmers during the work week; the labor force returned home on weekends. These transient workers in their humble temporary quarters represented the core of what would later become a thriving community, complete with all the amenities of churches, schools, commerce, and a decent way of life. In the coming years the Station would need a town to accommodate full-time workers, to educate children, and to provide liveable conditions for the thousands who would eventually be employed by the Station.[1]

Paradoxically, the very factors which made the site desirable for a proving ground made it undesirable as a place to live and work. Charles County had been in a period of economic doldrums since the Civil War, and by 1890 the county's population hit the lowest point ever recorded by a census. The 1890 population of 15,191 was more than 25 percent less than the population of 1790, when Charles County was still a bustling agricultural center.[2]

Maryland was no longer the breadbasket of the nation as it had been in the eighteenth century, and as railroads created a nationwide market for the vast production of the Midwest, Maryland's agricultural economy was further set back. This agricultural decline, through the last decades of the nineteenth century, caused a steady drain on the labor force of Charles County. The loss of slave labor was especially debilitating to Charles County, as tobacco, the county's main crop, was labor intensive. During this time the size of farms in Charles County had decreased, as had the overall productivity of the region.[3]

In a series of articles on Charles County, the Baltimore newspaper, *The Sun*, asked in 1910, "What have they done here in farming since the Civil War?" and answered, "Practically nothing but to cut down and sell trees and to raise on an insignificant corner of each of the immense farms a few acres of tobacco." The author also noted that a visitor would see "some of the scarcity of labor which prevails" in Charles County, indicating some of the difficulties that the Navy had in recruiting workers not only for construction but also for the work of the Powder Factory and Proving Ground. A railroad, which brought some activity to the area, opened through La Plata and Waldorf in the late 1870s, but the construction and operation of Indian Head provided the largest single industrial boost to Charles County in its history.[4]

To compensate for the meager local labor supply, the Navy had to rely on recruiting workers and technical staff not only from the local region but also from more distant locales. Yet the isolation and lack of social and economic amenities constantly worked against the efforts of early officers in charge to recruit and hold personnel. For Indian Head to develop into a successful naval station, Indian Head the town had to develop, with all the aspects of town life that were taken for granted elsewhere—housing, utilities, streets, stores, churches, schools, and, most importantly perhaps, friends and neighbors. In the period from 1890 through 1920, Indian Head had little to offer. Even peace and quiet, an abundant resource in the past, had been effectively destroyed by the loud guns and ordnance testing. After just ten years, the light frame buildings housing officers were no longer suitable due to "the constant jar incident to the discharge of heavy guns," testimony to the working and living conditions which reigned at Indian Head while the guns remained.[5]

Nor was Dashiell a great believer in physical comfort or psychic well-being. Off-duty he could be found drafting ordnance plans in his drafty fishing shanty, and as commanding officer he did not like to give paid holidays. However, even Dashiell requested base housing as it was a necessity for officers and permanent staff. Captain Newton Mason was much more in tune with the long-range social impact of conditions at Indian Head. As noted earlier, Mason requested that a doctor be sent to the Station and explained, in his annual report in 1893, that the nearest doctor was six miles away and "his professional work takes him all over the county, so that his attendance can not be depended upon within twelve hours after sending for him." Mason soon obtained the services of medical officers who rotated through service at Indian Head, including Dr. Carey Grayson, later to become Woodrow Wilson's personal physician, confidant, and fellow artillery target.[6]

Many of these first important steps in the development of social resources at Indian Head were necessarily taken by the Navy. If the naval base was to remain, however, the town would have to establish resources independent of the Station, which the Station could, in turn, rely upon in crisis situations. This chapter will examine how naval policy interacted with the growing, independent community, producing in the first fifty years of its existence a town called Indian Head which became a necessary adjunct to the naval facility called Indian Head. Out of the difficult conditions and the alternating cycles of neglect and attention from the federal government, a powerful sense of community identity and community loyalty would be forged. Those spiritual resources would become, in turn, assets which supported the naval facility through times of need.

INDIAN HEAD'S EARLIEST YEARS

In 1896 Francis E. Mattingly arrived in Indian Head and opened a general store a few steps from the front gate of the Station, and he remained active in the community for many years. Mattingly was the first postmaster, served on the first Board of Trustees of the first one-room Indian Head school, and was one of the original three commissioners when Indian Head was incorporated in 1920. When he first opened his store, he found a locale without schools, churches, or roads—or, for the most part, people.[7]

In the 1890s Glymont, about three miles east of the Station, was the nearest identifiable hamlet. Indian Head children walked to Glymont for school and were lucky on some days to receive a ride in a horse-drawn carriage provided by Lieutenant John Gering or Mattingly. Ursula Gray, a teacher at the colored school in Glymont, supervised the boys who split wood and started a fire each morning at 7:45 in a wood-burning stove in the center of the one-room schoolhouse. The students also performed janitorial duties according to a schedule posted every Friday. Two boys moved the desks and two girls swept every day; water was obtained from a neighbor's pump. The majority of the colored school's Parent-Teacher Association consisted of employees from the Indian Head Naval Station.[8]

The problem of education for the children of both civilians and uniformed personnel remained a difficult issue through the first decades of the Station. Two small buildings serving as the Glymont School for white children burned down in the 1890s, and a new one-room schoolhouse was built. A student from this time remembered that a tree stump was used as the front step into the school. This building also burned down and yet another school was erected. In 1898 George Gering, the four-year-old son of the paymaster at Indian Head, went to school a few months short of his fifth birthday to bring attendance to the minimum required to keep the school open. That same year the Charles County Board of Education in La Plata, the county seat, began to feel the burden of schooling Station children. The board had its secretary inquire "as to whether children residing at Indian Head, a Government reservation, were entitled to the benefits of the public schools." However, by 1902 the citizens of Indian Head had their own "Log House School" on Raymond Avenue, which was still being used in 1917.[9]

Mattingly's business relied upon the early workers at Indian Head as customers. Slowly, workers at Indian Head who had commuted from various scattered farms and hamlets in the area became permanent residents, as evidenced by the school. Aside from Mattingly, there were a few other employees and businessmen whose stories survived from the early years.

One such story was that of Captain Jansen, a riverboat pilot at Indian Head for over twenty-five years. Mr. Jansen was born in Denmark in 1868 and became a U.S. citizen as a young man. He had been working for the government when, in his mid-twenties, "at the request of Lieutenant Mason . . . he took command of the passenger vessel 'Barbara' at the proving grounds." He worked there, "when the difficulties in establishing communication with Washington or the surrounding country was dangerous because of dense fog and crushing ice," until his death in 1919.[10]

On days that threatened severe storms the employees at Indian Head depended upon Captain Jansen for safe boating. On 4 March 1909, William Howard Taft was inaugurated the twenty-seventh president of the United States, and on that day one of the most destructive snowstorms in local memory blew through Charles County. High winds or low visibility endangered travel and communications between Washington and the Station and dramatized the dependence of the Station upon the river pilots. An oft-repeated conversation on the shore of the Potomac, according to one source, captured that sense of personal dependence.

"Who is the pilot on the river tonight?" a waiting passenger would ask.

"It's Mr. Jansen," came the reply.

"Then we are safe," another would remark, confident that the trip to Washington would proceed according to schedule, despite conditions.[11]

Other early employees came from less distant areas. One, Perry Wright, was transferred from the Annapolis Experimental Battery in 1892. From then until his death in 1919 he worked continuously with the ordnance gangs. Men like Wright and Jansen made up the backbone of the town. Both of these men had families and depended upon Indian Head for their well-being.[12]

E. E. Evans, a storekeeper in Indian Head, was an active member of one of the earliest churches in Indian Head, the Episcopal Church. The Rev. Theophus Smoot served the Indian Head area from 1891 to 1900, and Mr. and Mrs. Evans took their turn in offering their house for the Sunday services. The Rev. George J. Graham ministered from 1900 to 1903, and his parish duties took him from Accokeek to Pomonkey and then to Indian Head. It was in 1903 that St. James Episcopal Church was founded in Indian Head and the Episcopalians had a permanent place of worship. The Evanses remained strong supporters of the church into the new century. The Reverend Bowyer Stewart served for many years, from about 1906 to 1915. He would come into town and spend Saturday night at the Evans's house, hold the service in the morning, be served breakfast, and then travel to Pomonkey and Accokeek for services.[13]

William Rainsford arrived in Indian Head in 1903 and noted that "at Indian Head, there was no fraternal organization, nor even a baseball team or other club." He helped organize the Ordnance Club which was dedicated to bringing a Masonic Lodge to Indian Head. The nearest Masonic Lodge was in La Plata, where some members of the Ordnance Club took their masonic degrees. This took a strong commitment as La Plata was fifteen miles away, and the road conditions forced the members to leave by 4:30 P.M. "It was always a lonely and dark trip coming home and in Summer it would be day light before Indian Head was reached," Rainsford remembered. In August 1907 ground was broken for a Masonic Lodge in Indian Head. The Ordnance Club dissolved upon the founding of the Lodge, having served its purpose to create a "strong, self-sustaining body." The Masonic Lodge and the Episcopal Church were two of the first of many religious organizations established in Indian Head which helped to promote Indian Head's sense of community.[14]

But for bachelors, the lack of female company and the isolated work at Indian Head represented a hardship that the officers in charge repeatedly addressed, often with considerable delicacy, in their reports. Bureau Chief Nathan C. Twining wrote to a prospective married employee that there would be "plenty of station society." In the same letter, however, Twining went on to describe a current employee at the station: "Michael wishes to leave because he is the only bachelor on the station, and he feels rather lonely."[15]

The hand of the Navy remained strongly involved in the advancement of the town. Joseph Strauss realized very early the importance of the social conditions on the station, and twice, in the period from 1900 to 1901, as commander of the station, and later as a high-ranking officer, he helped where he deemed best to improve the town:

> When I first went to Indian Head, in 1900, after looking over the ground I became convinced that there was a positive danger to what would be a large community eventually, in not taking every step toward making the community an entirely self-reliant one, and all my efforts since then have had that end in view in spite of any ill feeling that I would be subject to, in carrying out the scheme.

He pointed to the famous company town of Pullman, Illinois, headquarters of the firm which produced the sleeping car for railroads and the site of a protracted labor conflict, and to Essen, Germany, the paternalistic headquarters of the vast Krupp armaments works, as locations where the employer had developed "model establishments" for their employees. Yet, Strauss noted of these locations, "discontent has arisen among the men and their families, for one reason, probably, that the community ceases to be self-governing." Strauss wanted to avoid the transformation of Indian Head into a company town.[16]

In the business of explosives manufacturing, company towns had become rather common. In a major study of the subject, *The Company Town in the American*

West, James B. Allen defined a company town "simply as any community which is owned and controlled by a particular company." His leading and classic examples of the type were found in du Pont's company towns which, according to one company official, were formed out of "economic necessity" and industrial realities: "Explosive plants required isolation and therefore, a company village for all employees was an absolute necessity." Indian Head's isolation was not so great as to preclude private enterprise, but as Strauss knew, the Navy had to continually encourage the private sector and work to achieve self-government in Indian Head.[17]

Some prospective employees, married or not, balked at the thought of working in such an isolated spot. Walter Farnum wrote to Admiral Twining on 5 March 1912: "I have greatly objected to this transfer as I am forced to go to an undesirable location at no increase of pay and it is well known that the pay at Indian Head like the Phillipines has to be higher in order to get men to go there." Twining disagreed: "I do not think I would have directed the transfer had I felt that any great hardship was being worked upon you." Twining pointed out that Farnum would receive a partially furnished home and a job which "will be far more important than it has been at Newport . . . chemical work in connection with the manufacture of smokeless powder." However, Twining had received a letter from Indian Head less than two weeks before he heard from Farnum which had warned him of the bad conditions at Indian Head: "Unless something is done to make the better classes comfortable the Government will suffer by getting employees of an inferior class." At Newport, Farnum had headed up a laboratory, but one whose function had declined in importance after the opening of the Powder Factory at Indian Head and the transfer of most powder work from Newport. Again, important work at the station was pitted against the isolation of the town at Indian Head.[18]

While a college-graduate chemist with years of service for the Navy might find the conditions oppressive, for many of the local people who came to Indian Head the Powder Factory provided a chance to escape the "horrors" of rural poverty. The pre-World War I story of one such case of social mobility was related many years later by Mrs. McWilliams, the wife of a Charles County farmer:

> DeSales was getting a little discouraged on the farm as we didn't seem to be getting anywhere for all our hard work. Farm labor was high and farm prices very low. After giving Father his part of what little profit we made, there wasn't enough to start off the next year. Our main crop, tobacco, was only selling then for an average of five cents a pound. . . . We tried for two more years, and when we found that we were going back instead of forward, we made the change, and DeSales registered here at Indian Head for a job.[19]

The Powder Factory continued to draw employees from near and far. Even the farmers in the area felt the benefit of the Station, especially during World War I. According to one report "in the war as high as 6000 men had been employed there. It has always meant extensive employment for the southern part of the State and a ready market for certain kinds of farm produce."[20]

In late 1913 Officer in Charge Julius Hellweg wrote that further government housing was unnecessary: "The surrounding country is gradually building numerous houses, and there is therefore no reason for furnishing houses." In Captain Hellweg's view, Indian Head was approaching the self-reliant mode which Strauss had so hoped for. However, Hellweg would soon be proven wrong on the housing issue, as the Bureau forced the pace of powder production in the years 1914 to 1917.[21]

Also, in October 1914 a committee of disgruntled workers made objections to the Inspector's Office concerning the local stores. Some of the objections the committee had were that there was no fresh meat at Mattingly's store, and in fourteen years there had been no improvements and the store had deteriorated. The committee also complained that better food could be bought cheaper seven-to-ten miles away. The citizens of the town continued to rely on the Station to solve problems—the opposite of Strauss's hopes. However, the town had reached a size that made it difficult for some businesses to ignore. Banking in Indian Head had its origins in these years leading up to World War I. In 1916 directors of the Eastern Shore Trust Company and the Southern Maryland Bank of La Plata held talks with Indian Head citizens about establishing Indian Head branches. At the time the naval payroll was nearing one thousand dollars a week, and no bank was within fifteen miles.[22]

WORLD WAR I AND POSTWAR GROWTH

As the slow process of migration and unplanned expansion of housing proceeded, the need for social, economic, and other improvements increased. Workers across Mattawoman Creek, in Nanjemoy and Marbury, needed improved facilities to get to work. Until 1917 workers walked on a frail footbridge to a landing where rowboats or an awkward cable-ferry would haul them across to the Proving Ground. This crossing was especially hazardous at night, and several persons drowned in the attempt. The alternative, to cross the creek upstream, constituted a seven-mile walk or drive on difficult roads. The quicker footbridge and ferry system needed upgrading, but plans for a new bridge presented new problems. Barges brought coal up Mattawoman Creek, and a bridge would have to be high enough to allow the creek traffic to pass. An automobile bridge would be an expense too high for the Navy to bear, and the state would hardly fund such a special access bridge for the exclusive use of the Proving Ground.[23]

Congressman Sydney E. Mudd advocated a unique solution to be funded with federal monies. He pushed in 1916 for a footbridge to span Mattawoman Creek connecting the Naval Station to a point near Marbury. A rather unusual sight, the narrow bridge could be raised in the middle as a drawbridge. It may have been the only draw-footbridge in the state of Maryland; it was certainly the only one funded with federal money. Mudd said, "The purpose of this bridge is to afford direct and convenient communication to the government employees, for nearly 50% of them have homes on the Marbury side of the creek . . . for the past ten years the Government employees have been using a wooden bridge constructed by local labor at their expense." Mudd continued, stating the current arrangement was "hazardous," and he cited the drowned commuters. On 22 May

1916, the House Committee on Naval Affairs approved the bridge, and it was open for use the next year.[24]

At the outbreak of World War I Indian Head braced itself for further expansion of the plant. The labor shortage was partially circumvented by assigning enlisted men to the Station. In town, and in the surrounding countryside, families made room for extra workers in their homes, and the government provided more housing on Station for civilians as well as the servicemen. Capain H. E. Lackey was the chief inspector at Indian Head from 2 January 1917, to 15 April 1920. The U.S. participation in World War I, from April 1917 through November 1918, fell entirely within Lackey's term of office. He remembered those years:

> New construction was in progress all over even down in Cornwallis Neck where a large chemical plant was to be erected. New dwellings were built in the form of cottages, ready built and otherwise, a new post office was built at the south end of the Village Green, ten room houses, an hotel, dormitories, library, recreation hall, cooperative store, etc. New roads were built and others improved. Two new steamers were obtained for ferry purposes between Indian Head and Washington.

Between the contractors, the military, and the civilian employees the captain noted "there were all told nearly ten thousand people living or employed on the station."[25]

Charles County's scarce labor supply was stretched even further by these developments. Mrs. McWilliams remembered that she and her husband had moved to a house in Gering Row. Her brother and her two brothers-in-law moved in with her as they had all found jobs at the Station. In her household were four Station employees, each with a different schedule, and she soon found that she and her kitchen, like the Station, were expected to operate on triple shifts. Mrs. McWilliams recalled that "those were hard years for me since they all worked on different shifts, and I was cooking and serving meals at all hours of the day." The town was being put through its first important test as a necessary companion to the Naval Station. Not only did the factory have to scale up for wartime production but also the town had to support the wartime employees. The Station needed a town full of people; the town needed the Station for employment. Together they met the requirements of a wartime economy.[26]

The gross effect of the Proving Ground on Charles County's economy can be inferred from census records of the population. From 1890 to 1920 the only Charles County district which showed a constant growth in population was District 7, the Pomonkey district in which Indian Head's population was counted. From 1910 to 1920 the district realized its largest growth, nearly doubling from 1,589 to 3,124. That same decade, La Plata's district fell to 2,001 from 2,050 in population and Bryantown's to 2,058 from 2,216. This propelled Indian Head's district into the forefront of Charles County's population. Also in 1920, a new census district with a population of 1,392 was formed at Marbury, reflecting the significant growth in the area across the Mattawoman footbridge.[27]

This increased population, that was so critical for the wartime operation of the plant, put more pressure on local schools and churches. The public school at Indian Head had changed little since 1902, and the two small buildings were in disrepair. Attendance usually did not surpass thirty. Rear Admiral Lackey later remembered that the school situation was affecting his employees:

> In the summer of 1917, shortly after we had entered the first World War, one of my most valued civilian assistants said to me that he would have to ask for a transfer to some place where he could obtain proper schooling for his children. I could not oppose his leaving since we had nothing to offer at that time.

Lackey also remembered this incident when the U.S. Housing Corporation proposed building houses on Station. He agreed to the housing after ensuring that a school building would also be constructed.[28]

Indian Head residents had also made plans for a new school. Upon the governor's advice the residents contacted Walter Mitchell, a Democratic candidate for state senator. In return for votes, Mitchell promised to work for funds for a new Indian Head school. Lackey and the citizens realized that if they pooled their funds they could build a larger, better school. However, this mixing of funds was illegal. The residents transferred their funds to a school in Glasva and supported Lackey's plans with the U.S. Housing Corporation.[29]

In September 1919 the Station newsletter, *Proving Ground Fragments*, announced proudly that the new school would be ready for the fall term. A photograph of the school appeared on the front page and the article read, "It should certainly prove a success, as it is felt that this fine new building, with a total capacity of 280 pupils, should fill the needs of this community for some time to come." Both the community and the Navy had wanted the school, for they each recognized the important place it would hold in Indian Head for many years for both of their interests.[30]

One year later, the community named the school after Captain Lackey, the "War Inspector," who was remembered for his community service "amid overwhelming demands of official duties during war time and after." The first class to graduate from the Lackey School was in 1922. Lackey himself returned for the ceremony and felt great satisfaction in the work he had done for the community.[31]

The young town of Indian Head had good reason to be pleased with recent developments but also to be apprehensive about the future. In August 1919 the "new village" on Station, constructed with U.S. Housing Corporation funds, was nearly finished. By August 1919 fifty houses had been occupied and fifty others were nearing completion. *Proving Ground Fragments* reported at the time that "by next summer the whole village will have the appearance of one garden spot." The work of the Proving Ground and the Powder Factory was still essential to the U.S. Navy; however, yearly reductions in staff began to cause widespread discontent. The community of Indian Head was further threatened when in the summer of 1921 the proving ground was moved to Dahlgren.[32]

In July members of the House Naval Affairs Committee, Sen. Ovington Weller (who had unsuccessfully run for governor in 1916), and members of the Navy Department boarded the naval yacht *Sylph* and headed down the Potomac for Indian Head. Congressman Mudd led this expedition to southern Maryland to evaluate Indian Head in light of naval decisions to move the proving ground operations to Dahlgren, Virginia. High-angle gun testing had begun at Dahlgren during the war, and citizens of Indian Head were set against the complete transfer of work to Dahlgren and expressed their displeasure to Congress.[33]

On August 2, before the House Committee on Naval Affairs, Congressman Mudd protested the removal of the proving ground on behalf of five hundred civilian employees at Indian Head. Mudd's presentation took into account the danger of long-range work at Indian Head, and the petition from the employees requested "the retention of such work at Indian Head, where it can best be performed—that is, powder and ammunition, proof of tracers, the proving of all medium calibre guns, proof of such major calibre guns as do not require high angle fire, and experimental work in ordnance which cannot be carried on with large calibre." In effect, Mudd and the petitioners accepted the argument that the short range at Indian Head was too restrictive for the large 12- and 16-inch guns, but they hoped to retain all other armor and proof testing. The Navy disagreed, but the Indian Head advocates kept up the pressure.[34]

F. E. Mattingly wrote to *The Sun* to express his opinion. He stated that the government had spent nearly $4 million at the Indian Head Proving Grounds since the war, and such an investment should not be squandered. *The Sun*, in an editorial, called for a full investigation into Indian Head's recent history to determine if the money had been well spent or grossly mismanaged, and paralleled Mattingly's argument that the money spent meant either that Indian Head could support the Proving Ground or "the men responsible for those expenditures deserve prosecution."[35]

The Maryland Legislature also made a bid to revive naval proving at Indian Head. Early in 1922 they passed a resolution arguing that Indian Head had been adequate during the war and that some proving operations should be returned to the old location. The resolution pointed out that Charles County's economy had been affected by low prices and that the removal of the proving ground was an "overtaxing burden" that could be remedied by a simple policy decision.[36]

In April the issue was argued again in Congress. Led by Mudd, the House committee voted for an amendment to close Dahlgren, and representatives from Indian Head went to the Capitol to gain support for it. The delegation argued that Indian Head had ample housing and facilities for the workers: "More than 300 workmen from Indian Head had been compelled to leave their homes and go to Dahlgren to work." Some of them, Mudd claimed, still had their homes in Indian Head. However hopeless the situation was for Indian Head, the representatives left with assurances of support from the two Maryland senators.[37]

Despite the local efforts, the Senate appropriations committee struck Mudd's amendment, reasoning that "the conditions at Indian Head and the available range from that point make it impossible to fire projectiles at high elevation and long range." Thus, about a year after the last gun had been fired, the protests against the move finally died out. Indian Head's rapid wartime growth had been dramatically reversed in the postwar period by constant employee and output reductions.[38]

Officers, enlisted men, civilians and their families, now recognized that Indian Head would have to survive without the Proving Ground. The difficult years of building housing and then struggling to build a government and schools to go along with the housing had forged the beginnings of a sense of community pride. That pride took an interesting form in the early 1920s with specific activities, designed jointly by the town and the Navy, devoted to drawing attention to the life and attraction of the community.

INDIAN HEAD BETWEEN THE WARS

At 9:25 P.M. on 9 June 1920, Francis Mattingly, Thomas Norman, and Frederick Shaw convened the first meeting of the Indian Head town commissioners. This step put the community more firmly on the path away from company town towards self-government. The commissioners elected Mattingly president, appointed J. Milstead clerk and treasurer, placed bond at five hundred dollars, and adjourned at 9:55 P.M. All of this was done according to "AN ACT to incorporate the Town of Indian Head, in Charles County, Maryland," passed by the Maryland General Assembly on April 16.[39]

The incorporation of the town institutionalized the Indian Head tradition of civilian committees approaching the chief inspector of the Station for services. In the past citizens had asked the Station for schools, petitioned for boat transportation, and conferred for improvement in the community stores. On these and other occasions the Station had always lent an ear and usually provided substantial help. The incorporation of the town reflected the prior growth of the "Community Council," a form of self-government established by Captain Lackey for those living on Station property. The 1921 report from Indian Head reflected the fact that Lackey's Community Council had been a school for democracy. "The plan of civic organization by which an elective Community Council makes recommendations of various kinds to the Inspector concerning the welfare, living conditions, improvement, etc. of employees, has been in successful operation throughout the year."[40]

Even in self-government the community still relied heavily upon government largesse. In November 1920 the citizens of Indian Head presented their commissioners with a petition asking them to work out a plan with the Station to provide electric light for the town. Mattingly was authorized by the commissioners to correspond with the authorities at the Proving Ground regarding the light project. Although the town did not receive electricity until 1925, it is clear that the important work on Station was tied so closely to the well-being of the town that the Station was attentive to the town's needs. The system was working as planned.[41]

In the early and mid-1920s, the commanding officers and the townspeople worked closely together to provide some of the essentials of a modern lifestyle. The Station arranged for local civilians to buy such items as ice, coal, and heating oil from the Navy. As local suppliers of each commodity began to go into business, the Navy would gracefully phase out its supply operation so as not to compete with the private sector. In the meantime, however, the community of Indian Head could benefit in these practical ways from the cooperative attitude of Station commanders. Sometimes the provision of services could backfire. Nongovernment employees felt that employees who also ran businesses had received an unfair competitive advantage because such employees could bring goods for resale from Washington on the government launch free of charge. In a petition to the Secretary of the Navy, the independent merchants protested the practice.[42]

Despite persistent paternalism, a sense of community identification emerged. From 1920 through 1928 the town held an annual community fair. This fair had been using the government schoolhouse, and in 1922 its exhibits moved into a frame building erected during the war to house employees. The Indian Head Fair Association organized these events that included dances, an exhibit hall, and many other activities, including baking contests. One popular activity was a form of jousting, in which the horseman attempted to spear a series of small rings while at a full gallop. Perhaps the Indian Head Fair, held every September, began Indian Head's long tradition of celebrating town events in September.[43]

Signs of town growth occurred throughout the twenties. In November 1923 the Bishop of the Episcopal Diocese of Washington came to Indian Head when the cornerstone of St. James's Episcopal Church was laid. The old St. James Chapel, erected in 1903, was sold in 1920, as plans were made for a larger building. By 1928 Lackey School needed renovating at an estimated cost of $25,000. The school's unusual operating procedure—maintained by both the U.S. government and the county—necessitated meetings between Maryland congressmen and Brigadier General Herbert M. Lord, the federal government's Director of the Budget.[44]

In the twenties the Powder Factory sought to produce small amounts of powder as economically as possible. The need for bulk powder production declined from the wartime high, and there was genuine concern that the Powder Factory, like the Proving Ground, could be phased out. However, the Navy rationalized keeping the production line in operation, if only on a partial basis:

> It is realized that manufacturing such a small amount of powder [1,400,000] is far from economical; however, the bureau considers that in the long run this manufacture represents an economy, as it keeps the powder factory in good condition, ready for use, and at the same time it permits the retention of a small body of trained men upon which expansion could be made.[45]

The removal of the Proving Ground, and the operation of the Powder Factory on a lower level, provided Indian Head with years of apparent quiet. Behind the scenes the intensive work on tracers, illuminating shells, and other new formulations of explosives and propellants from the Experimental Ammunition Unit at the Washington Navy Yard provided an exciting agenda for the chemists working for George Patterson and Walter Farnum. But such activities did not employ the vast numbers that had been required for powder production during the war years. Thus, the Station took on a deceptively quiet appearance. Local tensions arose from political actions to help improve the town, or from rumors of the closing of the Station, as well as the actual removal of the Proving Ground facilities and the personal and work-place troubles arising from it.

Often through these quiet times, the most action on Station was provided by patrolling marines who rode between the residential and factory sections of the Station. One third of the houses were boarded up, as the number of factory employees had been reduced to the lowest safe operating level. Brady Bledsoe remembered patrolling the nearly empty streets of the Station for eight hours every day. Three marines would patrol in shifts from 8:00 A.M. until noon, noon until 4:00 P.M., and 4:00 to 8:00 P.M. Bledsoe recalled the quiet patrols when he would take along an extra horse for his girlfriend, whom he would later marry. Dorothy Artes recalled living on Station as a young girl and going to sleep to "the clop clop of the hooves of the marine's horses."[46]

Mrs. McWilliams walked to work at the Naval Station every day: "In those days everyone walked." She and her children would see a movie each weekend, and the social life of her family revolved around church and school affairs, altogether a clear mix of Station and town operations. Apparently, the "Roaring Twenties" left Indian Head when the last gun roared in the summer of 1921; the decade is more aptly remembered as a rustic idyll.[47]

The next decade was one of depression in the United States, but Indian Head escaped layoffs and scarce jobs. Indian Head had experienced its economic turbulence almost a decade earlier when the proving ground moved away and production at the Powder Factory was reduced to a minimum. However, elements of the Depression did seep in. Mrs. McWilliams remembered the times

> I have vivid memories of the Depression which did not affect the people who worked for the government like it did the farmers and people who had other occupations. No one was laid off, and we worked our usual five and one-half days per week without any reduction in pay. Our bank, however, failed, which did affect some of us. . . . After it opened again we all had letters telling us that one-third of what we had in the bank would be paid at once, another third in six months, and the balance would be paid in bank stock at ten dollars per share. This made some people mad as before the bank failed, its stock was being sold at one hundred dollars per share.

Although the failure of the bank in the thirties angered citizens, the town could still rely upon the government payroll to see them through.[48]

The programs of President Roosevelt, who visited Indian Head during his days as the Assistant Secretary of the Navy, also helped Indian Head. In 1933 the Navy Department announced it would allot Indian Head $71,000 for the production of smokeless powder. The funds were made available by the Public Works Administration, a relief program designed to raise consumer purchasing power. The Civilian Conservation Corps, another New Deal program, opened a camp in Indian Head. The Corps worked to improve the roads, plant new trees, and beautify the Station.[49]

President Roosevelt's interest in Indian Head continued. While on the Potomac, he observed the soil erosion on the bluffs along the riverat Indian Head. He directed the Station to take corrective action and suggested the use of Works Progress Administration funds and resources if necessary. Four months later Roosevelt followed up on the situation, and the chief of the Bureau of Ordnance, William R. Furlong, told Indian Head that the "President has personally indicated his desires" that the erosion be arrested.[50]

So business continued as usual in Indian Head, and that business, beginning in 1929, turned more and more to roads and cars. In September 1929 the town commissioners authorized a survey of road conditions in Indian Head. The commissioners made arrangements to get gravel for the roads and also inquired into obtaining signs to post a twenty mile per hour speed limit in Indian Head. This process continued unhindered by the depression, as the U.S. government kept Indian Head on an even economic keel. In June 1931 the town accepted the government's offer to scrape and furnish gravel for the roads at $25.25 per load.[51]

The work began in the summer of 1931 and by September the project was nearly wrapped up. According to minutes of the September meeting, "Commissioners McWilliams and Rison were authorized to finish graveling and oiling the road," and the commissioners "agreed that seven hundred $700 be burroughed [sic] from the Eastern Shore Trust Co. for completion of oiling and graveling the road." Dorothy Artes remembered that transportation by land to Washington was not reliable until World War II.[52]

The condition of Route 224, the road leading to the Station, was of particular concern to station officials. Route 224 was nearly unusable as late as the spring of 1940, just when the Station was gearing up for years of tremendous output. The commanding officer warned that the road would have to be closed due to the coming thaw combined with the increase of heavy freight shipments and employee passenger cars using this highway. The poor road conditions were due in part to neglect. The officer further reported residents of the "community and outlying districts depended largely upon boat transportation for their mode of commuting to and from the District of Columbia, where they were obligated to do most of their shopping for clothing, household supplies and other necessities, as well as a large portion of their groceries, since the source of supply at Indian Head was so limited."[53]

Dr. Frank Susan remembered coming to Indian Head in 1935 when it looked "just like a western town." He arrived with the Civilian Conservation Corps and found himself to be the only doctor in Indian Head. Before the Second World War, the commanding officer at Indian Head called Susan, who had a reserve commission with the Army, into his office for a meeting.

"I understand you're a reserve officer," the commander began.

"Yes, I am," Susan replied, not yet understanding the purpose of the meeting.

"I want you to resign your commission."

"Well, why is that?" Susan asked.

"You're the only doctor between here and Washington; war clouds are gathering and we might get involved." The commander was worried that the town might lose its only doctor. Susan wrote to the Army, cited the position of the Navy, and resigned his commission.[54]

Dating back to Mason's first pleas for medical attention for Indian Head, medical facilities and doctors had always been scarce. In 1936 Senator Millard Tydings of Maryland appealed to Claude A. Swanson, Secretary of the Navy, to reverse a ruling which prevented many government employees living off Station from receiving medical assistance from the dispensary. Tydings argued that the nearest hospital to Indian Head was thirty miles away in Washington. The ruling was reversed.[55]

In an odd twist of tradition, the construction of a modern sewer system for the town was instigated by the Station. In 1935 the commissioners held a town meeting to discuss sewage plans at which fifty residents and a representative from the state health department, Dr. Campbell, were in attendance. Campbell delivered an important message from the Station:

The commissioner related to those present the attitude which the Navy Department had taken with regard to the unsanitary condition existing in certain parts of the town which condition they claim affected the residents of the government reservation. Those taxpayers present, realizing the seriousness of the situation, agreed to have their property tax raised to fifty (50) cents on the hundred dollars.

Mr. McWilliams, chairman of the town commissioners, circulated the price estimate from Whitman Reynold & Smith of Baltimore among those present. A vote was taken and the proposal for the sewer system was unanimously approved. The meeting authorized the town commissioners to push ahead with the project. To pay for the project immediately, the commissioners approved a public offering of $10,500 in registered bonds, and announced that the Works Progress Administration would provide the remainder of the funding.[56]

The quick response by the town government and citizens to correct the situation indicates the relationship between the Navy and the town. The town depended upon the Navy for its jobs, for transportation to Washington, and for business at its local stores. Those present, "realizing the importance of the situation," immediately accepted new taxes, and the commissioners acted quickly to remedy the unsanitary conditions.[57]

Although Indian Head was southern Maryland's only industrial center, its effects on the rest of Charles County were surprisingly slight. Most of those who worked at Indian Head moved to the town or nearby Marbury, and the base was in a corner of the county closer to Washington than to many points in the county. As Sharon L. Camp said in "Modernization: Threat to Community Politics"

> The Navy, following the pattern of military enclaves elsewhere, built its own self-contained community with school, post office, dispensary and 100 government houses. The dominance of the base in the lives of area residents encouraged an inward looking community which kept itself somewhat apart from the rest of the county. The rest of the county, on the other hand, gladly reciprocated in kind. By way of illustration, oldtime plant employees still recall with some resentment the area's long exclusion from county political life and the hostility of many older county residents to newcomers attached to the base.[58]

When Indian Head began to expand in the late 1930s, it had more of an impact on other parts of the county. By 1940 nearly a thousand houses had been built to accommodate the influx of workers due to the expansion of the plant. One such impact involved medical facilities in La Plata. On 2 December 1941, the Federal Works Agency announced the "Presidential approval of a Defense Public Work project to provide additional hospital facilities for defense workers and their families at Indian Head, Maryland." This expansion was to take place at the Physicians' Memorial Hospital in La Plata that had opened in January 1939. The announcement stated that "employment at the Indian Head Naval Powder Factory has increased from 750 to 2400 and is expected to reach 5000 before long. Defense workers at Indian Head have brought their families and the county population has had a corresponding increase, so that present facilities are entirely inadequate." Despite the fact that Indian Head was beginning to break down its isolation from La Plata and the rest of the county, the issue of the relationship of the Powder Factory's town to the rest of the county would persist for decades.[59]

The idea that the county community and the community of Indian Head remained isolated from each other was later echoed by Jay Carsey, who became president of the Charles County Community College after having been a chemical engineer at Indian Head for several years. Carsey remembered that the division had a social as well as a political side. He made an effort, as college president, to hold social functions at his home, inviting representatives of the local social elite and Indian Head management, intending—with some success—to bridge the gulf.[60]

The bombing of Pearl Harbor brought Indian Head, along with the rest of the country, into a new age. Production at Indian Head continued to grow as the base went through a second period of rapid, war-induced expansion. Indian Head recruited workers from North and South Carolina and West Virginia as well as from Charles County and the rest of Maryland to meet this new challenge. New houses would have to be built and production facilities improved. After the war, new questions concerning the function of Indian Head, the naval facility—even its continued existence—would be discussed again, but the town, with fifty years of preparation, was strong and ready to play a role of political support.[61]

CHAPTER 6

World War II

Magazine No. 1
Building 67
1901

THE YEARS OF World War II brought rapid, unplanned growth to the facility at Indian Head. Some of the growth spawned other laboratories and facilities, which relocated elsewhere. Some of the new processes gave new capacities in facilities and personnel that would help shape the postwar identity of Indian Head. Indian Head saw the temporary housing of the Jet Propulsion Laboratory (JPL) there, a group which later formed the core of the establishment of the Naval Ordnance Test Station at Inyokern, California. Almost overnight, the Navy set up at Indian Head another facility, called the Explosives Investigation Laboratory, which like the Experimental Ammunition Unit (EAU) in the First

105

World War, did extensive examinations of captured enemy ordnance. A third major expansion came in the area of production. In cooperation with researchers at the California Institute of Technology (Cal Tech), in Pasadena, California, Indian Head put into operation an extrusion plant for pressing ballistite, or "Jet-Propulsion/Navy" (JPN) powder into rocket "grains," while continuing to produce smokeless powder and Explosive D. The expansion into production of extruded grains would reshape the destiny and mission of the facility.

In 1941, before American entry into the war, the Powder Factory took orders for the production of 1.3 million pounds of smokeless powder and 600,000 pounds of Explosive D, to be scheduled over the period April 1941 through June 1943. Even before U.S. entry into the war, additional plant capacity was added, with du Pont, as the major contractor, providing over five hundred people to build extensions to the smokeless powder plant under an appropriation of $4,464,000. Through November 1941, new distillation, storage, solvent recovery, and magazine buildings were accepted. Du Pont also oversaw the construction of new power facilities, steam lines, railroad tracks, and water lines. The contract was completed 23 January 1942. A new Explosive D plant went into operation on 15 April 1942.[1]

The Chemical Laboratory, founded by George Patterson, continued to play a major role for the Bureau of Ordnance in providing chemical expertise in the fields of explosives and propellants. After his retirement in 1940, Patterson's laboratories continued to grow. The development of laboratories, staff, and equipment to test the new products left Indian Head, at the end of the war, with a recognized research capacity, despite the transfer of specialists to other facilities and the shortage of replacements. Admittedly the research and the production facilities had grown "like Topsy," with little central coordination or long-range planning.

The proliferation of other facilities elsewhere in the United States during the war, sponsored by the National Defense Research Council (NDRC), the Army, the Navy, and corporate contractors, created an atmosphere of institutional competition in the new areas of rocket research, development, testing, and production. Production facilities at Radford, Virginia, and at Sunflower, Kansas, made rocket propellants. The Aerojet Corporation also focused on making new propellants. Aerojet had been founded by Cal Tech scientists engaged in the development of the first Jet Assist Takeoff Units (JATOs) and had started with a government-owned plant in Azusa near Pasadena. An NDRC facility, the Allegany Ballistics Laboratory in Cumberland, Maryland, was taken over by Hercules Powder and run as a contractor-operated facility, doing research and development. The test facility established by the Navy at Inyokern, California, the "China Lake" Naval Ordnance Test Station, became a major research and development center for naval rocketry at the end of the war. Older naval test facilities at Dahlgren, Virginia, and Chincoteague, Maryland, vied for a role. The Naval Ordnance Laboratory at White Oak, closer to the nation's capital, would clearly continue to be the Navy's major facility for ordnance research.[2]

With the rapid growth of new private and governmental facilities, factories, laboratories, and testing stations, the production and research facilities at Indian Head were faced with several potential challenges. Would Indian Head adapt from gun propellant manufacture to the new technologies of the rocket? And, having addressed that issue by moving in wartime into extrusion of rocket propellant, would Indian Head be able to maintain a significant role when private corporations, now flush with government contracts and government-provided facilities, continued as suppliers to the Navy in the postwar world?

The immediate postwar years would determine whether Indian Head would successfully adapt and find a niche in the new world of large-scale defense research and production. If it failed, in the worst scenarios imagined by its defenders, it could fall by the wayside as an outmoded factory, dedicated to a by-passed product. Its shrinkage or closure would create a local economic hardship, but from a national perspective, the argument for Charles County would have to compete with hundreds of other similarly affected regions throughout the country. If Indian Head succeeded in defining a place in the world of new propellants, its role would be assured over the coming decades as the Navy continued to modernize.

The war years would bring several assets which would help in the search for a strengthened role. The new highway connection would transform Indian Head's location from a disadvantage to an advantage, when the postwar relaxation of gasoline and tire rationing suddenly would bring it into the outer ring of Washington, D.C., suburbs. But more crucial were the assemblage of technical successes, new functions, and new equipment which came rapidly during the war years. The work in extruding propellants for rockets, testing rocket motors, developing new weapons and manufacturing techniques, and building production facilities for new explosives and propellants would show the potential for future expansion in certain particular directions.

INDIAN HEAD IN ROCKETS, WORLD WAR II

During the war, naval advocates of development of an air-to-air rocket established two centers to produce rocket propellant grains, one under Linus Pauling and Charles Lauritsen at the California Institute of Technology in Pasadena, and a second, on a larger scale, at Indian Head. In the postwar years, the Cal Tech work at Pasadena would be far more famous than the Indian Head contribution, but a close examination of the details of the extruded rocket story shows that Indian Head in fact played a larger role in total production and a significant role in much of the developmental engineering work.

Through 1941 and into early 1942 Lauritsen worked on acquiring a suitable propellant, settling on an experimental double-base formula from Radford Arsenal in Virginia and setting up extrusion presses to produce small rocket grains in Eaton Canyon, about two miles from the campus in the hills above residential Pasadena. The double-base powder was composed of both nitrocellulose and nitroglycerin. Radford would manufacture this ballistite propellant in sheets, then wrap it up in "carpet rolls" for shipping to the Cal Tech people. The propellent had the consistency of soft leather or, when warm, a child's modelling clay. At the Eaton Canyon facility, the carpet rolls would be extruded through a press into a cylinder of the appropriate diameter and length, machined to final dimensions. The cylindrical grains would be shipped elsewhere for loading into 1¾" and 2¾" rockets. By the end of February 1942, Lauritsen had two presses producing a total of two hundred pounds a day of experimental propellant grains.[3]

Although Lauritsen later complained that the Navy was slow to see the value of rockets, and historians have viewed the struggle to put the rocket into production as a fight against bureaucratic inertia, a close review of the record from 1942 to 1944 shows that the Bureau of Ordnance hastened to transform the small demonstration plant in Eaton Canyon into a model for large-scale production. Indian Head was at the heart of that effort, playing a role in scale-up from laboratory to industrial production which would become its trademark in later years.

In July 1942 two members of the Indian Head staff, Ensign T. F. Dixon and civilian chemist J. B. Nichols, traveled to Pasadena, California, to study Lauritsen's methods of extrusion and finishing of double-base, JPN powder grains. JPN was a nitroglycerin-nitrocellulose double-base propellant only slightly altered from earlier gun propellants. The two staff members spent five days consulting with the engineers at Cal Tech, taking both still and motion pictures in the bright California sunlight, and examining the presses, testing facilities, and storage magazines that had been set up on a small scale in Eaton Canyon. Nichols and Dixon reported in detail on the setup of the presses, safety features, installation of primers, testing, storage, and problems of designing an automatic cutter to slice the grains to specified lengths after extrusion from the press. The two men reported on methods developed at Cal Tech of wrapping the grains with cellulose-based tape, a material similar to cellophane. The acetate would be applied to the outside of the grains to regulate the burning rate of the propellant, serving as an "inhibitor."[4]

Both observers recommended several changes in operation before the system would be installed at Indian Head. In particular, a process by which the grains were shaved down on a lathe to conform to the specified weight and length struck them both as fraught with danger. Indeed, the lathe operation in a Cal Tech lab had resulted in several accidents and the death by burning of two technicians early in the project. In general, Dixon and Nichols found that the Eaton Canyon plant relied on highly skilled technicians, rather than on production workers, and they both sought ways to improve and simplify the process so the presses could be operated on constant, round-the-clock shifts by unskilled production staff. Dixon recommended that the automatic cutter be designed to cut the lengths within the desired tolerances to eliminate cutting by hand. The two men planned to design the equipment with automatic controls for safety shutdown and to staff the production equipment with women employees.[5]

The suggestions made by Nichols and Dixon were incorporated into the plans for an extrusion plant at Indian Head. Perhaps the most significant redesign of the extrusion equipment was to set up presses which worked on a horizontal axis, rather than a vertical line, as the Cal Tech presses were arranged. To set up Indian Head extrusion on a production, rather than an experimental basis, additional presses were ordered and a large scale of operation was planned. Dixon was put in charge of seeing that the procurement of equipment proceeded on schedule, and by November he reported to the Bureau of Ordnance that he "had to go from desk to desk most of the time" to get the orders through purchasing. "I've got their hair going gray over here, but I don't care, they are going to have to get the stuff."[6]

By mid-1943 the first Indian Head presses were at work. Two presses were put in operation by August, with three more ready to go if sufficient staff could be hired to run them. In order to gear up the production, Officer in Charge Captain Mark L. Hersey requested authorization to hire 87 men and 104 women, finding the civil service recruiting office at the Washington Navy Yard of little help.[7]

Small alterations in the extrusion press allowed for adapting the extruded grains to conform to modifications of the "specs." At this stage, the ballistite grains under production were labelled with a different mark number assigned to each diameter of grain and "mod," or modification, numbers issued to improved grains. Decades later, as the modification numbers climbed, representing later generations of improvements, a great variety of extruded rocket propellant grains could trace their "ancestry" back to these early products at Indian Head. Modifying the diameter of the major axial perforation in the center of the grain and minor changes in overall weight and length of the grain could be achieved with minor changes to the extrusion press die and automatic cutter settings.[8]

The extrusion presses turned out a steady flow of ballistite rocket grains, and the Army's Sunflower Ordnance Works in Kansas, along with Radford, began to provide carpet roll Jet Propellent (JP) powder for the presses. The Navy set the specifications of the carpet roll dimensions, receiving rolls just under 8 inches in diameter for the 8-inch press and just under 10½ inches for the larger press. By April 1944 the Indian Head presses were scheduled to produce over 150,000 pounds of grains a month, or about 119,000 individual grains. The grains, depending upon mark and modification, weighed from less than half a pound to about a pound and a half each. By June 1944 the Bureau hoped to get the eleven presses to produce over 900,000 pounds of grains a month, in a wide variety of marks and modifications, utilizing JP material provided from Army facilities. Actual production did not quite meet the planned level, but in November Indian Head could report production running about 670,000 pounds a month, in eight different marks of grains. Mark 1 grains—the large original grain—weighed about one and one-half pounds each and remained the major product in 1944, with over 100,000 grains produced a month in that single category.[9]

As the demands of the war increased with heated naval action in the Pacific through 1944, the Bureau continued to urge Indian Head to step up production to over 900,000 pounds a month. Early in 1945 the Bureau ordered the closing of three 8-inch presses and retooling of them in order to shift from Mark 1 production to Mark 13 and Mark 18 grains, which were larger and heavier than the earlier rocket grains.[10]

Through this period, the double-base JP powder had a chemical formula that consisted of about 50 percent nitrocellulose (of a 13.25 percent nitration rate), about 44 percent nitroglycerin, about 3 percent diethylphthalate as a plasticizing agent, and traces of potassium sulfate and ethyl centralite. The term "about" may seem inappropriate in describing such formulations, but the specifications allowed variation of up to 2 percent of the major two ingredients in the interests of maintaining a steady flow of powder and grains. The variations in the supplied raw material and variations in processing produced grains with widely varying performances.[11]

A major bottleneck developed as the application method of the ethyl cellulose inhibitor material by hand at naval amunition depots held up delivery of completed grains to the fleet. An experimental machine to apply the inhibitor at Indian Head went into production on Mark 13 grains in 1944. Indian Head conducted experiments on the resulting grains, testing them at various temperatures to evaluate the effect of different ratios of acetone, alcohol, and ethyl cellulose in the inhibitor mixture.[12]

F. C. Thames, the civilian chemist in charge of production, was selected as supervisor of the extrusion work and visited Eaton Canyon as well as the Army's Sunflower plant. In reporting on both facilities, Thames noted how variation in the slurry process of making the double-base propellant, together with variation in temperatures and pressures on the presses, produced a variable product. He lamented, as did Lauritsen at Cal Tech, the lack of time for a more scientific approach. As he discussed the absence of scientific data, the wide latitude in specifications, and the inability to gather and correlate information, Thames made it clear that he viewed rocket motor production as a shockingly imprecise business. He noted the wide deviations from specifications in every lot of produced grains. He complained that the whole operation, both at Indian Head and at Eaton Canyon, was run with too many variables. He was somewhat appalled at the attitude towards specifications which he encountered at Cal Tech but recognized that in the interests of producing large numbers of rockets, some variation had to be accepted.[13]

The Eaton Canyon experimental production facility continued to produce small quantities throughout the period, but the bulk of the standard production of rockets for combat use by the Navy came out of the extrusion plant established at Indian Head. By mid-1944 Thames constantly asked for more skilled assistants to replace some who he felt were unsuited in skills or intellect to handle the challenges of the job. Thames complained that his division had grown so fast that personnel would only be assigned long after the need for them had mounted to a nearly impossible level. He worked with Nichols, Harry Stern, and John Scott, all of whom he recognized as having important professional skills, although he hinted at a variety of personality difficulties. Scott was a "pusher," good at getting out production but a man who needed supervision in "desk work." Nichols was professional, but Thames alluded to difficulties he had had in his previous assignments, noting that he had found "a niche" in extrusion. He was even less complimentary about some of his other assistants, noting of one that he "gave him up as hopeless." When he found a good, promising young worker with knowledge of both factory and naval procedures, he agonized over promoting him above others because it would lead to discontent and "practically rebellious action." Despite all of Thames' difficulties in putting together a team on short notice, he was quite proud of their performance under pressure.[14]

Indian Head and Cal Tech worked closely together on the extrusion plants, with frequent cross-country visits to each other's facilities to study safety methods, inspection techniques, and ways of improving production rates. The frequent comparisons of techniques, management, and procedures were instructive to both facilities, and often the Indian Head representatives reported with some pride that their operation seemed superior in one respect or another. The Indian Head production facility was built with the Eaton Canyon experience behind it. In many ways it was not only bigger in production scale but more capable of accurate work. Stern, one of Thames's appointments and an associate chemist, visited the Eaton Canyon facility early in 1944. He noted problems with equipment and management there. "I was not at all impressed," he noted, "by the complicated set up at Cal Tech." He found it very inefficient to have different functions handled in separate departments rather than under a single administrator as at Indian Head. Cal Tech had separate staffs and directors for each unit: Magazine, Motor Loading, Static Testing, Calculating, and Inspecting. Thames especially noted the excessive reliance on skilled technicians in production at Cal Tech, several non-standard storage methods that struck him as dangerous, and other minor flaws in the hastily assembled Eaton Canyon plant.[15]

By January 1944 six firing bays had been constructed at Indian Head for the testing of extruded grains, and the commanding officer requested personnel to staff this "Ballistic Laboratory," as it was to be called. Captain James Glennon requested a total of 291 personnel slots to handle the work of testing the extruded grains on a three-shift basis. Glennon recommended that the total complement be made up of WAVES, and requested new barracks and housing for 25 officers and 266 enlisted women.[16]

Later in 1944 Thames set up in his inspection group a radiography unit to x-ray the produced grains to supplement the system of visual inspection and test firing or "destructive tests" of sample grains. When the chief of the Bureau of Ordnance ordered the construction of x-ray buildings to house the new equipment, he suggested that the project could result in savings, since fewer grains would have to be consumed in firing tests. He planned two buildings, each to handle about half of the extrusion press production, and each to be staffed by nine technicians to handle the equipment and to organize the flow of grains through the x-ray equipment. By July the Powder Factory had settled for a single, temporary x-ray facility, located in Dry House #58.[17]

Shortly after the war, early in 1946, Thames was asked to describe the contributions of the extrusion division to the war effort. In the hectic pace to meet production goals and in the generally classified environment of World War II production, there had been no opportunity to develop a public relations presentation of the achievements, and by the time Thames offered his assessment, the work was no longer "news." However, in retrospect, it is now clear that the rapid construction, from the ground up, of an extrusion plant, together with a host of specific contributions there, represented a relatively "unsung" success story which passed almost unnoticed in the world of naval rocketry. Like much of the incremental technological work done by Indian Head during the 1920s, the rapid and quiet innovation, essential to progress in a new area, got little public attention compared to the more spectacular innovations in armament elsewhere during the war. Thames listed several specific achievements:

◇ production of rocket propellant grains of quality in quantity

◇ installation of the first 15-inch press for rocket
◇ propellant in the United States and development of the dies for the press
◇ conducting of experiments in re-extrusion of scraps
◇ development of new inhibitors and new methods of inhibitor application
◇ development of improved methods of loading both Mark 13 and Mark 16 motors
◇ study of the reasons for failure of Mark 4 motors, and change of specifications in accord with Indian Head suggestions
◇ contributions by machinists Sergeant and Howard in developing end-facing and drilling machines for use on propellant grains[18]

Although Thames stressed the research and development aspects of the propellant work, the achievement which he mentioned first—the production of "quality grains in quantity"—was probably the most significant. Indian Head had taken a partially developed laboratory method and converted it into a factory-scale operation. The development work done by Thames's staff had been for the most part auxiliary to improving the production. That whole process of taking an innovation from the laboratory to the factory and then actually producing the new material for fleet use, in response to the needs of war, had become an established, if somewhat underplayed, tradition at Indian Head.

The pattern of rapid establishment of a production plant in response to a wartime need would have been familiar from earlier decades if the engineers and chemists at Indian Head had taken time from their busy schedules to dwell on their own history. Patterson had converted the smokeless powder process from the small Newport facility into a major factory in 1900, after the Spanish-American War had revealed the dire need for large-scale production. In World War I, Walter Farnum had set up the Explosive D works. And in the midst of World War II, Nichols and Thames had done a similar job, putting together a plant to produce hundreds of thousands of small double-base rocket grains. In each crisis, Indian Head had taken a laboratory-scale technology and turned it into a factory to supply the fleet directly. That pattern, apparent now in retrospect and mentioned by Thames in passing, was a major strength of the facility.

Later, that particular role—conversion from laboratory scale to production scale—would become the key to redefining the place of Indian Head in the scheme of naval weapons procurement. In 1946, however, Thames' comments were quietly filed away.

Other World War II achievements at Indian Head added to the track record in various aspects of research, development, testing, and evaluation. Two research facilities housed at Indian Head during the war years, the JPL and the Explosives Investigation Laboratory (EIL), made especially noteworthy contributions in ordnance. Although their work was part of the Indian Head success story, several factors mitigated against full recognition of that fact. The work was classified and hence received little contemporary publicity. Furthermore, neither facility was directly under the Powder Factory administration. The forward-looking work at the JPL, like that at the EAU and Dahlgren in earlier decades, would be sheared away and relocated. The EIL, although dramatic, was a temporary, ad hoc group under an independent administration. Its successor organization remained located at Indian Head, with a continuing separate existence as the Explosive Ordnance Disposal Technical Center and the Explosive Ordnance Disposal School.

THE JPL AT INDIAN HEAD

The Jet Propulsion Laboratory at Indian Head had been set up in 1942 as a joint activity of the National Defense Research Council and the Powder Factory, with the factory contributing funding for equipment, administrators, and nontechnical labor, but less than 5 percent of the budget covering college-trained technical workers. The organization was changed 1 January 1944, when it was set up with six naval officers, entirely as a Navy-supported unit. When Captain Hersey's successor as officer in charge, Captain Glennon, established the reconstituted JPL, he specifically asked for an increase of another eleven naval officers with skills in mechanical engineering, machine design, electrical engineering, physics, electronics, chemistry, chemical engineering, and report writing. The group was to continue rocket research, augment the rocket work of the NDRC, and provide experienced personnel to the extrusion plant when difficulties arose in putting new marks and modifications into production.[19]

The JPL at Indian Head continued to do experimental work through early 1944, and it constituted a research and development laboratory specializing in small rockets. Staffed with naval officer engineers and chemists, the JPL was independent of the extrusion production facility but worked closely with Thames, using the extruded small rocket grains as propellants for new devices. The officers occasionally served as consultants to the extrusion plant.

Thames and his machinists made innovations in production methods; the JPL concentrated on inventing specialized new applications of small rockets in a wide range of weapons and illuminating devices. The Bureau asked the JPL to work on a photoflash rocket for night photography, using the 3.25-inch "CIT" rocket as the propellant. Another project involved a rocket flare for post-attack illumination, utilizing a 1.25-inch rocket motor. Again, the Indian Head JPL worked on the project, using an experimental rocket produced at Cal Tech. Another rocket designed to eject "windows," a code word for anti-radar chaff, was designed at the JPL under the direction of Lieutenant C. C. VanderWall. The request for the development came in December 1943, and by April a successful run of four hundred samples of the window rocket had been produced for testing of the anti-radar aspects by the Naval Research Laboratory. The JPL worked on several flare and illuminating rocket projects, target rockets, and an Armor-Piercing Rocket Bomb with a 1,250-pound warhead. During its stay at Indian Head, the JPL, under Lieutenant Commander R. A. Appleton, could report progress on some seventeen rocket projects, as listed in Table 9, and issued memorandum reports on most of the projects.[20]

Lieutenant Commander Appleton reported on progress on all of the seventeen rocket projects by July 1944, and the group issued regular reports on the work. As the Navy concentrated its rocket work at Inyokern, California, establishing the Naval Ordnance Test Station (NOTS) there, the whole JPL group was

TABLE 9
Indian Head JPL Projects, July 1944

1. Window Rocket
2. Rocket Bomb; 12-inch motor to increase velocity of the Mk 33 bomb
3. Flare, Rocket
4. Rocket Bomb, Mk 50-12-inch Bomb (1,250 pound)
5. Flare, Rocket (PT Boat)
6. Flare, Rocket (Aircraft) Mk 7 motor with 4-inch Flare
7. Flare, Rocket (Aircraft) Mk 7 Motor with 5-inch Flare
8. Post-attack Rocket Flare (Aircraft) 1.25-inch Mk 2 motor with 3-inch Flare
9. Photo-flash Rocket Bomb (Aircraft) with Mk 7 Motor with 4.5-inch Bomb
10. Flare, Rocket, 3.25-inch Mk 12 or Mk 14 Motor with 4-inch Flare
11. Flare, Rocket, 3.25-inch Mk 5 Motor with 4-inch Flare
12. Tandem Barrage Rocket—two 2.25-inch Mk 9 motors in Series
13. Black Powder Rotating Rocket Flares
14. Dale Target Rocket (Black Powder)
15. Ejector Body
16. Rocket Flare Mk 1
17. Study of 4.5-inch Barrage Rocket Stability

Source: Lieutenant Commander R. A. Appleton, Memorandum Report, Jet Propulsion Laboratory, Indian Head, 4 July 1944, Record Group 74, Entry 25, NP8/A9, National Archives and Records Administration, Washington, D.C.

shifted from Indian Head to the new facility in the late summer of 1944. Thus, just as Indian Head had been the parent to the Naval Ordnance Laboratory eventually established at White Oak and the Naval Surface Weapons facility at Dahlgren, so it was one of the parents of the NOTS/Inyokern facility, along with the Cal Tech specialists.[21]

EXPLOSIVES INVESTIGATION LABORATORY

In June 1942 the chief of the Bureau of Ordnance authorized the establishment of an "Ordnance Investigation Station" at Indian Head. The history of this unit during the war years would make an interesting book in itself, and only a fraction of the work accomplished by the unit can be described here. Set up in July, the formal name of the new laboratory was the Explosives Investigation Laboratory (EIL), and it was established on the Stump Neck section of the Indian Head reservation. Several quite hazardous types of work were delegated to the new laboratory from its inception:

◇ stripping and disassembly of recovered enemy ordnance, including mines, torpedoes, and bombs
◇ researching methods of location, identification, and disposal of enemy and "our own" ordnance
◇ demonstrating, for instructional purposes, ordnance disposal techniques[22]

To avoid overlapping jurisdiction with other research facilities, the new laboratory would only conduct research on specific approval from the Bureau. The Bureau planned to funnel requests for specialized studies from other facilities including the Mine Disposal Unit and the Bomb Disposal Unit, both under the office of the Chief of Naval Operations and the Naval Ordnance Laboratory, and to work out demonstrations for the Advanced Mine School and the Bomb Disposal School. The Bureau planned to move the two schools into the Stump Neck facility as soon as possible. A separate "Inspector of Ordnance" would be in charge of the new laboratory, and he would work out liaisons with the various laboratory facilities under the jurisdiction of the Powder Factory. The unit was planned with three divisions: Research, with a staff of thirty; Investigation, to study enemy ordnance, with two officers and ten enlisted men; and Operations, a unit to conduct demonstration and instruction, staffed with thirty men, mostly graduates of the existing mine and bomb schools.[23]

By the end of July 1942 the new laboratory was at work. One of its first research and development projects was to work out techniques for cutting mine and bomb cases by means of special explosives, using the "Munroe Effect." Charles Munroe, who had headed the smokeless powder work at Newport in the 1890s, was internationally famous for identifying the concept of focused explosive force through shaped charges. The Bureau hoped to use the concept in opening and defusing captured ordnance. The new laboratory was to develop a method of using small charges to open mines or bombs without detonating the main charge in the device.[24]

Lieutenant D. Klein was appointed as officer in charge of the EIL in August 1942, about a month after the founding of the laboratory. Klein was provided with the yacht *Elsie Fenimore*, commissioned as the *John M. Howard*, to conduct experiments in the river. Klein's staff incorporated men previously assigned to other units, including six enlisted men of the Mine Investigation Unit of the Powder Factory. In addition, he was assigned some twenty-two officers, twelve seamen, and about forty petty officers, bringing the complement a few men over the originally planned size.[25]

Captain Hersey enthusiastically approved the appointment of Lieutenant Klein to head the new laboratory. Klein was technically capable in the area and Hersey recognized that fact, approving the concept that the young lieutenant should have considerable independence. He suggested that Klein deal directly on technical matters with the Bureau of Ordnance, the Naval Ordnance Laboratory, the two Disposal Schools, and directly with the Vice Chief of Naval Operations on matters relating to mine and bomb disposal, only keeping Hersey himself informed.[26]

Through the period 1942–45, the EIL produced reports on a wide variety of specialized ordnance disposal problems, published as "Ordnance Investigation Memoranda (OIM)" and "Explosives Investigation Laboratory (EIL) Reports." While a project was still in process, it would be designated by an OIM number, and when the work was completed, an EIL report would be issued. The topics ranged over an interesting spectrum of ordnance issues stimulated by the rapid technology transfer of war itself. Several were derived from the early work on using explosives to cut mine cases, investigating several applications of shaped charges to "countermine" various enemy ordnance, and studies of jet propagation from shaped charges. A variety of shaped charges in strips, applied as an inverted "V," would allow for cutting shaped openings in metal and also would produce straight line shearing in the metal cases. A number of specialized investigations came out of the shaped charge work, including methods for opening a variety of bomb and projectile cases and for cutting chains. The unit conducted studies of x-ray techniques, steaming-out techniques, and techniques for igniting floating oil on water, as well as studies of demolition charges and the effects of underwater explosions. In cooperation with British researchers, the group conducted x-ray examinations of British high explosive 15-inch shell, British depth charges, and Canadian warheads.[27]

The EIL reports, complete with photographs, were typically brief, specific, and to the point, spelling out research procedures, results, and recommendations. Lieutenant Klein directly participated in some of the research work as well as in the administration of the overall unit.[28]

Another small category of projects were assigned "DIM," or Demolition Investigation Memoranda, numbers, with a range of special studies on both American and British demolition charges. Included in the DIM projects were such studies as the sensitivity of blasting caps to nearby explosions; reasons for the failure of half-pound TNT-tetryl blocks; and tests of modifications of demolition outfits provided by the Naval Ordnance Laboratory. Special pentolite demolition charges developed at the Naval Ordnance Laboratory for the purpose of cutting chain and cable were tested in this category.[29]

The EIL worked on literally thousands of samples of captured enemy equipment and would assign "CEE" numbers to the weapons. The EIL's role would be to open safely and disassemble the device, label the parts, and then refer the materials to other laboratories in order to produce reports detailing chemical composition. The EIL itself would issue brief descriptions of mechanical design and methods of disassembly and demolition. Some of the chemical analysis work on the more important types of captured enemy equipment was performed by the Powder Factory's chemistry laboratory, headed by Farnum through the war years, in cooperation with the EIL. The chief of the Bureau noted that the EIL was "the only unit established for this purpose" and gave a blanket order specifically empowering the laboratory to proceed with the disassembly and inspection of all captured enemy ordnance, "without prior direction or approval."[30]

By November 1944 the CEE series was in the seven thousand range, with hundreds more coming in each month. The work ranged over every type of Japanese, German, and Italian ordnance device, including anti-aircraft projectiles, bombs, mortar projectiles, artillery projectiles, primers, hand grenades, smoke devices, fuzes, mines, float lights, Very pistol cartridges, torpedo warheads, rockets, and box mines. Both naval and army ordnance would be studied, including army bombs and projectiles and anti-tank mines as well as naval equipment. A few items from allied nations were examined under the same procedure, including British small arms ammunition and fuzes. A study of boosters from Russian bombs showed one booster made of tetryl with a paraffin coating and a second made of pressed picric acid covered with a resin-coated paper.[31]

By 1944 more non-ordnance items were included in the studies, including equipment such as recording devices, clocks, pistols and machine guns, electric drive shafts and torpedo motors, voltage regulators, tools, parachutes, and other equipment auxiliary to ordnance. One piece of equipment was apparently the famous V-1 "Buzz bomb," listed as CEE No. 7470 "German Robot (buzz) bomb." The orders to examine that device were specific and separate from the more general marching orders to examine all captured material. Still another famous device tested at the EIL was the solid propellant rocket grain from the "BAKA" rocket-driven kamikaze plane utilized early in 1945 by the Japanese.[32]

Most of the EIL work was classified as "Confidential," with occasional radiography projects on captured equipment classified as either "Secret" or "Top Secret." As an all-uniformed outfit used to hazardous work and staffed by an eager and highly trained group of specialists in explosives, the small team of less than one hundred men turned out a spectacular array of studies, investigations, and developmental research. It is clear that the bureau chief and the Vice Chief of Naval Operations could call on this unit to work with anomalous foreign equipment, with experimental modifications of ordnance produced by researchers at the Naval Ordnance Laboratory, and with massive numbers of captured bombs, shells, and torpedoes, and could ask them to study the most advanced and classified devices which came into the Navy's possession, including the V-1. Hazardous studies of the effect of underwater explosions upon divers and other high-risk work with sensitive, unknown explosive devices added to the unit's reputation for both courage and a can-do attitude.[33]

This quick-response group could be called upon to assist in the research and development carried on by other laboratories and units, and the chief took advantage of this capacity frequently. An early project studied a combination shaped charge and incendiary device to be used in igniting storage tanks of liquid fuels, including diesel, which would be useful to commandos, paratroopers, resistance fighters, and others operating behind enemy lines. Klein personally reported on a suggested improvement which used a shielded flare set several feet from the tank, combined with a ring of plastic explosive to blast a hole in the tank. Such specialized studies, produced on specific orders, suggest the range and variety of the unit's work, along with its more routine study of the thousands of pieces of captured equipment which made up the bulk of the workload.[34]

In the postwar years, the EIL would be transformed into the Explosive Ordnance Disposal Technical Center (EODTC), operating the Navy's Explosive Ordnance Disposal School. That organization, under a separate officer in charge, retained the EIL facilities. Thousands of experts would be trained there over the coming decades, each spending a few weeks at Indian Head. In an indirect fashion, the presence of the training school would become an asset to Indian Head as its graduates went on in naval careers. Many of them would be important friends to their former Indian Head associates in the vast web of naval political networking. However, in the postwar decades, the EOD Technical Center, the EOD School, and their predecessor, the EIL, were perceived organizationally as "tenant activities," not directly part of the institution descended from the early work of Robert Dashiell, George Patterson, and Joseph Strauss.[35]

POWDER FACTORY RESEARCH AND DEVELOPMENT DURING THE WAR

The EIL was a separate, all-military unit, with its own command and somewhat separate location on the Stump Neck section of the Powder Factory grounds. The JPL, also an all-military unit, was shifted away to Inyokern in 1944. While both of these highly active research units were organized separately from the Powder Factory, the expansion of production of smokeless powder, of Explosive D, and the rapid adaptation of the Cal Tech extrusion process to production at Indian Head had all given direct proof of the strengths in propellant research capacity of the Powder Factory.

Even more central to the story of Indian head was the Powder Factory's own laboratory, the Chemical Investigation and Development Laboratory. This was the laboratory once headed by Patterson under the jurisdiction of Farnum during the war years. It continued to provide a steady output of explosive and propellant related chemical experimentation. The work that the Bureau of Ordnance sent to Farnum and his laboratory showed that the Bureau regarded the lab as its own repository for expert knowledge and expert opinion on matters of explosive and propellant chemistry.

Although the United States did not enter the war until 7 December 1941, increase in the fleet and expansion of research had created pressures on the small scientific staff as early as January 1940. In that month Patterson recommended an increase in the chemical staff, and Captain Preston B. Haines, the officer in charge, endorsed and forwarded the request. In addition to the senior chemist, Farnum, and Patterson himself, whose rank was the special "Powder Expert," the chemical laboratory had eight chemists and seven lab technicians or "Science Aides." Haines and Patterson requested two new chemists and two new lab assistants. The level of employment had not changed since 1936, but the work in routine analyses and tests had increased 44 percent, from about seven thousand routine jobs per year in 1936 to over ten thousand in 1939. Not only greater powder factory production but increased special analyses and tests for other naval activities in connection with powder and explosives had increased the pressure.[36]

Patterson finally retired in 1940 at age seventy-two, and Farnum (who had assisted him for twenty-eight years) then took over full charge of the chemical laboratory. The Navy could call on Farnum and his staff to play the role once played by Patterson, as the resident expert on powder and explosives. With the proliferation of new explosives and propellants during the war, partly brought on by the hectic arms race, partly stimulated by shortages of materials among the combatants, and further stimulated by the application of civilian advances in plastics to the field, such expertise was frequently required. In early 1941 Farnum attended a conference in New York with Drs. Chadwell and Kistiakowsky of the National Defense Research Council to discuss with British and Canadian representatives their methods for manufacturing cyclonite, or, as the British called it, RDX. The conference also discussed other explosives, including "Pentolite," a mixture of 50 percent TNT and 50 percent PETN. The RDX was superior to other explosives and was particularly useful as a plastic explosive for demolition charges and as shaped charges for anti-tank weapons, for armor penetration shells, and for demolition of steel rails and the like. Farnum reviewed the methods suggested by the Canadians and the British in order to assist in the planning of a pilot plant to produce RDX.[37]

The Powder Factory laboratory continued to test powder produced by other laboratories to evaluate whether or not it met naval specifications. For example, it evaluated smokeless powder produced by the Tennessee Powder Company and ammonium picrate produced by Maumelle Ordnance Works. In 1944 the laboratory produced an analysis of two varieties of flashless powder, EX5020 and EX5021, double-base powders of nitroguanidine and DINA, or diethanol nitramine dinitrate, as well as nitrocellulose. The method of analysis included not only standard chemical analytic methods but several innovative spectrophotometric and chromatographic methods, checking against a similar analysis conducted by researchers at the California Institute of Technology. In 1944 such methods of analysis were on the cutting edge of technological progress in chemical work; they held the promise of allowing rapid confirmation of, or substitution for, the more tedious, traditional "wet chemistry" methods of analysis. Researcher H. A. Arbit reported that the methods employed at Indian Head independently confirmed an analysis conducted by Cal Tech.[38]

Farnum's lab reported on an extensive method for analyzing the double-base propellant used by Thames in the extrusion plant, to determine if material supplied to the plant met the naval specifications. R. H. Kray, a senior chemist working under Farnum, represented Indian Head at a conference held at Bruceton, Pennsylvania, at the NDRC's laboratory there, regarding apparent discrepancies in studying the effect of water on the three different explosives, torpex, DBX, and minol.[39]

Farnum's laboratory continued to produce specification tests and analyses of a wide variety of explosives and other materials through the war. One study of pyrotechnic chemicals included analyses of barium nitrate, aluminum flake powder, magnesium powder, potassium nitrate, and paraffin wax. His lab worked on all sorts of auxiliary materials as it had in the past, including such items as "luting," or a petroleum-based plastic used for sealing threads in fuzes, acids used in cartridge case cleaning solutions, shellacs, adhesives, cellulose acetate cartridge cases, and rubber. Farnum's laboratory had responsibility for rewriting the Joint Army Navy (JAN) specifications for ammonium picrate, or Explosive D. The laboratory produced some ten to twelve technical reports a month through the war years, often analyzing foreign explosives, propellants, and pyrotechnic mixtures, detailing methods of testing, examining corrosion of explosives, and reporting on samples of a wide range of nonexplosive auxiliary materials used in ordnance.[40]

Captain Hersey gave a detailed summary of existing work and planned new work for the chemical laboratory in 1943, describing both the heavy current load and the proposed new work as justifications for adding new laboratory space. In the category of older, continuing work he included: analysis of enemy munitions; analyses and tests of ordnance materials, including explosives, propellants, solvents, rubber, salt, paints, metals, wax, and chemicals used in ordnance; and work with the new methods of spectroscopy. In the proposed research he included: study of the properties of new propellants and explosives, such as RDX; improvement of stability testing; improvements in manufacture methods; use of micro-crystallographic analysis and x-ray methods for rapid analysis and identification of explosives; study of new plastics and synthetic materials to be used in explosives; physical testing of electrical conductivity of materials; an experimental propellant plant and an experimental high explosive plant; studies of extruded propellant; and special studies of accidents and defective explosives. With the exception of the ideas for experimental production plants, which were postponed until well after the war, Hersey's plans formed the base for the development of new research capacities for Farnum's laboratory.[41]

During the war years, the Powder Factory Chemistry Laboratory divided into two sections, both reporting to Farnum as "Chief Chemist." Section A, the "powder laboratory," worked more closely with analyses and older problems related to factory production, while Section B, with the newer explosives research outlined by Hersey, included several projects related to the JPL-designed rockets and other projects as assigned from the Bureau of Ordnance. A few months after the end of the war, a full listing of the research projects in Section B of the laboratory showed twenty-three projects under way. Designated by this point as the Naval Powder Factory Explosives Research Laboratory, Section B was headed by W. C. Cagle. Cagle reported on a rich agenda, including rocket powder surveillance, studies of JP stabilizers, and responses to Bureau of Ordnance inquiries on a variety of rocket motors. The division into "A" and "B" laboratories administratively divided the more research-oriented work from the production-oriented work. The "B" laboratory later became the Research and Development chemistry laboratory at Indian Head from the postwar years through the 1960s.[42]

With the research strength of the chemical laboratory under Farnum, the production of double-base JP grains under Thames, the advanced research and development in the Jet Propulsion Laboratory before it moved to Inyokern, and with the continuing high-risk work of the Stump Neck group under Lieutenant Klein, the Indian Head facility proved a key element in the Navy's ordnance research during the war. Although the depth and variety of the work was not publicized, a review of the actual schedules and achievements shows that the term "Powder Factory" hardly conveyed the variety of innovative and scientific work proceeding at the Indian Head facility during the hectic years 1941–45. The quiet evolution to a far more complex and sophisticated role in the world of naval ordnance was well under way by the end of the war.

Ensign Robert Brooke Dashiell was Indian Head's first Inspector in Charge of Ordnance from 1890 to 1893. He was also the inventor of an improved rapid-fire breech. The destroyer USS Dashiell was sponsored by his widow and commissioned 20 March 1943.

The valley test site during World War I received guns and materials from the Washington Navy Yard upriver along the Potomac. The Wharf handled boat traffic but railroad tracks along the dock allowed for heavy guns to be placed on flatbeds and easily transported to different areas of the test site. In the background stands the North Battery test range. (circa WWI)

The valley at the Indian Head Proving Ground. Here, officers tested guns against armor produced by Bethlehem, Carnegie and other companies.

Testing all Navy guns was the mission of the Naval Proving Ground for thirty-one years. Here a 6-inch Wire Wound gun shows considerable damage after the 45th round of testing. (circa 1910)

Example of armor testing in the valley. This Carnegie armor plate was tested 1 September 1910 and struck by projectiles from a 10-inch gun. Examples of previous test firings are shown on the background structure.

Indian Head laborer stands beside 14-inch and 16-inch shells, circa 1911.

A bomb-proof shelter in use at the Naval Proving Ground. (1911)

Opened in 1917, the drawbridge crossing the Mattawoman Creek replaced the rickety and dangerous footbridge that for years was the only passage for foot traffic between the community of Marbury and the Station.

Captain Henry E. Lackey directed the Naval Proving Ground as a Commander from 1917 to 1920 and for his efforts was awarded the Navy Cross. Through his civic involvement, he initiated many community improvements. He retired with the rank of Rear Admiral.

Proving Ground employees and military personnel pose in the valley along the North Battery wall where guns were usually placed for testing. (circa 1918)

Franklin D. Roosevelt was Assistant Secretary of the Navy from 1913 to 1920. He supported a yardstick role for Indian Head, using Indian Head costs to determine the fairness of private industry charges. Here FDR visits the Proving Ground in 1919 to do some target practice.

Aerial view of Naval Powder Factory in 6 September 1919.

High explosives magazine located in wooded area away from structures, circa 1920.

Walter W. Farnum, second row, third from right, was one of the Navy's leading experts on powder production and Dr. George W. Patterson's assistant at the Powder Factory. (1920)

Dr. Robert H. Goddard, known as the father of American rocketry, came to Indian Head in 1920 to initiate research on rocket ordnance.

Valley employees came to memorialize the last armor plate (taken from the German battleship Ostriesland) test firing at Indian Head, 21 July 1921. That year all gun and armor testing was moved to Dahlgren, Virginia and the Proving Ground was renamed the Naval Powder Factory.

Dr. George W. Patterson served at Indian Head from 1900 to 1940. He directed both powder production and chemical research for most of his 40-year career with the Station. (circa 1940)

Early example of a static test firing at Indian Head. (1945)

IF YOU ARE IN ESSENTIAL WAR WORK - DO NOT APPLY

Cover of a recruitment brochure for the Naval Powder Factory. World War II demands overseas had drained stateside factories of large numbers of employees. Replacement workers were enticed to Indian Head with the availability of on-base or near-base housing, a good community life, and attractive wages and benefits.

Dr. Francis C. Thames organized and directed the first Research and Development Department at Indian Head in 1947.

Indian Head has been the site of explosives ordnance training since 1942 when the Explosives Investigation Laboratory was initially set-up. Later, in the postwar years, EIL was transformed into the Explosive Ordnance Disposal Technical Center and Explosive Ordnance Disposal School. Pictured is the graduating Class of 1947.

Dr. Sol Skolnik directed the Research and Development Department of the Naval Propellant Plant from 1953 to 1958.

The Cast Plant, opened in the mid-1950s, produced propellant grains which were loaded into motor cases. The Plant produced grains for the first and second stage of the Terrier weapon.

A pilot in a high-altitude flight suit stands beside the Sidewinder missile in October of 1956. Indian Head's work on the Sidewinder and other missiles and rockets helped define the Station's mission during the 1950s.

A Polaris motor is put into vertical position in May 1964. Indian Head scientists produced and tested one-third scale-size Polaris motors starting in 1959.

The Station and the Town of Indian Head had no borders for many years. With the exception of the Cyclone fence put up in 1917 around the Station's perimeter, pedestrian and vehicle traffic had virtually unobstructed access to the reservation. Local businesses and the Navy resided "cheek and jowl" with one another. (1962)

An AA4E Aircraft attack squadron armed with Bullpup rockets in June 1964. The Bullpup is an air-to-surface rocket for which Indian Head produced cast propellants and conducted surveillance tests.

Zuni rockets are mounted on the fuselage of the F8 Crusader aboard USS Ticonderoga CVA-64 in August 1964. Indian Head produced rocket grains for the Zuni and proposed an upgrade in the quality of the grain which received funding in the early 1970s.

The Advisory Board, consisting of prominent individuals from the field of propellants, advised Indian Head on management and technical issues. Seated, Dr. L. T. E. Thompson, Rear Adm. Rawson Bennett II, USN (Ret.), Dr. John F. Kincaid, Chairman of the Advisory Board and Technical Director Joe L. Browning. Standing, Capt. Oscar F. Dreyer, NPP Commanding Officer, Rear Adm. F. S. Withington, USN (Ret.), Capt. W. J. Corcoran, USN (Ret.) and Mr. Robert William Haigh.

Aerial view of the industrial area of the Station looking toward to Mattawoman Creek, the Stump Neck facility, and the community of Marbury across the Creek.

Aerial view of the Patterson Pilot Plant along the shores of the Potomac River.

A destroyer launches the Navy's Anti-Submarine ASROC rocket in January 1965. Indian Head scientists helped develop and produce the fuel that propels the ASROC.

Employee in the Patterson Pilot Plant prepares a rocket motor for casting. Additional rocket motors are shown on table and stand. (1965)

Indian Head Commanding Officer Captain Oscar F. Dreyer congratulates Dr. Otto Reitlinger on receiving a patent award. Dr. Reitlinger developed at Indian Head Otto Fuel II used in the MK 46 Torpedo.

Joe L. Browning served as Indian Head's first civilian Technical Director from 1962 to 1974. (1966)

2.75-inch line throwing rocket launcher with rocket motor in place ready for test firing at Indian Head. (1966)

The Dispenser Pit Tank for the Inert Diluent Process Plant being lowered into position in March of 1967. The Inert Diluent Plant provided a safer method for mixing energetic materials.

Swiss scientist Dr. Mario Biazzi (center) was the designer of the Station's Nitroglycerin Plant during Captain Leslie R. Olsen's tenure. Standing right is Walter A. Carr, one of the Station's nitration experts. (1967)

The Gunfighter Program was an important element of Joe L. Browning's plan to rejuvenate Indian Head with young professionals and new projects. Shown is a disassembled Gunfighter projectile.

150 gallon horizontal Sigma Blade Mixer. Building 1026.

The Biazzi Nitrating System Control Panel in the Nitroglycerin Plant. The Plant opened in 1954 to produce the chemical for double base rocket propellant. The Plant represents an important step in Indian Head's transformation from smokeless powder to rocket propellant production. (1970)

David Eugene Lee was Technical Director of the Naval Ordnance Station from 1975 to 1985.

A roar and jet of flame come from a Talos booster being test fired. Data recorded during this static test enables engineers to predict the useable life of this type of missile motor. (1975–80)

Major plant areas.

Twin Screw Extruder. This versatile machine allowed researchers to process new and exist-ing formulations of propellants and explosives, safely and flexibly taking them from the laboratory bench through pre-production engineering in the manufacturing technology ("mantech") program.

Roger Smith. Technical Director 1989 to 1999, Smith coupled his congenial style of management with leadership in refocusing the mission of Indian Head as the National Center for Energetics during a period of defense downsizing.

Mary Lacey. Director 1999–2002, Lacey emphasized networking and a staff mentoring program. She sought to engage the technical experts at Indian Head in a concerted effort to bring their knowledge and skills to bear on issues of military preparedness and homeland security.

CHAPTER 7

Postwar Patterns

Magazine No. 2
Building 65
1901

Mᴜᴄʜ ᴏf ᴛʜᴇ work begun during World War II set the patterns for the immediate postwar years at Indian Head. By the end of the war, the Powder Factory had become much more than a factory for manufacturing smokeless powder. The new work in extrusion and the work in testing and evaluating new weapons, as well as the extensive program in examining captured enemy equipment, had changed the nature of the facility. Personnel had changed as well. George Patterson was gone, although the laboratory which he had developed and turned over to Walter Farnum continued in operation. This laboratory had divided into two sections in the war years, with the "B" section doing research.

Much of the war work at Indian Head, as elsewhere in military laboratories and factories, went on behind a shield of wartime-induced security. Much of the work at the facility was classified "Confidential," and most of it was automatically downgraded to unclassified twelve years later. As a consequence, very few of the projects or the research could receive immediate publicity; when the classification was lifted more than a decade later, the accomplishments and participation of Indian Head no longer seemed like news. For such reasons, the crucial role played by the facility in a number of major weapons developments at the end of World War II and through the late 1940s simply went unnoticed by the broader public and, indeed, by many in the naval establishment. The name itself, "Naval Powder Factory," became more and more misleading, since casual outside observers, including many naval officers unfamiliar with the nature of the work, would assume that Indian Head simply continued in its traditional role of manufacturing smokeless powder.

The end of the war years and the immediate postwar years saw the gradual development of a new institutional identity behind the double screen of security and the increasingly outdated name.

EARLY WORK ON JATOS

In 1944, the Bureau of Ordnance gained control of the development of "Jet Assist Takeoff Units" (JATOs) which had been designed by the Bureau of Aeronautics. Nine of the completed units, together with some preliminary study of the propellant done by "Air Corps Jet Propulsion Research" at the California Institute of Technology (Cal Tech), arrived at Indian Head in June of that year for study.[1]

The Aeronautics-designed propellant was called "GALCIT 61-C" and consisted of a petroleum-based material. The composition was 16.6 percent Texaco No. 16 asphalt, 7.2 percent Pure Penn Oil SAE 10 weight, and 76 percent potassium perchlorate. This primitive, "greek-fire" type propellant was insulated from the walls of the cylinder by a rubbery liner made up of vegetable pitch and 10-weight motor oil. Like the acetate on solid grain propellants, the liner served to restrict the burning of the propellant as well as to cement the propellant to the walls of the cylinder. The takeoff units were quite unreliable. At low temperatures, the propellant became brittle and cracked, leading to irregular burning; at high temperatures, the propellant became soft and flowed. The Bureau requested, as a start on the research on the newly acquired units, that the Powder Factory conduct preliminary tests over a temperature range.[2]

The commanding officer requested that the preliminary study of these JATOs be conducted by Lieutenant D. Klein's group on Stump Neck. Klein borrowed equipment from the Powder Factory and set up a system of cold-boxes, chilled by dry ice, to study the effect of cold temperatures on the propellant by x-ray of the units at ten-degree intervals down to minus 30 degrees Fahrenheit. The points at which the propellant cracked and broke away from the container walls were noted.[3]

In March 1945 the Bureau established "a loaded rocket motor surveillance program" at Indian Head to determine the "safe life" of jet propellant (JP) grains and other rocket propellants. From these beginnings, Indian Head emerged as the testing station for both small and large solid-fuel naval rockets.[4]

In 1946 the group testing small rockets, the Ballistics Laboratory headed by W. J. Moore, continued work on a group of designated problems, including several new projects assigned by the Bureau:

◇ B-506 Rocket Propellant Surveillance
◇ B-545 Temperature Conditioning of Mark 19 Grains (4.25″)
◇ B-564 Stability Tests of Plastic Cartridge Cases
◇ B-570 Smokeless Powder for Airplane Starter Cartridges
◇ B-571 5.0″ Rocket Motors Testing
◇ B-574 5.0″ Experimental Aluminum Motors

The methods and experiments which Moore reported followed precedents from the war years, measuring the effect of storage at different temperatures and testing thrust and burning time in static runs. He installed and used various new pieces of equipment, including a set of high-precision ovens for temperature tests and a six-channel Cathode Ray Oscilloscope for recording various temperature points on the rocket motors.[5]

Meanwhile, testing of larger JATO motors produced by Aerojet Engineering Corporation in Azusa, California, was established at Indian Head, with funding provided from the Bureau of Aeronautics. The first shipment of fifty JATO units, together with sixty igniters, arrived by truck in the fall of 1946. A standardized system of visual inspection, static firing at different temperatures, and a system of recording test data on cards for later statistical analysis were established. One of the purposes of these early tests was to determine whether faulty units could be detected by visual tests.[6]

A wide variety of rocket motors were designated "JATO" whether they were actually intended to "assist" airplane takeoff or whether they were designed to propel a warhead. In order to clarify terminology, a joint Army-Navy aeronautical bulletin in July 1948 defined JATO as "an auxiliary rocket for applying thrust to some structure or apparatus. A JATO is defined further as being a complete, self-contained unit having a definite burning time and a fixed thrust," and included liquid-, plastic-, and solid-propelled rockets. JATOs intended for aircraft were designated "Aircraft JATO." Over the next six years, as a variety of solid-fuel missiles were developed, the test unit at Indian Head evaluated JATOs developed at Aerojet, at the Allegany Ballistics Laboratory (ABL), and at the Naval Ordnance Test Station (NOTS) at Inyokern. The Indian Head testing group conducted surveillance checks to determine deterioration in storage as well as inspection checks to determine whether the motors met specifications.[7]

THE POSTWAR SEARCH FOR A RESEARCH AND DEVELOPMENT MISSION

Whether or not the research strengths developed at Indian Head during the war would form the basis of the facility's postwar identity remained an issue. In the period 1946–49, the commanding officers of the station worked with civilian researchers and with the Bureau of Ordnance in attempting to get the official mission of the Indian Head Powder Factory modified to reflect the new strengths. The effort was a conscious and concerted one through the immediate postwar period.

In 1946 a Research and Development Division was informally established at Indian Head. However, at first the division could not staff a productive research program, and a number of promising projects proceeded without notable achievement. Limitations on budget, difficulties in recruiting and holding sufficiently trained chemists, and the fact that many other facilities were engaged in various aspects of naval propellant research all mitigated against Indian Head establishing a well-known role in the emerging world of rocket propulsion which followed after the war. More spectacular projects at White Sands, at Inyokern, and at the Naval Ordnance Laboratory at White Oak threatened to eclipse the strengths of the older parent institution at Indian Head. The lack of a strong central personality in the research area after the retirement of Patterson in 1940 also made it difficult to transform the factory into the lead propellant "R and D" facility.

Records from the war and postwar years were replete with efforts to bring leadership and order to the research tasks at the center. Bureau chiefs recognized the potential of Indian Head, and their support for a research emphasis at Indian Head was clear and strong through the postwar decade. The difficulty did not lie with the inclination of the Bureau to identify and support the facility but with organizing and building a team with the skills and morale to meet the Bureau's expectation.

In November 1945 Bureau Chief George F. Hussey, Jr., gave to Captain James Glennon, the commanding officer at Indian Head, an eleven-point research program, couched in "general terms" for planning purposes. Six of the projects were based on wartime-developed strengths in propellant work:

◇ 549. Improved Propellant Analysis Techniques
◇ 550. Correlation of Ballistic Properties from Closed Bomb Tests with Tests in Guns
◇ 551. Study of Various Nitration Rates with Nitrocellulose
◇ 552. Study of Decomposition of Propellants
◇ 553. Kinetics of Propellant Burning
◇ 554. Solventless Powder and Procedure in Manufacture

The other five projects were even broader, including surveillance and stability studies; new stabilizers; explosive plasticizers for cool propellants; control of burning rates by surface inhibitors; theory of decoppering of guns; and surveillance of full-scale rocket "grains" or motors.[8]

Through 1946 Indian Head used the Bureau of Ordnance research and development program as the overall authority for continuing research in those eleven areas. The months after the war were hectic, and the first monthly report of the Research and Development Department reflected the difficulties: "The Powder Factory has been greatly handicapped, however, by reduction in force, transfer of employees and the red-tape connected with allotments, attempts to procure new personell [sic], etc. Stability tests studies, including Taliani, were proceeding well until all eight employees had to be released leaving one employee, transferred from another section to carry on the work." The contrast with Cal Tech, which had twenty full-time employees working on similar projects, was frustrating.[9]

Over five thousand rocket motors returned from storage in the South Pacific, which had been manufactured in 1944 and 1945 at Indian Head, at the Sunflower Ordnance Works, and at Cal Tech, were sorted and selectively tested for flaws. That work could fall under the general rubric of surveillance work, one of the eleven general tasks in the bureau chief's orders. Similarly, after a literature survey on gun-coppering, a series of tests on that project were scheduled for Dahlgren, representing progress on another of the eleven tasks.[10]

Eighteen months after the first Bureau order, on 30 June 1947, the chief of the Bureau of Ordnance set out to define more clearly the major research and development assignments in the explosive and propellant area, assigning various aspects of research to different laboratories. "Broadly stated," the chief said, "the research and development program at the Powder Factory is aimed at the invention of ideal solid propellants for all military purposes, but especially for U.S. Navy high and ultra-high performance guns and for catapult systems." The Bureau assigned fourteen specific tasks to Indian Head, a slight expansion over the eleven assigned a year and half before.

1. Closed Bomb Tests
2. Cool Propellants (with du Pont)
3. Investigation of Organic Preparations for Propellants
4. Properties of Nitrocellulose
5. Physical Properties of Propellants
6. Analytical Methods
7. Propellant Decomposition
8. Gun Type Catapult Systems
9. Surveillance of Loaded Rockets
10. Preparation of New Multibase Rocket Propellants
11. Nitrocellulose Pilot Plant
12. Construction of Nitroglycerin Pilot Plant
13. Construction of Solvent-Solventless Propellant Pilot Plant
14. Construction of Cast Propellant Pilot Plant[11]

There were ample precedents for all of the fourteen task assignments. Setting up the propellant extrusion plant during the war had established a place for Indian Head in the burgeoning field of small rockets, especially in production, quality control, and surveillance. Commodore Mark L. Hersey had included pilot plant work in his 1943 recommendation. The closed bomb work and the analytic and testing work reflected the strengths of Farnum's expanded laboratory, the Ballistics Laboratory, and the test and evaluation section of F. C. Thames' extrusion plant.

Over the next decade, the Bureau of Ordnance letter of 30 June 1947, was treated as an original authorizing document, a sort of biblical text used to justify establishment of the R and D Department, further assignments of tasks, claims for funding, and internal reorganizations to achieve the various goals. The agenda held great promise as a rich research assignment, and the Bureau had carefully

TABLE 10
Categories of R and D Projects, December 1947

Category	Number of Projects
Properties of Explosives	2
Installation of Equipment	2
Synthesizing	4
Surveillance of Smokeless Powder	5
Burning Studies	4
Nitrocellulose Studies	12 (in two groups)
Chemical Analysis	8
Rocket Motor Work	6
Chemical & Physical Properties, Explosives	2
Total	45

Source: B. Hall Hanlon to Chief of the Bureau, 12 December 1947, Record Group 74, Acc. 5595, NP8/ Re2-L5, 1947, Washington National Records Center, Suitland, Md.

delineated these areas as appropriate for Indian Head. Through 1947, some progress could be reported on the specific research tasks, and planning went forward on the proposed pilot plants.

With the more structured program came a more structured fiscal accounting system, with work assigned to budgeted "tasks." In order to transfer over to the new system, various research tasks already "on hand" were grouped together and then the Bureau developed "task assignments" with a budget for each. In December 1947, the commanding officer, now Captain Byron H. Hanlon, could report to the bureau chief with a complete structured system of monthly and quarterly reports, with all work assigned and billed against numbered work tasks. Hanlon reported various degrees of progress on over forty experiments and tests relating to burning rates for propellants, polyvinyl nitrate in propellants, new manufacturing specifications, microphotography, heat tests of rocket motors, establishment of pilot plants, experiments on reduction of smoke, experiments with plastic polymers in propellants, and installation tests on new equipment. The highly structured and organized reporting system which Hanlon set up was perhaps more impressive on paper than the actual degree of accomplishment. In several cases, Hanlon had to report no progress for lack of trained personnel or delays in installation of new equipment.[12]

Altogether, the Bureau converted forty-five tasks already in progress at Indian Head into the formalized, "assigned" research and development agenda in 1947. Table 10 lists the categories as of December of that year. Each category, for the most part, represented the work of one or another group already established at the facility.

Many of the projects to which a separate budget line was assigned were in reality a long-range, repetitive series of experiments or tests. Others were one-time studies which would result in a report. Several of the continuing projects, such as analyses of sample powders, represented continuation of work which had begun under Patterson in the first decade of the century; others, such as the various tests of rocket motors, represented types of work added during the war years.

In early 1948 Captain Clarence E. Voegeli continued the monthly and quarterly reporting system, with Thames, formerly head of the new extrusion plant during the war years, serving as director of Research and Development. Thames reported progress, with less attention to the accounting system of budget lines and hours than Captain Hanlon had utilized, but giving details of chemical formulas and indicating progress with several of the polymer investigations.[13]

The laboratory was part of a wide network of Army and Navy facilities. Members of the staff attended meetings of the Solid Propellant Advisory Group (SPAG), an interlaboratory association of Army, Navy, and contractor scientists working on solid propellants. SPAG met through these years, sometimes in conjunction with the American Chemical Society and at other times at one or another member laboratory. Copies of the monthly and quarterly research and development reports from Indian Head through this period were sent to a wide distribution list of SPAG members, including major contract laboratories such as the ABL run by Hercules in Cumberland, Maryland, other naval facilities such as NOTS Inyokern and the Naval Ordnance Test Station at Chincoteague, Maryland, and a wide variety of corporate, university, and government laboratories across the country. As Indian Head experimented with new sample lots of explosives, propellants, and other materials, the R and D Department would send samples to other labs for work. In order to obtain data, sample lots of polyvinyl nitrate manufactured at Indian Head were sent to NOTS Inyokern, Aerojet, Cornell, and Picatinny Arsenal.[14]

To make best use of the solid propellant capacity of the Navy, the Bureau of Ordnance would assign specific tasks to each of the laboratories. The tasks assigned to Indian Head during the late 1940s tended to be a continuation and extension of the earlier projects. While this division of work had a certain logic to it, in that assignments were built upon recognized capacity, the procedure did not have the flexibility to use the abilities of newly arrived younger researchers, and it tended to "freeze" the agenda along certain lines. Such an atmosphere was not always conducive to innovation and progress.

PILOT PLANTS AND MANUFACTURING EXPANSION

Although the Powder Factory had taken on an R and D program, the larger scale employment at Indian Head depended, as it always had, upon the production facilities. With the decline in smokeless powder consumption and with the layoff of personnel at Indian Head in the postwar months, congressmen and senators from Maryland soon came to the defense of the facility, urging the Bureau of Ordnance to utilize the plant facilities there on a larger scale and to rebuild to meet new needs. Perhaps the most effective support came from Sen. Millard Tydings (D-Md). Working with Congressman Lansdale Sasscer (D-Md.) and with Sen. Herbert O'Conor (D-Md.), Tydings asked acting Bureau Chief Admiral Malcolm F. Schoeffel whether the Indian Head plant could be shifted over to the production of "rocket powder" in the light of declining consumption of smokeless powder.[15]

Tydings and Schoeffel discussed the matter in a phone conversation early in 1947. Schoeffel at first argued that the Navy planned to purchase powder from the Army, "except for some experimental jobs there at the Powder Factory."

"Wouldn't it be possible," Tydings asked, "and economical, with this plant as large and modern as it is, and with whatever small additions are necessary, to utilize the part that won't be needed now for the making of the heretofore used Naval powder to turn that over to the manufacture of the rocket powder?"

Tydings went on to explain that he looked to the future.

"I'm afraid," he said, "that with the advance of the rocket, some of this powder business may drop to a smaller proportion . . . in the overall picture."

Schoeffel said that he would have the matter studied from that viewpoint.[16]

Within weeks, Bureau Chief Hussey wrote to Congressman Sasscer, blandly making the noncommittal statement that the Powder Factory was "absorbing the full Naval load of work that is within its capabilities." On a more promising note, Hussey said that the question of the "feasibility of establishing a rocket powder production line" was under study—a study stimulated by the pressure from Tydings and Sasscer himself. Hussey also spoke on the telephone with Tydings, stating that he had initiated the study of whether to go into rocket powder production in response to Tydings' earlier request. Nevertheless, Hussey had to report that the total number of civilian employees at Indian Head would be cut from 1,233 to about 1,000 under current plans.[17]

The political pressure was only one of several factors at work in the Navy's plans for Indian Head. Nevertheless, it was shortly after Tydings' pressures that the Bureau finally agreed to the establishment of a set of pilot plants at Indian Head which would have the capacity to produce experimental new propellants for naval research use. A key event in this decision was a conference held 27 March 1947, at Indian Head, involving local staff members and several representatives of the Bureau. The conference recommended construction of several pilot plants, after discussing a variety of objections that might be raised on safety, on economics, and on the grounds that other facilities might be better suited to the work.[18]

Whether or not the pilot plants were planned as a response to the political pressure, they certainly provided evidence to quiet the political uproar over the threat to Indian Head. When Tydings and the Maryland State Senate protested to Secretary of the Navy James Forrestal in June 1947 that it still appeared the Navy planned to close Indian Head, Forrestal responded by pointing to the pilot plant plans: "As further evidence that the Navy plans to keep the Naval Powder Factory in the forefront of research and powder manufacturing development, a project is being sponsored at this time to construct additional pilot plant facilities for the manufacture of double base powder. . . . [The project] will include additional facilities for a nitroglycerine plant, a pilot plant for rolled sheet ballistite, and a cast powder pilot plant."[19]

Over the next six years, after much delay and change of plans, Indian Head gradually built up its "pilot plant" capacity, building on the strength in experimental propellants which Tydings had suggested. In this step-by-step fashion, a change and clarification of the role of Indian Head took place, building on the tradition of taking laboratory-scale production to industrial-scale, as in the case of Patterson's smokeless powder, the Explosive D plant, and, in World War II, under Thames, the extrusion work. The explicit concept of pilot plants, coming at a time of crisis, began to reshape and redefine the mission of Indian Head. The survival of the station required redefinition, which slowly but surely emerged.

Immediately after Tydings' protests, the chief of the Bureau of Ordnance authorized the establishment of a special pilot plant for the production of "variable" nitrocellulose. The product of this particular pilot plant would be experimental lots of nitrocellulose. The lots would have varying degrees of nitration, viscosity, and molecular structure. In order to produce improved propellants in which nitrocellulose was a main ingredient, large sample lots would be required. The purposes of the pilot plant were to produce sufficiently large samples for gun testing and to obtain information on manufacturing techniques.[20]

Using existing buildings, the pilot plant was planned with a budget of $67,453, and Captain Hanlon went to work to make the changes. After some backing and filling over which line in the budget would cover the modest remodelling expenditure, Bureau Chief Hussey authorized the amount out of the 1947 Ordnance and Ordnance Stores Appropriation.[21]

Hanlon also recommended establishment of another pilot plant to produce the double-base propellant, which would require both a plant for the production of nitroglycerin and another for mixing nitrocellulose and nitroglycerin. By the summer of 1947 the Bureau had accepted the plan to construct four pilot plants altogether: a nitroglycerin pilot plant; a plant to produce varied nitrogen-content nitrocellulose; a plant for mixing and rolling experimental lots of solvent and solventless propellant; and a fourth plant for experimental production of cast propulsion units. It was this plan which Forrestal was able to use to quiet the insistent rumors of the imminent closing of Indian Head. Hanlon's successor, Captain Voegeli, pushed in 1948 for the completion of the nitroglycerin plant as the first priority, followed in order by the nitrocellulose plant, the mixing and rolling plant, and a casting plant. The R and D Department also proposed setting up a plant for production of fine grade nitroguanidine, to be consumed in the propellant pilot plant.[22]

Before proceeding with the nitroglycerin plant, the Bureau of Ordnance discussed the exact function of that plant and debated whether or not to proceed with a batch or a continuous flow process. At least part of the delay in completing the construction of the nitroglycerin plant occurred because of changing views on whether to build a batch process system or one based on a continuous process. Plans to install a continuous process nitroglycerin manufacturing facility, using the most modern European designs, took years to implement. The "Biazzi Plant," following a Swiss design, opened in 1954—fully seven years after the commitment to Tydings.

As early as 1947, both Farnum and Thames supported the plan for a continuous flow plant, sending one of Thames' assistants, Arthur Mayer, to visit Britain to study the continuous production methods there. However, while Mayer was in Europe, the Bureau decided that the plant should act as a source not only of nitroglycerin but of other nitrated materials, which would require that a specialized flow-process nitroglycerin plant be abandoned in favor of a batch process which could be more readily adapted to different explosives. After his trip in

September 1947 to both Britain and Germany, Mayer submitted a detailed report comparing point by point the Schmidt and Biazzi systems, finding the Biazzi system simpler and smaller and recommending it. But while he had been gone, officials at the Bureau had already decided to go with a batch plant instead of either the Schmidt or Biazzi continuous flow systems.[23]

Chief of the Bureau Albert G. Noble indicated that a continuous flow process had some advantages and that it might be considered in the long run. However, he ordered the Powder Factory to proceed with a batch process plan. The nitroglycerin batch plant made good progress through early 1948, utilizing refrigeration equipment transferred from Camp Davis, North Carolina, and installed by the York Corporation. However, construction of the cast plant and the nitrocellulose pilot plant was suspended while work proceeded on the nitroglycerin plant and the mixing and rolling plant. Slow progress was made on acquiring and installing equipment. A small mixing and rolling plant for experimental batches of Cordite-N propellants, together with modification of three of the eleven extrusion presses established for rocket grains in the war, were scheduled for completion in April 1948. The nitroglycerin batch process was in production by 1951; it could produce thirteen batches every twenty-four hours, with each batch containing 435 pounds of material.[24]

By 1951 plans for the continuous production system for nitroglycerin were revived. In that year, under a contract with the Biazzi company, Bureau Chief Schoeffel ordered the installation of continuous nitration equipment, and through 1952 he debated the terms of the contract with the company. Resident Officer in Charge of Construction D. H Murrell worked with the Biazzi company at Indian Head and with the contractor which installed the equipment, the Leeds and Northrup Company, working out modifications to the plant incorporating a system of pH meters. The bureau chief had further disputes with the Biazzi firm over whether the procurement included drowning tanks.[25]

Despite the clinching of the contract with Biazzi, Commanding Officer Voegeli at first opposed the Biazzi plans. He advocated expansion of the American-built batch process plant, with a back-up in case of equipment destruction by accidental explosion. Voegeli objected to the idea that European replacement equipment for the continuous process would have to be purchased. Bureau Chief Schoeffel patiently insisted that the Powder Factory proceed with the Biazzi system, and that continuous production facilities, in contrast to batch production plants, were accident-free. The provision for back-up equipment in case of accident was characteristic of batch type plants, he argued. There had been a "firm decision" in 1948 to proceed with a continuous type production plant, he claimed, although in fact the recommendation for continuous production had been shelved for several years in favor of the batch process. Schoeffel indicated that part of the plant be constructed using captured German Schmidt equipment. Schoeffel "directed" Voegeli to proceed and to support the completion of the Biazzi system.[26]

The coming of the Korean War in 1950 had provided a needed stimulus to increased explosive and propellant production. Funding implementing the 1947 plans for the cast double-base facilities and for the completion of the nitroglycerin plant finally came through in 1952. Bureau Chief Schoeffel noted, with somewhat circular logic, that "the delay attendant upon the obtaining of authority to use the funds provided herein for the indicated purpose has caused serious interference with the programmed production from these facilities." Schoeffel approved over $1 million for the cast propellant facilities and for finishing the construction of the Biazzi nitroglycerin plant. The Biazzi plant finally went into production with its first run on April 23, 1954.[27]

In 1949 the Powder Factory had obtained over $1.3 million to rehabilitate the extrusion facilities in order to produce 66,000 grains of ballistite for the 5-inch HVAR (High Velocity Aircraft Rocket) Mark 18. Detailed plans in cooperation with Radford Arsenal and with the Naval Ammunition Depots in Crane, Indiana, and in McAlester, Oklahoma, got the system under way.[28]

The demands of the Korean War required increasing production of extruded grains. The "Mark 31," 2.75-inch, folding-fin aircraft rocket, or FFAR, required an expansion of the extrusion plant facilities, which were limited to 25,000 grains a month because of the bottleneck on machining and wrapping the grains with inhibitor. With a modest increase of machining, curing, and inhibiting facilities, totalling under $400,000, the production was increased to 80,000 grains a month.[29]

RESEARCH AND DEVELOPMENT: PROBLEMS WITH IMPLEMENTING THE MISSION

The Research and Development Department had been formally established on 30 June 1947, with fifty-eight professional positions allotted to the department. However, well into 1948, there remained nineteen vacancies which could not be filled, and an inspection revealed that from June through December of that year the "personnel situation was indeterminate." A naval inspector, Captain E. C. Craig, reported in a quite matter-of-fact tone in 1948 that "it is apparent that Navy civilian procurement methods do not allow expeditious procurement of desired personnel." He recommended upgrading the laboratory director and in general recruiting at a higher federal pay grade. His suggestions were not implemented.[30]

The problems of administration of scientific personnel hinted at in the Craig report and in the complaints of Indian Head administrators were widespread in other military laboratories through this period. Aside from the problem of attracting and holding highly trained researchers within the civil service grades, other difficulties affected morale at laboratory after laboratory. The by-the-numbers approach adopted by some naval officers often irritated civilian scientists. Furthermore, young scientists fresh from academic or fast-paced industrial settings frequently developed disdain for supervisory personnel, either in or out of uniform, who were not knowledgeable regarding technical specialties. It was difficult to work for administrators whose own scientific knowledge was limited or nil and who could not identify talent among subordinates. With the agenda frozen by Bureau decision, including and excluding certain research paths, there was little or no flexibility to adapt to the ideas or talents of newly recruited people. Such rigidities of structure were hardly appropriate to the rapidly changing technology of rocket propellants.

Doubtless, the older supervisors regarded some self-confident junior chemists and chemical engineers as immature, callow, and disrespectful. In all such settings, differences of personality were magnified and occasionally became an additional source of managerial problems. The organized assignment of tasks was not up for discussion, and certainly not by those at the bottom of the bureaucratic decision-making process. Junior people did not attend conferences between the laboratories which helped sort out priorities, such as those held under SPAG regularly through the late 1940s. Bright young researchers resented being assigned to work which they saw as a dead end, or whose long-range function was not apparent.

Indian Head, like other military laboratories in the period, suffered from many such causes of poor morale. The dry report of Captain Craig suggested how widespread the difficulty was. Thames had earlier reported a series of personality conflicts in his internal comments to Farnum, suggesting the problems which arose when talented but inexperienced younger staff members were promoted over the heads of those with more seniority.

Some hint of the difficulties in these years of adjustment was conveyed in an "Industrial Survey" conducted by outside experts in 1951. The group recommended fifty-one changes to increase safety, improve morale, and deal with the question of research management and planning. Early in 1952, the acting commanding officer, Captain Francis W. Scanland, Jr., reported on the progress or lack of progress in implementing the recommendations. Of the fifty-one recommendations, seventeen dealt with minor alterations of equipment for convenience or safety purposes, and thirty-four were long-range reforms affecting personnel or management. Of the seventeen simple recommendations, twelve were either implemented or largely accomplished within a few months. But more significantly, of the thirty-four managerial reforms, substantial progress could be reported on only seven; the rest were either postponed, stalled, or awaiting a decision in the Bureau of Ordnance.

The reforms which addressed the problems of stimulating research, improving researcher morale, and clearing the bottlenecks of work had little chance of success. For example, the inspection had recommended several moves to recruit new personnel and to upgrade the pay scales of those already aboard; a personnel "freeze" prevented any action on those points. Dr. Sol Skolnik was appointed to the position of assistant director of Research and Development. Shortly thereafter, he was promoted to replace Thames as Director of R and D, who returned to head Production.

It should be noted that at this time there was no civilian technical director of the whole facility, and division of Production and Research and Development meant that there were two co-equal civilian administrators, with the commanding officer in a supervisory role over both of them. As in the past, the civilian tenure would outlast the military tours of duty. Officers in charge in the postwar years are shown in Table 11.

While many of the administrative reforms could not be implemented, another recommendation that a more liberal policy be set for attendance at professional meetings had some minor impact. Some shifting of space allotments also resulted from the study, providing more room for administration and instruction. In case

TABLE 11
Postwar Commanding Officers

Years	Commanding Officer	Rank at Indian Head
1943–1946	Glennon, James	CAPT
1946–1948	Hanlon, Byron H.	CAPT-RADM*
1948–1948	Gallery, Philip D.	CAPT
1948–1952	Voegeli, Clarence E.	CAPT
1952–1952	Scanland, F. W., Jr.	CDR
1952–1955	Benson, William H.	CAPT

Source: "Inspectors of Ordnance in Charge and Commanding Officers," Indian Head Technical Library, Naval Ordnance Station, Indian Head, Md.
*Hanlon was promoted while at Indian Head.

after case, however, the recommended reforms were postponed, reported as under study, or more or less politely rejected as inappropriate or unachievable.[31]

Still another indication of the difficulties surfaced when Scanland's successor, Captain William H. Benson, reported later in 1952 upon the manpower requirements at the Powder Factory. Although increases in production had brought the total civilian employment in September to 2,891 against a ceiling of 2,907, problems in recruiting scientific personnel continued to plague the facility. Benson noted that "the Research and Development workload at the Naval Powder Factory is seventy per cent greater than the capacity of the presently authorized personnel (while still being well within the capacity of present facilities)." He remarked that the Powder Factory had continued to attempt contracting out the research, "but little relief is anticipated from this source." Even if the ceilings were raised, Benson indicated, the local supply of labor would make it difficult to increase the staff.[32]

Early in the 1950s, two promising research and development projects remained relatively stymied, and debates over them reflected the deeper problems of management and morale. Plans to develop an aircraft catapult propellant had been on the agenda since 1947, but the project did not move forward, partly because of low priorities and partly because of an imposed technical limitation. Indian Head was "tasked" to work on low-pressure propellants—those which would generate less than 125 pounds per square inch. To several of the younger researchers, including Joe Browning, Eugene (Gene) Roberts, Todd Braunstein, Robert Michaels, and Jack Crooks, this idea seemed patently absurd. Propellant was most efficient at higher pressures, and some system utilizing high-pressure propellant, with venting to a low-pressure chamber, would be more effective. However, when the group suggested such a move, they were told by Director of Research Dr. Sol Skolnik, who had replaced Thames during this period, that such an approach was "forbidden." China Lake was to work on high-pressure propellants and high-low systems, and Indian Head was to work on low pressure. However, the young researchers believed that China Lake had stopped "dead in the water" after running through some five or six propellants. Consequently, Browning went ahead, "in rebellion," developing over one hundred new catapult propellants. In 1954 the Navy decided to move away from propellant-actuated aircraft launching catapults. In the meantime, however, Indian Head, at the urging of some of the younger researchers, had proven its capacity to do research and development in new propellants.[33]

Looked at from an objective point of view, Dr. Skolnik had failed to get the younger researchers engaged in the project by explaining to their satisfaction the role that Indian Head had obtained, and they had failed to promote their different approach sufficiently to win his endorsement and to get administrative backing for doing things their way. What Browning would remember years later as a function of personality was, from a management point of view, largely a failure of communication.

Similar problems of communication inhibited work on Weapon A, a planned antisubmarine rocket. Although Skolnik was a good research chemist, he was viewed by some of the younger researchers as particularly unskilled when it came to propellant testing. Preliminary cast grains produced by the Allegany Ballistics Laboratory did not meet the specification as to range. Skolnik believed that substituting propellant N-4 for JPN (jet propellant, Navy) held great promise. Browning later complained that Skolnik had "mixed up the samples," leading to confusion in record keeping, ineffective test runs, and unsafe tests because the rocket nozzle sizes were not properly adjusted to the different propellants.[34]

Browning and the other young researchers did not fear that Indian Head would be closed. Rather, they viewed the central problem as one of poor management. Although some of their views may have been technically correct, the ideas of these younger researchers simply flew in the face of the designated assignment of problems to the various rocket and propellant research facilities. Clearly, the structured decision-making process, the assignment of tasks based upon 1945–46 capacities, and the placement of a research chemist like Skolnik in charge of a development program which was not his specialty could mitigate against innovation, good work, and productive use of the new staff members' talents. Later, Browning would admit that he made little personal progress in his career at this time and that management viewed him more as a trouble-maker than as a rising talent.[35]

With no civilian in a management role over both the Production Department, headed by Thames, and the Research and Development Department, headed by Skolnik, personal friction between the two produced what Browning later remembered as a "battle royal." Browning transferred out of the Research and Development Department to the Production Department to manage that department's laboratory, which served the newly constructed production plants. Skolnik assumed Browning would soon resign, because any "good R and D man," in Skolnik's opinion, would find production boring.[36]

Browning, however, would later claim that he found more challenges in the first month in the Production Department than in the entire time he had previously been in R and D. He was able to take over responsibility for seeing that the newly constructed pilot plants went into operation. When "shaking down" a new plant, he would go to the plant with the head of the Gun Division, Canfield Jenkins, from the Bureau of Ordnance. Together, they would live at the plant until they got the process working.[37]

Browning took a research attitude towards the opening of the new plants. When Browning and Jenkins opened the Cordite-N plant to produce a triple-base powder made of nitroglycerin, nitrocellulose, and nitroguanidine, they found that the extrusion presses operated at under three hundred pounds pressure rather than the specified twelve hundred pounds. Jenkins said, "Shut it down." Browning said, "Let's run it and see what we get." The resulting product met the specifications with more consistency than any batch produced elsewhere in the Navy. Browning recognized that the different crystalline structure of the nitroguanidine used by his presses was the factor which allowed for different extrusion pressures. When he started he did not know what he would get, but he saw no reason to simply throw out the product without evaluating it.[38]

However, Browning found little tolerance on the part of the administration for such experimentation. If a plant did not operate like another one, it would be regarded as "wrong." Over and over he viewed such situations as opportunities to conduct research. For about two years, Browning headed the Production Department's laboratory (the successor to the wartime-designated "A" laboratory), simply taking over the task of putting the plants in production. He built a reputation as a competent researcher, a man who would make his own job, and a bit of a character with explicit contempt for those whose work he regarded as sloppy, unquestioning, or pretentious. The production workers fondly put up a sign over a piece of equipment he had used for research in the Cordite plant, designating it as "Old Joe's Mixer" because he had lived with it so much. At that time, he was about twenty-five years old.[39]

Despite continuing difficulties through the early 1950s, it was clear that Indian Head had passed a major crisis in survival when the Bureau decided to set up the pilot plants there. The Research and Development Department, with an officially assigned and potentially expandable agenda, addressed a designated segment of the Navy's rocket research problems. Four pilot plants to produce experimental quantities of new explosives and propellants were functioning, and others went into production, including the nitroguanidine and Cordite-N plants.

At this stage the Navy hoped to procure new rockets by a structured process. The Naval Ordnance Test Station or the Allegany Ballistics Laboratory would develop new formulations. Indian Head would put the new product in pilot-scale production, learning the details and the know-how required. Then production could be put out for competitive bid, and the manufacturers would have detailed information from Indian Head about manufacturing processes and production engineering.[40]

By the mid-1950s Indian Head was in a position to redefine itself from Powder Factory to a research and development facility in propellants and to branch out in the expanding world of naval rockets. What was required was an infusion of leadership and a resolution of the haunting problems of management and morale. Commanding Officer Scanland addressed several of these issues to the chief of the Bureau in 1952. He found that the term "Powder Factory" itself "implied a plant primarily engaged in the manufacture of gun powder." While the facility's name had been correct in the 1940s, Scanland argued, "it is no longer correct, and becoming more and more incorrect." He recommended that the name be changed to the "Naval Propellant Development Station." The station, he claimed, "is ready, willing and anxious to assume the roll [sic] of the Navy's and the Bureau's solid propellant development activity; in fact, it feels that it has already achieved this status." Scanland went on to develop a comprehensive restatement of the station's "mission," which he proposed be adopted by the Bureau. He recommended a research and development and pilot plant focus, and continued housing of the Explosive Ordnance Disposal School.[41]

The opening of the cast plant in 1952, followed by the nitroglycerin Biazzi plant, the Cordite-N plant, and the nitroguanidine plant, had set the course away from "powder." The full and formal transition to a new role and a wider recognition of the change from powder production to rocket propellent development work took much longer, however. The changes through the mid-1950s, as the new role became more firmly established, are the subject of the next chapter.

CHAPTER 8

Years of Transition, 1955–1963

Valley Storehouse
Building 54
1908

T HE PERSONAL IMPACT of Technical Director Joe Browning on the direction of Indian Head was so pronounced that it was difficult for many there to perceive that the changes associated with him derived in part from larger trends and tendencies. However, many of the changes he was able to initiate in the 1960s grew out of forces already at work at the facility in the preceding decades. Others came from deeper shifts in the larger context of naval research and development, and out of nationwide revisions in management style inside and outside of the Defense Department. The period from the mid-1950s through the early 1960s was one of transition at Indian Head, in both management and in propellant technology. The modernization of its research management emerged out of several parallel developments. In 1956 Browning resigned from Indian

Head and took an appointment at the Special Projects Office (SPO), which oversaw the development of the Polaris missile, a long-range ballistic missile designed for launch from submarines below the surface. He stayed with the SPO for about two years before returning to Indian Head. The position not only built his personal reputation but exposed him to a variety of new and exciting management concepts. When he returned to Indian Head, he brought with him an invigorated sense of the possibilities of conducting research and development in a government program. He and others were also exposed to a variety of new management ideas coming out of the reforms in management theory which swept through the private sector and through management schools in the mid- and late 1950s. Browning and his contemporaries adapted the ideas of management experts and innovators in a conscious and explicit way at Indian Head.

The reforms in management being discussed and applied in that period had as a goal the breaking down of the traditional top-down authoritarian model of management that had dominated corporations and the military services for generations. A variety of means for tapping the decision-making and creative potential of staff members lower in the hierarchy characterized this management revolution, and Browning and Indian Head were in an ideal spot to implement some of the new concepts.

During the 1950s and early 1960s, the command structure of the Navy began to recognize the problems which grew in the large civilian-staffed laboratories and engineering centers, and addressed the issues with a series of reforms instituted through nationwide policies. Some of these Navy-wide reforms were designed to provide firmer internal civilian management and to establish more clearly the mission of each of the facilities.

In this regard, what happened in this period of transition at Indian Head can be viewed as a historical case study in the growing problems and emerging solutions at naval research establishments during the Eisenhower, Kennedy, and Johnson administrations. From 1953 through 1969, the secretaries of defense were drawn from business. Charles Wilson from General Motors and Robert McNamara from Ford Motor Company consciously sought to bring business concepts to the sprawling government-owned research enterprises. At Indian Head, the ideas took shape in personnel changes, in new programs, and in new ways of conducting work.[1]

The late 1950s and early 1960s saw a burgeoning of new types of missiles and propellants. The long-range Polaris missile was the Navy's most notable achievement in this era. When developed and deployed, the Polaris missile would give the Navy a major role in the nation's strategic deterrent force. However, smaller weapons, including new varieties of anti-aircraft, air-to-air, and air-to-surface missiles, together with other specialized applications of propellant such as pilot-ejection capsules, depth-charge launching, and "gas generators" to operate devices within missiles, all presented new challenges. The gas generators served as power packages, used to pressurize hydraulic accumulators, to provide hot gases for running turbines, or to activate pneumatic controls.[2]

As the second Eisenhower administration and the Kennedy administration sought to close the supposed "missile gap" revealed by the 8 October 1957, launching of Sputnik-1 by the Soviet Union, missile innovation, production, and testing became a major enterprise. Interservice rivalry, multimillion-dollar private contracts, and the growth of a variety of government-owned testing and development laboratories created a hectic and fast-changing market environment. Indian Head was caught in the midst of these developments, attempting internal reform and adjustment to external pressures simultaneously.[3]

CONTINUING PROBLEMS: PERSONNEL, SAFETY, AND EQUIPMENT

Through the 1950s, Indian Head commanding officers had to address a range of management difficulties, most of which were common to other naval and government laboratories. One crucial problem at Indian Head was the identification, recruiting, and retention of talent in government service. The difficulty was manifold. First, government work had a reputation for being rather controlled and uninviting to the brightest talents. Secondly, the government paid less for similar training than the private sector. Thirdly, older, high-ranking researchers with seniority often grew increasingly out of touch with new developments, and as they moved up to positions of branch and division heads, their supervision of younger researchers did not always sit well. Government civil service rules made it difficult to weed out deadwood, with "tenure" of nonproductive individuals representing the sort of problem affecting the academic world as well. In such an environment, the brightest might not be recruited. If recruited, they might soon become disillusioned with the pay, conditions, and quality of leadership. If they continued to perform well despite such handicaps, their chances of moving away to a more attractive position would increase and they would be lost to the local effort.

Another problem was the difficulty of producing innovation on order. Science, by its nature, produces results intermittently, often after long periods of apparently unproductive research in pursuit of alternate paths. Ever since the 1890s, the Navy, like other federal and private institutions, had struggled to convert the creative research process into technological progress by command. By contrast with science, engineering and development can more often be scheduled, but unexpected delays, discoveries of unworkable solutions, and the tedious, careful, repeated testing of materials and compounds very frequently take unpredictable amounts of time. For such reasons, research and development managers found it difficult to plan results and to solve problems as fast as the client (or the budget manager) would like. The Navy and the Defense Department would have preferred to procure innovation in the same way that standardized products were procured—on a schedule, in a predictable budget year, at a predictable cost. Some new management concepts tried out at the SPO addressed these issues.

These problems took a somewhat exaggerated and unique shape at Indian Head, growing out of its location, its mission, and the changes in naval weapons themselves. The relative remoteness of Indian Head from Washington continued to pose difficulties, despite the growth during the interwar years of a community with some of the social and economic aspects of a small company town. Faced with irregular bus service from Washington, junior researchers were confronted with a choice of moving to Indian Head, with its less than cosmopolitan atmosphere, or commuting by automobile from the capital. In the mid-fifties, the expense of an automobile for a low-ranking government employee was still a severe burden. Thus, the continued isolation, although ameliorated by the highway connection, tended to add to the difficulties of recruiting and holding new talent at all such laboratories.

Another problem for managers at Indian Head was the continuing issue of safe handling of energetic materials. Whether working on small batches of propellant or on the production of nitroglycerin or nitrocellulose, the tasks remained hazardous. The danger of fire and explosion was always apparent, and the intermittent accidents that claimed lives added to the management problems in several ways. Safety could only be achieved by careful delineation of procedures and strict adherence to them. During this period the number of "SJPs," or Standard Job Procedures, expanded. By its nature, an SJP could be confining and not conducive to innovation. Ways to reconcile the restrictions of safety with the need to find better, faster, cheaper, and more efficient ways of making propellants would eventually be worked out, but the two approaches often ran head on in the 1950s.[4]

The contrast between the innovative style and the safety style came to the fore in many ways. Browning thought it ironic that he was appointed to be the "safety officer" for Polaris, when he was known for taking risks. Commander William J. Corcoran, who recruited Browning for Polaris, told him that he was being recruited for his reputation for making his own job and for taking a fresh look at problems, rather than for adherence to rules. At a deeper and less personal level, the tension between safety and innovation helped define the divisions between production and research, between chemical engineers and chemists. Specific conflicts and arguments over small points, which seemed matters of personality to the participants, frequently sprang from different approaches to the problem of reconciling procedure and inventiveness, adherence to rules and willingness to think independently.

A second aspect of the safety problem was that the occasional accidents had a devastating effect on morale. Tragic and sudden deaths disrupted families, depressed co-workers, required assessments of blame and guilt, and sometimes resulted in a blot on the careers of surviving supervisory personnel. In the period before 1925, fatal accident investigations often concluded with a stock phrase: the "death was accidental and due to the action of the unfortunate victim." However, by the 1950s such an assessment could no longer be accepted, and investigations often revealed the need for tightened procedures and closer supervision.[5]

In the 1950s Indian Head faced specific problems with its obsolete or worn-out equipment. Facilities built in a catch-as-catch-can fashion under wartime conditions were no longer adequate to the emerging needs of rocket propulsion. Just as in the area of personnel, holdover and outmoded equipment had a long-established place at the facility. Buildings and machines, based on 1940s needs and technology, grew more and more inappropriate as each year passed. In some cases, it was a matter of simple wear and tear and delayed maintenance; at other times, devices for testing, production, mixing, laboratory work, or data recording were not up to the specific demands of the new projects. To gear up for a new task would often require a capital investment, and program managers at head-quarters would prefer to place a job where the buildings, equipment, and, of course, the personnel were already in place to avoid the start-up costs. Thus, program opportunities would be limited by the "tool kit." A new task, once assigned, provided an opportunity to get new tools. As in all such enterprises, a bit of marketing skill would be required to deal with the paradox that one needed projects to get resources, and resources to get projects.

THE SEARCH FOR A MISSION

Personnel, safety, generally restricted budgets, and outmoded equipment all contributed to the problem of finding, defining, and aggressively holding to a clear mission. For years, Indian Head's agenda of work had expanded through a piecemeal and apparently random process of accumulating tasks and budget items. To convert the "task list" into a clearly defined mission was no simple procedure. Without good equipment and people, one could not take on a new set of jobs; without the jobs, it was difficult to lay claim to the resources for people and equipment. Only a particularly adept manager, with vision and leadership, could overcome the problem of bootstrap-lifting.

Officers and senior civilians sought to clarify Indian Head's role as the Navy shifted to increasing use of larger rockets. Through the early 1950s, the development of large, cast-propellant rockets, in which the propellant would be cast directly in the casing, suggested the need for technical changes. At first, Indian Head was slow to respond, but researchers and production people at the facility recognized that if Indian Head were to adjust and survive, it would need new equipment and new programs to deal with the fundamental changes in rocket propulsion. Indian Head's background in extruded rockets in the 1940s and its addition of a casting plant in the mid-1950s suggested that it could lay claim to a solid role in the new development. But to define and grasp that new role in the face of competition from other naval facilities and from the private sector would require decisiveness and strong leadership.

Exactly what role should the old "powder factory" play in a world in which powder production appeared less and less vital, and in which guns might be replaced by rockets? Certainly a change in name from "Naval Powder Factory" would be an appropriate step in the right direction, as Captain Francis W. Scanland, Jr., had advocated as early as 1952.[6]

The name change was finally achieved on 14 August 1958, with the official change in designation to "Naval Propellant Plant" under Captain Griswold T. Atkins. But the new name would only convey a hint of the sector of the naval weapons business which Indian Head hoped to take on. Would the "plant" have a research or testing role? Would pilot production of rockets in peacetime provide the scale of employment once required for smokeless powder production?[7]

141

TABLE 12
Civilian Employment, Indian Head, 1950–1962

Year	Total Civilians	
1950	1,100	
1951	1,700*	
1952	2,900	
1953	3,000	(Korean War Peak)
1954	2,700*	
1955	2,500*	
1956	2,300*	(Post-Korean War Drop)
1957	2,100	(Sputnik Launched)
1958	1,400	(Polaris Work Began)
1959	1,700	
1960	2,200	
1961	2,700	
1962	3,300	

Source: RG 181, Acc. 71A-7407, Box 45, Advisory Committee, Washington National Records Center, Suitland, Md., and Office of the Commanding Officer, Advisory Board Correspondence, Naval Ordnance Station, Indian Head, Md.
Note: Figures rounded to nearest 100.
*Estimated.

Employment at Indian Head had risen from about 1,100 in 1940 to a peak of 3,500 during the war years, to fall back and then rise again during the Korean War, reaching a high of 3,000 in 1953. But the facility declined through the mid-1950s almost to the pre-World War II level. Some researchers and managers felt it was on the verge of being closed, despite the opening of the new plants during the Korean War. Table 12 is a listing of civilian employment figures through the period.

For many of the production workers, employment at Indian Head had become a family matter. By the 1950s hundreds of "blue-collar" employees in Production and in Public Works (the building and maintenance department) could speak with pride of their fathers' and even their grandfathers' work at Indian Head, in the Proving Ground and early Powder Factory days. To some, it seemed impossible to believe that "Washington" would decide to close the facility; "there will always be an Indian Head," they told the worried managers and researchers. But a realistic view of the politics of technical research suggested to the naval officers in command and to the senior civilians in the mid-1950s that something had to be done to clarify mission and adapt to the changes.[8]

The mission of Indian Head in the Polaris program and its difficulty in finding a place in the scheme of missile development became a matter of a congressional investigation in mid-1958, just before the name change from Powder Factory to Propellant Plant. In hearings stretching over two days, a subcommittee of the House Armed Services Committee headed by Congressman F. E. Hebert questioned Director of the Special Projects Office Admiral William F. Raborn, Chief of the Bureau of Ordnance Admiral Paul Stroop, and former Secretary of the Navy Dan A. Kimball, then serving as president of Aerojet-General, as to why a Polaris research and development contract had been placed with Aerojet in Sacramento, California, and not with Indian Head.

Raborn, Stroop, and Kimball all testified that Aerojet was in a better position to do research than Indian Head because of personnel in place and because of facility investment; that once the research work was concluded, the production could be taken over by Indian Head to supply fleet needs. Unlike gun ammunition, the missiles would not be consumed in target practice once the development phase was past. Clearly, there was more profit in the research and development phase than in the long-term production phase. This fundamental difference between rockets and guns, spelled out in the testimony, helped explain why so many private firms had become interested in aspects of development and were willing to abdicate the production for replacement purposes back to the government. It was quite a change from the days when Indian Head, the Experimental Ammunition Unit, and Dahlgren concentrated on gun and gun shell research and development, and then turned the bulk production over to the private sector.[9]

Hebert concluded that Aerojet had superior staff and facilities for the research work, but that the concern over declining employment at Indian Head should not yield a picture of "hopelessness." Rather, he pointed out that the Navy had promised a production line for Polaris powder there and that the future was "encouraging." Indian Head, Hebert remarked, was "an assuredly potentially successful contender for participation in the new weapons program." However, Hebert did not spell out exactly what aspects of that new weapons work would come to Indian Head. The Propellant Plant's mission continued to evolve.[10]

INTERNAL DIVISIONS

In a classic causal cycle, the lack of a clear mission exacerbated internal differences; the internal strife made the definition of the mission more difficult. The definition and evolution of the mission itself contributed to internal friction, as different individuals would bring a different emphasis to the balance between "foundational," or "pure," research and more applied aspects of development and engineering. Getting tasks added to the agenda required that headquarters be convinced of capacity and talent. Sometimes a capacity could be developed quietly and then claimed as part of the competence of the facility, as Browning and others had pushed in the early 1950s for a role in formulating experimental propellants. Since those hoping for a more developmental role had to do so against the inclinations of some of the supervisory personnel, it seemed like expansion by conspiracy to the participants. Browning, for one, remembered it in exactly those terms.[11]

The search for a clear mission was hampered by severe internal divisions among civilian staff over approach to research and over controlling scarce resources of money, equipment, and people. The combined problems of safety, personnel, equipment, and mission contributed to the internal strife and emphasized the need for good management.

In the early 1950s the friction between departments over such issues surfaced in several conflicts. In 1954 Commander Corcoran was appointed to Indian Head to head up the new Quality Surveillance Department. Corcoran worked out a schedule, in conferences with the Research and Development (R and D) Department, for the use of personnel until the new Quality Surveillance Department could recruit sufficiently to take over the functions of administration and local technical supervision. Corcoran planned on internal borrowing of personnel as a temporary expedient. He also planned to utilize R and D personnel in the future to avoid duplication. On this second, long-term use of R and D personnel, a degree of friction developed. Dr. Sol Skolnik, as director of R and D, made it clear that he would prefer to have any such work assigned through regular channels from headquarters and to report back through channels. Corcoran balked at such layering of authority and announced "it is my plan to work as rapidly as possible to relieve Research and Development of all or most of the locally assigned work load under the quality surveillance program." Skolnik, needless to say, did not appreciate the suggestion that his work as well as his personnel be taken from him.[12]

When Corcoran ordered changes in static firing equipment "to permit Inspection to conduct acceptance test firings in accordance with specifications," Skolnik complained to Corcoran about the change. He inferred that either inspection had not been in conformity with specifications or that specifications had been altered. "In either case," he went on, "R & D has not been so informed. . . ." Skolnik held up the equipment work order for sixty days.[13]

Friction between Production and Research also surfaced during the period. Skolnik complained regarding a proposed plan for pilot production of "Gimlet," a modified 2.75-inch rocket, that "no consideration" was being given by the Production Department to the participation of R and D in the planning for the new plant. Such conflicts seemed to have a petty character, but they gave evidence of the serious difficulty of managing three roughly co-equal divisions without a powerful supervisor over all three.[14]

The lack of a strong civilian leader such as Corcoran with responsibilities for Research and Development, Production, and new activities left the departments—from a management point of view—inevitably in conflict as separate "fiefs." The commanding officer, while nominally in charge, found his ability to make peace among the departments hampered by several factors. Since 1900, when George Patterson set up the factory and organized powder production, civilians had emerged as the true managers of the day-to-day factory operation and, later, the research activities. The structure of the civilians' careers kept them at Indian Head for longer periods than the commanding officers.[15]

Although Patterson's official title had been "Powder Expert" in the 1930s, he played a role much like that of a modern civilian technical director, with a degree of supervisory coordination between research and production. On Patterson's retirement in 1940, Walter Farnum had succeeded to the post briefly, but after World War II, there had been no single civilian who could manage the whole operation on the basis of prestige alone. Unlike officers, the civilians could not be ordered to another post; civil service regulations prevented their dismissal on grounds that appeared to be matters of personality conflict.

When, from time to time, a uniformed expert like Commander Corcoran was given direct responsibility for a technical program, entrenched civilian specialists were in a position to obstruct his work if they chose to do so. If they felt that his role would tend to undermine their own, their full cooperation would be difficult to obtain. But even between the civilian-managed departments, cooperation was arranged by the politics of co-equals, rather than by management from above. Thus, favors, obligations, "log-rolling," and jealous protection of one's turf would naturally be the norm.

These problems were in no way peculiar to the Indian Head facility. Such difficulties existed throughout the defense establishment laboratories and would be addressed in several stages by the Defense Department in the early 1960s. In 1961 Secretary of Defense Robert McNamara ordered that "Task Groups" investigate 120 questions he had regarding defense management. Question #97 was: "What must be done in order to enhance the capability of our in-house research and development laboratories?" Task Group 97, headed by Deputy Director of Defense Research and Engineering Eugene Fubini, produced a study of the laboratories which pinpointed the question of short tours of duty for commanding officers as one cause for poor morale at the laboratories. Fubini, a physicist who later became a director of Texas Instruments Corporation, identified other severe problems, including poor salaries, lack of clarity of missions, and complex budgeting and programming procedures that had been patched together over the years. Another report from David Bell, President Kennedy's Director of the Bureau of the Budget, focused on the "lack of clarity" in the relationship between civilian technical staff and military officers as one of the causes of poor morale. Over the period 1961–64, Congress approved "Public Law 313" or "supergrade" positions for civilian technical directors of the major military-operated research and development laboratories. These technical directors would manage the civilian scientific or technological side of the operation, while the commanding officers would retain responsibility for military personnel and the facility itself. As the technical directors were appointed, the exact relationship would evolve somewhat differently at each establishment. However, until such a position became available at Indian Head, the old methods and old problems would persist.[16]

For the researchers and officers at Indian Head in the early and mid-1950s, the difficulties identified a few years later in the Fubini and Bell reports made day-to-day management a sequence of turf-fights, continuing attempts to redefine the mission and to add tasks, and constant struggles to recruit and retain talented personnel. In the mid-1950s Fred Zihlman moved over from Research and Development to work for Corcoran in the Quality Surveillance Department, as part of the buildup of Corcoran's division. While in Quality Surveillance, Zihlman wrote a proposal for Indian Head to take a much larger role in the overall surveillance of propellant. After his proposal was considered and rejected, he departed Indian Head to work first for the Bureau of Weapons in the Research-Missiles, Missile-Propulsion (RMMP) office and then for the National Aeronautics and Space Administration on the Gemini program. These steps in Zihlman's career reflected three difficulties of management typical of the era: internal personnel recruiting among divisions, attempts at mission redefinition, and departure of talented individuals from the institution for more exciting and responsible jobs.[17]

Zihlman was one of several younger, energetic researchers who moved on to take advantage of opportunities elsewhere in this period. The whole rocket industry was expanding, and some of the best sources for such people were the in-house government laboratories. Others who departed in these years included Joe Browning's colleague Gene Roberts, who took a position with United Technology. John Murrin, who served in Indian Head's Research and Development Department in the early and mid-1950s, followed Zihlman in the move to the RMMP office in the Bureau of Weapons.[18]

AD HOC MISSION ADD-ONS: CAD, TERRIER, WEAPON A

Several of the projects and areas of research developed in this period would set precedents for long-range areas of excellence for the station. Such piecemeal additions contributed to refinement of the mission. In 1956 the commanding officer, Captain George E. King, arranged a conference of representatives from the Naval Ordnance Laboratory (NOL) and from the Bureau of Ordnance (BuOrd) to discuss progress on a Pilot Ejection Catapult Cartridge (PECC). Much of the initial work on this project had been conducted at the Philadelphia Air Material Center, and King worked out plans for Indian Head to become engaged in testing of the cartridges and to introduce experimental grains developed at Indian Head, variations of the N-5 propellant, for testing as aircraft pilot ejection catapult propellants.[19]

Later that year, a full program was developed and presented at another conference, outlining the kind of work in which Indian Head could engage to develop pilot ejection seat catapults, and suggesting such a role in the development of "cartridge activated devices," or CADs, more generally. This conference, which involved representatives from the Indian Head, NOL, and BuOrd, was held on 23 October 1956, and appeared to be the earliest official move of Indian Head into this activity. Earlier work in preparation of propellant specifications and in evaluation of propellant performance data, worked up in earlier studies of experimental propellants for the airplane launching catapult and drone systems, gave the R and D Department a good claim for pertinent background for this work. A paper detailing work on catapult propellants, prepared by Gene Roberts and H. N. Sternberg and presented at the Joint Army Navy Air Force (JANAF) eighth specification conference in 1956, also helped establish the credibility of Indian Head in this area.[20]

The opening of the casting plant in the mid-1950s led Indian Head into production in a new area—larger propellant grains which would be cast into beakers and then loaded into motor cases, rather than extruded through a die. Two of the first cast products were the grains for Weapon A (also designated Weapon Alpha), a device for launching depth charges at ranges up to 950 yards, and for the "booster," or first stage, and "sustainer," or second stage, of the Terrier weapon. The Terrier was a large, short-range surface-to-air weapon. In the period from 1954 to 1956, planning conferences dedicated to these specific weapons were held at Indian Head in attempts to sort out cooperative plans with the developing laboratory, Allegany Ballistics Laboratory (ABL) at Cumberland, Maryland, as well as the internal lines of authority between the Research and Development, Production, Quality Surveillance, and Inspection Departments at Indian Head. Representatives from the ABL, Naval Ordnance Test Station (NOTS) at China Lake, Indian Head, and BuOrd attempted to work out detailed specifications. In an effort to agree amicably on drawings and specifications, the group would thoroughly discuss lines of responsibility, "interfacing at all levels." Production and Research people from Indian Head would help guide the development at the ABL, identifying problems and making suggestions during the early phases. Indian Head had grown to expect that it would be actively engaged in every naval program which was sent to the ABL for development.[21]

In the fall of 1955, the Powder Factory produced Terrier booster grains in NPF Lot 14 and, early in 1956, sustainer grains, NPF Lot 17. Using these grains, inspection records yielded base data while the grains were held in ambient storage. In March 1957 the program of Terrier manufacture was approved by BuOrd, and a task assignment for fiscal year 1958 was issued. Under the task assignment, Indian Head conducted type-life surveillance tests which involved putting several Terrier motors through a cycle of controlled temperature storage.[22]

Other task assignments for surveillance work covered storage and periodic testing of Weapon A, the Sidewinder Propulsion Unit (SPU), and a Rocket-Assisted-Torpedo, affectionately designated "RAT." Together with the surveillance on Terrier sustainers and boosters, the surveillance of the five large-scale rockets entailed a system of ovens and a program of pretest conditioning, static firing, data assessment, and equipment recovery. Periodic chemical and physical tests on the stored motors would add to the data. Since the surveillance program would last ten years, a twenty-four hour security watch represented over $800,000 in labor cost over the course of the five surveillance projects. Captain Atkins indicated over two-thirds of the cost could be saved with the installation of a safety and warning system in the conditioning ovens.[23]

In 1954, working on the assumption that the ABL's Weapon A grain would be put into pilot production, representatives from the ABL met with F. C. Thames as head of the Production Department and Skolnik as head of Research and Development. Original planning was based on the assumption that about 1,550 rounds would be manufactured. The conference drew up initial plans for testing the grains and for long-term, controlled temperature surveillance.[24]

A later "Weapon A conference," in October 1955, chaired by the representative from Allegany, settled on production plans for Weapon A. Although the ABL prepared the final drawings and plans to send to BuOrd, suggestions, recommendations, and changes were proposed and agreed upon by the ABL representative after closely reviewing the preliminary plans with both Production and R and D personnel from Indian Head.[25]

Later weapon conferences, through 1956, followed a more formal pattern. Commander King brought in BuOrd representatives to review thoroughly further problems related to the Terrier rocket. A sustainer grain from a specific lot, NPF-13, had blown up during a qualification static firing. At the December 1955 meeting, Dr. Loren Morey from the ABL focused on that specific malfunction. A review of existing manufacturing data and of inspection data gave no clue as to the problem. The conference worked out a plan for a close examination of production information to determine the exact cause of the malfunction. In order to trace the difficulty, a program of sixty-four tests was planned on the same lot, to be held at Indian Head, while the ABL held a series of tests in which they deliberately changed the casting process in an attempt to identify the problem. The conference concluded with the relatively vague guess that the cause of the blowup was probably an excessive burning surface. Skolnik suggested a possible causal sequence, stemming from a grain flaw. Even though the Production Department had not been able to identify any such flaw in its testing program, R and D laid the blame there. ABL and Indian Head staff agreed to examine further the remaining grains from the lot, using a wide variety of tests, before deciding on its disposition. The results of the meeting were hardly remarkable. What was clear from the minutes of this and other conferences was that differences among Indian Head personnel could not remain private; the problems would have to be resolved in cooperation with ABL staff and BuOrd personnel.[26]

As Indian Head became more involved, through the weapons conferences, with other laboratories and production facilities, any internal divisions over responsibility, planning, or minor management issues became publicly aired. If Indian Head was to move effectively at the inter-institutional level, it needed to present a united front and to resolve its internal difficulties amicably before such meetings. Otherwise, awkward incidents of publicly aired squabbles could mar the institution's standing. In such ways, the need for strong leadership, capable of directing and resolving the issues "at home," became more and more apparent through the 1950s.

Similar meetings were held on malfunctions of the 2.75-inch folding-fin aircraft rocket (FFAR). Representatives from BuOrd, NOTS, and the quality assurance and research groups at Indian Head met to review results from a set of tests on 300 rounds selected from 30 different grain lots involved in a program to rework the propellant. The results were compared to 300 test firings at NOTS on one ammunition lot. The conference produced recommendations for the reworking of 1.5 million rounds at the Naval Ammunition Depot at Shumaker, at Camden, Arkansas. While the cause of earlier malfunctions had not been pinpointed, the conference agreed that "exudate" from older grains had led to uneven burning. A few procedures regarding disassembly and reassembly of the motors were agreed to, including replacing rubber o-rings which might have been damaged by exudate.[27]

The extruded 2.75-inch rocket, one of the standards of the World War II era, was taken out of production in 1957. However, eight years later, the 2.75 would be revived and redesigned, and would go into mass production once again for the increased needs of the Vietnam War, when it became the "bread and butter" round, especially from helicopter gunships. In the mid-1950s, however, problems with the inhibitor wrapping for the 2.75 and the rocket's unreliability at extreme temperatures led to its being dropped from the Navy's arsenal.[28]

By the mid-1950s, specific weapon conferences resolved a wide variety of interfacility issues. A March 1956 conference reviewed various aspects of the Weapon A grain, particularly focusing on the fact that inhibitor appeared to deteriorate on rockets stored for over six months at 130 degrees Fahrenheit. At such warm storage, the inhibitor tended to be pressed tightly against the motor case and flow over the end of the grain. The ABL and NOTS disagreed over the seriousness of the problem, with the ABL arguing that it was unlikely that such temperatures would occur in normal storage, and NOTS arguing that the figures could be extrapolated to standard temperatures and indicated a deterioration problem. In the long run, it turned out that in this case NOTS was correct. The large-scale "conferences" over Weapon A with representatives from Indian Head, NOTS, ABL, and BuOrd led to more local, ABL-Indian Head meetings to discuss even finer details of production and testing. Representatives of the Production and Research Departments at Indian Head went to the ABL to reciprocate and work out further details of impulse measurement, plan for "round robin" comparative tests, and discuss the reasons for variation in Weapon A grain lots.[29]

MISSION REDEFINITION BY PLANNED REMODELLING

Despite such promising cooperative work, some of the younger researchers and engineers, colleagues of Joe Browning, including Gene Roberts, Todd Braunstein, and Frank Worcester, were quick to point out that the dated nature of the equipment placed severe limits on what Indian Head could do. Roberts, as associate head of the Engineering Division, prepared an extensive "justification" of an experimental propellant processing facility in 1956. He noted that experimental equipment "is scattered, misfitted spacewise, incomplete, inadequate, and uneconomical to operate. That it has accomplished what it has thus far is a tribute to the ingenuity of the people who operate it."[30]

The next year, a frank and even more thorough review of the testing equipment revealed that the layout was outdated and based upon 1940s rocket technology. Worcester, who worked under Braunstein in the Research Department, agreed with and went further than the complaints of Roberts. Worcester began with a brief review of the history of the equipment. Much of it, he pointed out, had grown without much planning from the test facilities for the extruded rockets in World War II and from gear installed over the years by different groups. He gave a brief history of the changing jurisdiction over rocket testing, by an Inspection Department, a Quality Evaluation Department, and then a Quality Control Department that handled both acceptance inspection functions and evaluation of fleet inventories. The shifting lines of authority and reorganizations, he implied, had helped account for the poorly planned and badly distributed equipment.[31]

Another facility for testing larger rockets had been built as a service to the cast propellant plant. That test ground consisted of static firing bays, loading facilities, and a recording room with capabilities of handling rockets up to 200,000 pounds thrust. Practical difficulties ranged over the gamut, including inadequate cooling of equipment, crowding, use of temporary quarters, inflexible x-ray equipment, inadequate x-ray safety shielding, old and scattered conditioning ovens for heating and cooling rocket grains, cramped clothing change houses and office space, and a wide variety of "compromises" with good safety practice. Some of the equipment required manual control, where simple thermostatic equipment could vastly reduce labor costs.[32]

The Worcester report on the test facility recognized that the Navy was at a turning point in rocket development. In the very period when Browning had left to work on Polaris at the Special Projects Office, Worcester made some predictions and proposals for improvement of the large-rocket testing program at Indian Head that would prove essential to its future role in the Polaris program.

He pointed to two "trends" in rocket development under common discussion in the missile community at the time: decreased demand for mass-produced rockets, and a growing complexity in rocket design which made assembly at the producer's plant desirable. He concluded that these changes could lead to a greater percentage of overall procurement from the Indian Head facility. Instead of routine acceptance inspection, the Navy would need "developmental evaluation" as evidenced by the work done at Indian Head on Weapon A and at NOTS on the "SPU," or Sidewinder. "We are now faced with the requirement not only to produce but to test and evaluate the device produced," he claimed. "In order to accomplish this, we must develop test facilities to meet the demands of our future development and production programs."[33]

Worcester noted the prediction of others that the "specific impulse" barrier would soon be broken. Researchers expected that new composite propellants, using light metals and fluorine compounds, would far exceed the propulsive power of the double-base propellants. Furthermore, problems in aerodynamic heating from air friction of missiles carried at sonic and supersonic speeds would require conditioning ovens capable of much higher temperatures. Other tests would require conditioning propellants at lower air pressures characteristic of high altitudes. He recommended flexibility to take on a variety of air-breathing and liquid monopropellant systems, as well as solid-fuel rockets. He also recommended work on systems for testing underwater rockets and deep-running torpedoes, using solid propellant gas generators. As a consequence, he spelled out plans for a new ballistic research facility, to include a separate x-ray building, a preparation building, holding bays, firing and test bays including a high-pressure testing tank, a low-pressure testing chamber, a centrifuge for simulating acceleration forces, static test bays for handling medium sized rockets, a centrally located recording building with closed circuit television, and a set of buildings capable of conditioning the rocket motors in the range of −80 degrees Fahrenheit to 800 degrees Fahrenheit. He worked into the plan several already existing buildings, including darkroom facilities, test bays, and a recording building.[34]

His recommendations had to await an infusion of money from the SPO engaged in Polaris. However, such mid-fifties planning indicated that various researchers, including some rather low-ranking in the organization, kept their ears to the ground and recognized the changes which would come about with new cast propellants. For example, the addition of aluminum as one of the ingredients in propellant powder was still several years away when Worcester noted that it was under study.[35]

THE SHAPING OF A DOMINANT PERSONALITY:
JOE L. BROWNING

In the period 1956–58, Joe Browning left Indian Head and worked for the SPO on the beginnings of the Polaris missile project. For Browning personally, the experience was rewarding. He later remembered, as one of the most exciting periods of his life, a weekend early in 1956 when he met with a small group to plan the basic, solid-fuel concept which was at the core of Polaris. Early plans had centered on a modification of the existing Army Jupiter, for use aboard submarines. Naval officers objected to a liquid-fueled rocket on safety grounds. By looking ahead to the planned warhead weight, and by predicating the propulsive force required on anticipated future weight reductions of the warhead, the planning group worked out—in three intense days—the basic size and requirements which became the solid-fueled Polaris. In the week after their preliminary planning session, the idea was approved by Admiral William F. Raborn, by Secretary of Defense Charles Wilson, and by President Eisenhower.[36]

For the first three years of the SPO work, Browning watched the team grow from three to over thirty technical people, with consequent layering of management and what he perceived as the slowing down of responsiveness in the organization. However, he was intrigued by "PERT," the program tracking system put in place under Captain Levering Smith and Admiral Raborn. This system later became famous throughout the Navy and more widely through the government as a model for structured planning around objectives based upon continually revised, expert-derived estimates of completion dates. He later claimed that the only time he was tempted to leave research entirely and to move into management was when he considered an offer to take over and run the PERT system.[37]

By 1958, as Polaris was well under way and plans for manufacturing propellant and testing it were being shaped, Browning accepted an offer to return to Indian Head. The three years with the Polaris project had exposed Browning to a number of new developments in technical management. Browning belonged to the American Management Association and took considerable interest in developments in the private sector at the time. However, the innovations of PERT and the excitement of working on the high-priority and high-visibility Polaris project also contributed to ideas he had about tapping the human potential among researchers and engineers at such facilities as Indian Head.[38]

The success of Polaris itself, with a pad-launch in 1959 and with a proven underwater firing on 20 July 1960, demonstrated that a new proposal could be developed rapidly and successfully if assumptions were challenged. Browning liked the sense that a fresh start with a new approach would overturn established and incorrect thinking. Technical innovation, such as the concept of building a long-range missile using solid propellant instead of liquid propellant, fascinated Browning. For a man of his personality this was no mere technical solution to an engineering question: how to move a certain weight a certain distance from a submerged submarine. Rather, it took on the quality of a crusade; Polaris became a means of demonstrating that hidebound ideas, defended by people in power, could be destroyed by force of intellect and diligent work, tapping the talents of a select few. Technological innovation could demonstrate the vindication of brains over power. It is such a personal, even passionate drive which, when communicated to others, that constitutes the elusive quality of charisma. When he returned, he would bring his new ideas and his personal drive to bear on the local problems.[39]

MANAGEMENT PROBLEMS ADDRESSED

Among the suggestions recommended in 1952 by the Industrial Survey Board to address management issues at Indian Head had been the institution of an advisory board, made up of prominent individuals from the field of propellants, to convene at the facility. From 1952 through 1960 the board met regularly, usually twice and sometimes three times a year. Most commonly through these years the board would meet in April and September. The early board consisted of Dr. A. M. Ball from Hercules Powder, Dr. Wendell F. Jackson from du Pont, Dr. E. T. McBee from Purdue University, and O. H. Loeffler from Rohm and Haas. In addition, F. F. Dick, a senior chemist from the Bureau of Ordnance who had served under Patterson at the early EAU in 1922, continued to serve on the board through the 1950s.[40]

The visits of the board were occasions for considerable anxiety, as a tight schedule of meetings, social occasions, and presentations by management required coordination between the commanding officer and various division heads. The reports of the Advisory Board would be forwarded to Bureau headquarters, and the commanding officer at Indian Head would have an opportunity to respond to specific suggestions and comments made by the board. On occasion, if one or another department was criticized by the Advisory Board, the commanding officer would allow the department head to participate in the response. Several of the commanding officers were able to take advantage of the board's presence and its ability to make independent, and sometimes critical, judgments on management, to implement internal changes.[41]

In particular, through the late 1950s, the board in several of its comments suggested more emphasis on "exploratory and foundational research," supporting the position of Dr. Skolnik that the Bureau did not provide enough funds for such work. However, at the next meeting, when the board praised the work of the Production Department in reducing overhead costs, Skolnik complained that due recognition for Research and Development steps in that direction had not been made. Debates over exactly how to phrase the Indian Head response to the board's comments and criticisms tended to reflect the internal divisions and conflicts, which, by 1957 and 1958, reached a difficult level.[42]

Early in 1958, as Commander King was retiring, the Advisory Board complimented King on several aspects of his management. The board particularly noted the establishment of committees to work out specifications and plans for Zuni, Sidewinder, and Talos, which continued the work of the earlier planning groups around Weapon A and Terrier. These committees cut across the traditional bureaucratic lines and appeared to reflect the parallel development in industry of "matrix management" and "product management". Such reforms were intended to address the problems which came from the cumbersome functional-department model.[43]

Although the Advisory Board did not make any effort to tie their observations to larger developments in administrative theory, or to use the language of management in describing the committees, they clearly took the position that a more flexible response was suitable. Such observations did not always sit well with the various administrative heads at Indian Head who could see such cross-divisional committees as undermining their own authority.[44]

In 1958 the problem of internal supervision was somewhat addressed by the appointment of Commander Jim Dodgen as technical director. As at other naval laboratories, the role of technical director was filled for a period from the late 1950s through the early 1960s by a technically trained naval officer rather than by a civilian. Although such a move did put a technical person in charge of the various functional departments, the fact that Dodgen was appointed on a tour of duty which would be likely to end in three years still represented a limitation on his authority and control.

King's successor, Captain Atkins, working with the board, instituted a series of rapid changes in 1958, hoping to address locally some of the typical management problems plaguing Indian Head and other laboratories. He recruited Browning, bringing him back from the Special Projects Office in the summer of 1958. Although there was no slot for a civilian technical director, he at first placed Browning in a quasi-supervisory position as head of a staff group designated Planning and Management. Atkins obtained the approval of the Advisory Board at its next meeting for the new position held by Browning. He obtained the change in name from Powder Factory to Propellant Plant in August 1958.[45]

At the meeting of the Advisory Board in September 1958, Skolnik, Browning, and Dodgen reported on the various projects in hand, while Atkins reported on the management reforms that he was instituting. The main tasks included work on Talos, Sidewinder, and Zuni, planning for testing and preliminary work on Polaris, and work on new propellants. In response to the new emphasis brought by Atkins, the Advisory Board dropped its recommendation to increase the "foundational" research, and concentrated on reflecting the need for increased engineering. Immediately after that meeting, Skolnik resigned, and Captain Atkins appointed Browning to direct the Research and Development Department.[46]

Browning not only brought the contacts and commitment of Polaris with him to Indian Head, he also brought some of its business as well. While still at the SPO, he arranged that some of the testing of Polaris motors be conducted on a one-third scale size at the Indian Head test facilities.[47]

As Browning remembered it later, Captain Atkins expected Browning to replace Commander Dodgen as technical director, as soon as the position was changed by Congress from a military "billet" to a Public Law position in the civil service. Browning later recalled that the original plan, which he expected to go into effect in 1958 or 1959, was delayed until 1962. He traced the cause of the delay to the fact that Atkins retired, and after two short-term interim commanding officers, Captain Otis A. Wesche became the commanding officer at Indian Head. Wesche conducted a national search for a technical director, and only after repeated recommendations from a wide variety of specialists that he put Browning in the post, and a threat from the command structure to withdraw the position unless he did so, Wesche finally appointed Browning to the Public Law 313 position in July 1962.[48]

In the intervening years from 1958 through mid-1962, Browning served at Indian Head briefly as director of the Research and Development Department. In that period, he began to implement several of his new management concepts within R and D, as Indian Head continued to adjust to its role as a center for propellant development.[49]

Like Atkins, Captain Wesche made use of the Advisory Board, obtaining the board's support for the recommendation that a civilian technical director at a "supergrade" level be appointed at the end of Dodgen's term. Wesche also noted with approval the Advisory Board's recommendation that Research and Development explore more closely problems related to hazards in production.[50]

BROWNING'S MANAGEMENT IDEAS

In 1958 and 1959 Browning felt he had to revise the direction of the Research and Development Department, and he eased out many of the scientists whom he regarded as less productive by encouraging them to resign with a positive recommendation. Within a year he had begun a rapid process of recruiting, filling thirty-six positions with new chemists and chemical engineers, many of them new graduates. By September 1959 the Advisory Board had approved the recruiting effort, although remarking on the lack of experience of the new personnel in the field of propellants.[51]

Even though the government positions did not pay well by comparison to private industry, Browning felt he could encourage young graduates to come to Indian Head by showing them opportunities and responsibilities for new research. After a few months or a year of work, he would send the recently hired individuals out to assist in recruiting more new college graduates. Many of the individuals he recruited between 1958 and 1963, who had gained their degrees in those years, stayed on at the facility, rising to positions of leadership and management over the next generation. By the late 1980s, the core of the facility's management was largely drawn from the cohort he recruited in those years.[52]

Browning solved many problems of inefficiency by simple and direct phone calls, and inserted a memorandum of the call in the file as a record to confirm authorization. By this simple expedient, he was able to cut much of the red tape that always plagued naval operations. For example, the "Terrier Type Life Program" ran on a different temperature schedule from the other rockets. Browning was able to obtain telephone authorization to convert the Terrier storage temperatures to the "NPP standard" used for the other rockets, thereby reducing manpower and space needs. Other telephone memoranda recorded how Browning obtained approval for contracts, for changes in equipment, and for expending funds in short order.[53]

During the late 1950s and early 1960s, Browning instituted an innovative management idea which he said was based on "multiple management," and which he consciously borrowed from a plan instituted at McCormack Tea Company in Baltimore. McCormack Tea had published a work called *The Power of People*, which described a procedure the company had instituted in the early 1930s—an Assistant Management Board. In order to harness the talents of younger staff members, Browning set up a similar board. His concept was that each branch head within his Research and Development Department should select one talented younger employee to serve on the board. The term of office on the board was set at six months; the board would select its own leadership and would create vacancies by mutual self-ranking, dropping the lowest-ranked from membership. Browning presented the board with his special concerns as Research and Development Department head and would implement their suggestions if they were made unanimously.[54]

This Assistant Management Board succeeded in motivating many of the younger staff members by giving them access to senior management and a sense of being singled out as members of an elite group. The actual agenda items and studies prepared by the boards through these years were obviously of less crucial import to the young chemists and engineers than the experience itself. Browning explicitly used the board to motivate and tap the energies of the young researchers; his more senior branch heads, he believed, would stifle any creativity the new recruits had among them. Browning also used the board to identify talented younger staff members, whom he would keep in mind for difficult or challenging assignments later.[55]

While the Assistant Management Board made up of young staff members flourished, the far more senior and distinguished Advisory Board of outside specialists in propellants began to languish in these years. Although the Advisory Board had proven a useful device for bringing difficulties to the surface and for clarifying the need for the very reforms which Browning advocated, as director of Research and Development he found its suggestions intrinsically limiting. He held that the board put him in a "damned if you do, damned if you don't" dilemma. If he proposed taking on a new piece of research which no one else handled, the board would ask, why, if it was a good idea, it was not being done at other facilities which had more staff. If he proposed doing something that was already being done elsewhere, he would be asked why the other laboratory couldn't do it.[56]

Browning later recalled that Captain Wesche summoned him to his office and informed him that the Advisory Board had recommended that he dismiss him. Instead, Wesche told Browning, he had decided to dismiss the Advisory Board.[57]

The combination of Browning's intense drive and personal charisma, together with his structured management devices for recruiting and motivating new talent, soon developed a dedicated following which grew in concentric circles, all committed to excellence and productivity. He would later utilize the same techniques when he moved to the position of technical director.

POLARIS WORK

Early in 1958 Indian Head began to propose Polaris-related jobs. In February Todd Braunstein, then working in the Research and Development Department under Skolnik, arranged by telephone to expend some preliminary SPO funds on design studies for a Polaris missile fire-suppression system. Meanwhile, the Production Department, after working with Aerojet and sending a visitor to the plant in Azusa, developed a production facility to manufacture 2-2-dinitroproponal (DNPOH), a chemical used in synthesizing nitropolymers and nitroplasticizers for use in Polaris propellants. By April and May of 1958, Indian Head was supplying small amounts of DNPOH, in the range of several hundred pounds, to Aerojet.[58]

Close cooperation between Research and Development and Production would be required if the recently renamed Naval Propellant Plant was to move into Polaris grain production and meet the stringent scheduling demands of the SPO. A plan for such cooperation was spelled out during the first few months of Browning's tenure as the new director of R and D.[59]

Through early 1959 the Powder Plant worked on plans for producing and testing Polaris motors at one-third scale. The planning meetings, coordinated by "K"—that is, the head of the seven-person local Polaris program management office, E. L. Watt—showed much less evidence of the earlier frictions over turf, equipment, and style of operation than had surfaced a few years before in the face of other coordination efforts. Factors that accounted for that new spirit of cooperation included the excitement of working on the high-visibility Polaris missile project, the firm management of Watt and Dodgen, and the drive and commitment of Browning.[60]

Planning for testing fifteen one-third scale motors between March and November and for nine full-scale motors between November 1959 and July 1960, required that Research and Quality Control cooperate closely to produce a "Problem Statement" in the form of a joint memorandum. Watt asked for Research and Production to prepare also some joint "investigation plans" and to exchange data sheets.[61]

A few minor hitches showed up. J. B. Nichols, in Research, was alarmed when he discovered that planning required three buildings, numbers 888, 889, and 890, be vacated for Polaris work. The buildings had been obtained he noted, "only after long and difficult justification and re-justification," and he made it clear to Watt that Research expected the space to be replaced.[62]

Further evidence of the push for coordinated work because of the Polaris project came in the nature of the management controls systems. Commander Dodgen submitted to Captain Atkins the first "Milestone" charts, numbered "0, 1, 2, 3, and 4," early in 1959. These charts, familiar throughout the government in later years, appear to have been the first of their type employed at Indian Head. Dodgen showed "Action Milestones" in each of four areas, with expected completion dates for specific tasks. The four areas were Propellant Characterization, Scale-Model Development, Full Scale work, and Construction of Facilities for fifteen-unit-per-month production. The milestones or anticipated completion dates for the various tasks in each of the four areas fell between June 1959 and June 1960. Watt called a meeting in January 1959 for a presentation on the milestone system itself by representatives from the Office of Naval Material and from the SPO. The management milestone system, called at the time "Line of Balance," was explained for the first time to all the department heads. In this concrete and explicit fashion, the management concepts developed in the SPO influenced and shaped the situation at Indian Head.[63]

During the fiscal year from July 1959 to June 1960, the Propellant Plant operated on a task assignment from the SPO, which provided funds and authorization to do four things. First, the Propellant Plant was to continue to work with the ABL in pilot processing and process development of one-third scale, second-stage motors. Second, it was to develop a batch-type base grain processing facility with a capacity for 30,000 pounds a month, to go into operation by June 1960. Third, the plant was to review all specifications, safety rules, and process designs for the rocket grain production. Fourth, the plant was to continue planning of improved instrumentation for a test layout for firing the motors. Separate task assignments covered the production of the base stock DNPOH and the testing of produced motors.[64]

By November 1959 the Propellant Plant had made and fired six one-third scale Polaris motors. All of the motors were static fired without any major "malperformances," although there were minor ignition difficulties on the last three motors. A wide variety of measurements and data were gathered from the six test firings, including blast stream temperatures, internal surface temperatures of the nozzle exit, internal pressure, char and erosion rates, and nozzle and shock wave position and motion. Pressure results were particularly difficult to obtain, as deposits clogged the pressure lines, resulting in clearly erroneous data.[65]

The plugged pressure lines and other testing problems required redesign of the instrumentation of the test stands. Problems such as means of cooling the center core, developing safety shielding, the location of strain gages for best consistency in readings, and positioning fixtures for radiography tests kept the Design Division in the Research Department, headed by E. Y. C. Tsao, occupied with constant redesign and rebuilding of the test equipment.[66]

While Indian Head had responsibility for pilot plant processing and for process improvement, for testing and for test equipment improvement, the research on new propellants went forward at the Allegany Ballistics Laboratory. Levering Smith, the technical director of the SPO, expected the ABL to develop a higher-impulse grain for the next modification of Polaris motors in mid-1960; for that reason, he ordered that Indian Head not proceed with full-scale motor production using the current propellant but stick with the scale model (SM-1) work. From December 1959 through mid-May 1960, the Propellant Plant had processed forty-nine of the scale model motors and conducted a series of about seventy-five tests on the motors. As in the World War I period, when multiple tests for different variables had seemed the way to economize on time and staff, the SPO was able to combine six or more of the ten different elements being tested on one static firing. Polaris base grain manufacturing facilities for the propellant opened at Indian Head in 1960, and production of the nitroplasticizers for Polaris began in 1961.[67]

Scale model work, involving casting various blends of ABL-formulated powders into casings and testing the rocket motors, continued through 1961. Other aspects of the Polaris work included production of nitroplasticizers, long-term surveillance tests of scale model motors stored at 110 degrees Fahrenheit, and experiments to find a substitute for aluminum as an ingredient in the propellant. Formulation "2056D," a standard mix incorporating the aluminum ingredient which was adopted through 1960 and 1961, produced several accidents in production when the base grain material deflagrated in a press during a pressure cycle.[68]

In 1962 Browning, now serving as technical director, helped organize a "round robin" test procedure to correlate processing and testing techniques among the three Hercules-operated facilities and the Propellant Plant. The three Hercules plants were ABL, Radford, and a facility at Bacchus, Utah; each engaged in production of standard forty-pound charges to be machined and tested. Each facility would make twelve charges, retain four as spares, test two, and then send two to each of the other three facilities. Those received would be subjected to chemical, physical, and firing tests, and the data exchanged. Indian Head had its samples ready and waiting while Bacchus geared up for production of the round robin samples in June 1962.[69]

The round robin procedure was not the first for Indian Head, and the Powder Factory had participated in a few such exercises in the early 1950s. A common technique in the private sector for ironing out differences between different companies' standards and procedures, the method had been widely employed by the petroleum business in the 1930s in standardizing octane testing procedures and in setting standards for motor oil viscosity. Professional associations such as the Society for Automotive Engineers organized such intercompany testing procedures, and the Bureau of Mines had assisted with independent and neutral data collection.

The fact that the round robin method was utilized in the Polaris program demonstrates several aspects of the burgeoning rocket enterprise in the early 1960s. Although Hercules operated the ABL as a Navy contractor and ran its other facilities with contract funds, the multiple laboratories and firms involved soon evolved specific differences in terminology, method, equipment, and procedures. Coordination by conference, as on Terrier and Weapon-A, was one way to insure that all participants were aware of each other's work. Another method was the "distribution list" for reports, which constantly expanded to include other laboratories and contractors. By late 1962, regular weekly reports, in addition to seventeen copies internally, would go to four individuals in the Special Projects Office, four at the Bureau of Weapons (including former Indian Head researchers Murrin and Zihlman), two at the Hercules Powder Company, sixteen individuals at the ABL, one at the Naval Research Laboratory, and one at the Lockheed Missile Space Company. The list continued to expand, and individuals at Aerojet-General as well as more internal offices got later copies of reports. The fact that by the early 1960s three Hercules-operated facilities and the Propellant Plant utilized the round robin procedure indicates how government production and testing of one complex product could rapidly come to resemble the variegated world of private enterprise. The methods of the private sector would be required to bring the diversity into mutually intelligible language, if not conformity.[70]

TABLE 13
Partial List of Rocket and Missile Systems, circa 1960

Name	Description
Asroc	Ship-carried, antisub, torpedo or depth charge
Bullpup	SR, air-to-surface, liquid or solid motors
Chaffroc	Modified Zuni for dispensing chaff
FFAR	Folding-Fin Aircraft Rocket, 2.75", air-to-air, ground
Shrike	Air-to-surface, antiradar, short-duration solid motor
Sidewinder	Air-to-air, single-stage, solid motor
Sparrow	Air-to-air, radar-guided, solid motor
Subroc	Subsurface-to-air-to-subsurface, cast composite motor
Talos	LR, surface-to-air, solid motor, ramjet 2d stage
Tartar	SR, surface-to-air, dual-thrust, single stage
Terrier	MR, surface-to-air or -to-surface, two solid stages
Zuni	Air-to-surface, folding-fin, unguided, solid motor

Source: Atkins to Chief of the Bureau of Ordnance, 10 August 1959, RG 181, Acc. 71A-7407, Box 45, Correspondence, Washington National Records Center, Suitland, Md., and Naval Ordnance Station, "Introduction to Ordnance Technology," IHSP 76-129, Indian Head Technical Library, Naval Ordnance Station, Indian Head, Md.
SR = Short Range; MR = Medium Range; LR = Long Range

Some of the Special Projects Office Polaris-funded work wound down in 1965, as the Hercules Bacchus plant picked up the next phase of production of new Polaris propellants. However, Indian Head continued to engage in surveillance, nitroplasticizer production, testing of burning rates, studies of casting process improvement, and the design of new equipment for extruded propellants.[71]

Through the 1950s the types and variety of rocket grains produced by both the casting method and the extrusion method had multiplied. Indian Head had a role in the manufacture and testing of many of the rocket propellants, as well as PADs—or propellant-actuated-devices—and gas-generators, propellants which burned to produce gas used in devices for other than propelling the rocket. By the 1960s the complexity and number of missile motors and gas generators had increased. A partial list of the systems either manufactured, regrained, or serviced at Indian Head during the early 1960s is presented in Table 13.

CHALLENGE AND RESPONSE

Finding, asserting, and holding a viable role in the ever-changing technology of rocket production would require a flair for aggressive management. Such managerial risk-taking would be uncharacteristic of the safety-conscious and conservative style required for the continued safe operation of the facilities. This difficulty of propellant research management could be resolved with the right mix of personality among the leaders, but finding and holding together that mix was not easy.

Over the period from 1962 to 1974, Joe Browning, as technical director, would work to expand the mission of Indian Head, to tap the talents of the younger researchers, and to address the continuing issues of safety and internal divisiveness which he inherited. Although the Polaris program represented an infusion of funds for equipment and production, it, like the earlier programs, had been added almost accidentally. The question of finding and holding a clear mission would persist. Browning brought an enterprising and energetic approach to the search for new tasks which could be used to sustain the operation, provide funds, and open new opportunities. His approach added a wide variety of new programs and activities, many of which grew out of the precedents established by 1961.

The next decade, particularly from 1965 to 1969, would see the increase in need for weapons in Vietnam. That outside demand, coupled with the energized administrative style of Browning, brought a period of creative and productive change to Indian Head.

CHAPTER 9

From Naval Propellant Plant to Naval Ordnance Station

Solvent Recovery House "D"
Building 302
1918

W HEN JOE BROWNING took over as the technical director of the Naval Propellant Plant in 1962, he worked to salvage the facility from its difficulties. Over the next twelve years, the station went through a cycle of recovery, expansion, and then decline in workload, personnel, and morale. Browning retired to private life in 1974. Through his early years as technical director, the unresolved question of a mission for the station remained. The mission continued to be shaped by the accrual of tasks and projects. Browning succeeded by bringing a personal entrepreneurial style to the effort. As he remembered it, the system of weapons procurement in the 1960s was hardly organized, and it was a "dog-eat-dog" scramble among the laboratories for resources, funding, and projects.[1]

161

Browning began to dedicate every Thursday to visiting different offices in the Pentagon and at the systems commands, explaining the facilities and services available at Indian Head and looking around for projects which might be brought to the station. To keep the budget over $100 million a year, he calculated, he had to raise an average of $2 million a week. At first the Polaris projects helped sustain the station, and as Browning won more and more contracts and tasks, the base expanded into new products and services. After the Gulf of Tonkin Incident in 1964 and the vast increase in American troop commitment in Vietnam over the years 1965–68, several wartime needs proved to be boons to the Indian Head facility.[2]

In October 1966 the name was changed again, from Naval Propellant Plant to Naval Ordnance Station (NOS), Indian Head, reflecting the diversification from propellants into related fields of chemistry, engineering, and production contract management. Then, with the beginnings of cutbacks in 1969, the Station faced a crisis of reduced funding and personnel layoffs. A ray of hope could be seen as several of the projects which had begun in the mid-sixties continued to flourish, and a new emphasis on the gun as a weapon brought vigor to the Station.

The sixties, however, would be well remembered at Indian Head as a period of dynamic leadership, expansion and successes, and exciting technical breakthroughs. As in previous decades, however, the precise definition of the facility's mission remained an open question, as each new category of work at once provided employment and contributed to the *raison d'être*.

MISSION VISIONS, MARKET REALITIES

Before he resigned in 1958, Sol Skolnik wrote an article presenting the concept that somewhere there should be the equivalent of the National Bureau of Standards or the National Institutes of Health for the rocket propellant industry. "The creation of such organizations," he claimed, "is dictated by the basic laws of organized society." Multiple producers would generate complete chaos without a central authority for standardization, and Skolnik regarded the need as "urgent."[3]

Skolnik spelled out how Indian Head had played such a role in the early twentieth century. After the Korean War, he noted, the number of companies engaged in rocket production vastly increased. Each concentrated on different propellants and different, modified formulations, attempting to beat the competition. Small variations in ingredients produced very different effects. Destructive tests of large-scale rockets such as Polaris would be extremely expensive. The solution to the problem had to be development of process control procedures, rather than postproduction destructive testing. Stability was a further problem, with deterioration in storage producing cracks and crevices, declines in potential energy, and changes in burning rates, all of which would result in different performances. Such variation could not be allowed in ordnance.[4]

The need for a common scale of measurement similar to horsepower in the automobile industry had produced the concept of "specific impulse": the total impulse divided by the mass of the propellant contributing to the thrust. But specific impulse would vary with the operating conditions. To get a standard, one had to compare specific impulse under "standard conditions," usually agreed to be 1000-psi chamber pressure and one-atmosphere exit pressure. But a wide variety of methods of adjusting data to the standard had evolved in different companies. In Skolnik's view, the proposed propellant "Bureau of Standards" should set a "truly standard method" for adjusting for varying conditions, in order to make specific impulse measurements fully comparable.[5]

Skolnik's article and his idea of a "Bureau of Standards" for the propellant industry struck a responsive chord among his colleagues at the station. Browning found the "Bureau of Standards" concept compelling and later discussed it with management theorist and consultant Peter Drucker.[6]

Even though he liked the "Bureau" concept, Browning's operative vision of the role of Indian Head was much more practical, wide-ranging, and competitive. His strategy was simple and pragmatic: he would try to identify young and productive people, often through the device of the Assistant Management Board, give them opportunities and recognition, and then put them to work on challenging ideas. He would then go door-to-door in the Pentagon, rounding up business for the staff. Utilizing funds from retained earnings, Browning was able to put some locally generated money into research. He established a liaison office in the Bureau of Weapons, staffing it with good advocates of the station, such as Jay Carsey, who served at once as contact points, "ombudsmen," and an in-town sales force.[7]

In these rather straightforward ways, Browning took a more organized approach than had been taken before to the problem of finding work for the station which matched its capabilities and staff. Increasingly through the 1950s, program managers asked Indian Head to bring into production chemicals, propellants, and motors researched and developed either at Hercules-Allegany Ballistics Laboratory (ABL) or at China Lake. By the 1960s, however, more than thirty large Government-Owned Contractor-Operated laboratories (GOCOs) participated in the manufacture of propellants, systems components, or completed rockets. Some of the contractors operating the government facilities had their own plants as well, including such corporate giants as du Pont, Hercules, Thiokol, Aerojet, and Rocketdyne.[8]

As the propellant industry grew more complex and the number of programs and sources of funds multiplied, Browning realized that he needed to explain the capabilities of the facility to the changing program officers, treating them as potential "customers." The Department of Defense reorganized naval procurement on 1 May 1966, by splitting the Bureau of Weapons into Naval Ordnance Systems Command (NAVORD) and Naval Air Systems Command (NAVAIR), which divided some of Indian Head's previous sponsoring groups. The Naval Ordnance Test Station (NOTS) in China Lake, which had previously worked closely with Indian Head, now drifted into more work for NAVAIR, while Indian Head remained largely an Ordnance-funded facility. These changes in personnel and responsibilities required that Indian Head continually re-introduce its talents to those with money.[9]

Although Browning, in his direct fashion, realized that he was engaged in "sales" and "marketing," the more formal, approved term for his activities in naval circles was "workload development." In order to develop workload, a realistic appraisal of the capacity and capabilities of the station was required. As he proceeded, Browning showed that he had mastered the central "lifting by the bootstraps" problem of needing resources to take on tasks, and needing tasks to justify resources. He consciously put the best light on the situation, depending upon his audience: to one funding official with a project, he would point to existing resources and claim he needed the work; to another, who held resources, he would point to the projects and claim he needed the resources. In the hectic world of propellant procurement in the early and mid-sixties, the process worked. Indian Head's products and its resources, despite earlier cutbacks and difficulties, did indeed make a good argument for more work.[10]

PRODUCTS, PLANTS, AND PROCESSES IN 1960

Several of the long-standing products of Indian Head continued to be utilized by the Navy through the early 1960s, providing a basis for employment in the production department. Although the 2.75-inch rocket had been out of production since 1957, producing other products, such as the basic pilot ejection rocket catapult Mark 1, Mod 1, and the propellant for the Sidewinder missile, together with a host of smaller gas generators and other missile grains, kept the plant busy.

About 1960 the Naval Propellant Plant produced two components of the Sidewinder missile. The products were two extruded grains, the propulsion grain Mark 29, Mod 1, made from N-4 propellant, and the auxiliary power grain, Mark 1, Mod 0, made from X-9 propellant. Both of the grains were extruded from carpet roll supplies purchased from Radford Arsenal, Radford, Virginia, itself a Hercules-operated GOCO. The rolls were about four inches thick, about fifteen inches in diameter, and made from sheets about seven-tenths of an inch thick. The basic process was very much like the World War II extrusion work.[11]

Safety and conformity to standards required not only testing and inspection but a whole set of procedural standards. When the testing, inspection, and procedures were considered together, they represented the "process control" to which Skolnik had referred. The controls had been worked out to ensure standard product with regard to internal defects, dimensional tolerances, and performance values. The propulsion units were examined, fluoroscoped, and measured. The gas generator units were measured carefully, x-rayed, and examined by light for fissures and foreign materials. Potassium sulfate filler and potting resin were carefully analyzed for conformity to specifications. The generator grains were "canned" before shipment, and the cans were checked for leaks and proper weight and dimensions. Altogether there were seven process control points for the propulsion grain and ten process control points for the generator grain.[12]

The method of heating the carpet rolls of propellant prior to extrusion was little changed since September 1944, when the group under F. C. Thames had set up a system of dielectric radio-frequency heating to replace an older system of hot-air oven heating. In "dielectric heating" the propellant, as a nonconductive substance, was placed between two electrodes, and a charge was "broadcast" through the material with rapid alternation of the polarity of the two electrodes. Dielectric heat would develop from the induced molecular motion in the propellant itself, a process familiar to a later generation as "microwave cooking." This method was safe since the heat could be raised with little danger of igniting the propellant. After some experimentation, the optimum frequency was set between 8.4 and 8.6 megacycles, which produced a three degree per minute rise in temperature.[13]

There was some danger that foreign material or poor loading could lead to ignition, but between 1949 and 1960, over fifty million pounds had been preheated with only one ignition. A detailed "SOP," or standard operating procedure, with thirteen procedural steps in the preparation of the carpet roll stage and twenty-four steps in the heating stage, together with extensive lists of safety precautions, had reduced the danger with the materials. The SOP for extrusion itself gave twenty procedural steps. Similar extensive procedures for milling, machining, potting, x-ray of grains, canning the generator units, and inhibiting the propulsion grains with ethyl cellulose tape all helped ensure conformity with specification and safety. The experience with stringent process control proved valuable in arguing later for new exacting tasks, like those in Polaris, and for those which required reliability in great volume, such as mass-produced rockets for the Vietnam conflict a few years later.[14]

A new chemical product, plastisol nitrocellulose, had been developed at NOTS. Prior to 1958, the nitroguanidine and ammonium picrate plant at Indian Head was converted at a cost of $10,000 to produce the new chemical, for use in small experimental motors. Because the nitrocellulose particles in the "plastisol" form were spherical, rather than fibrous, the new form had higher density than regular nitrocellulose. Because of the higher density, it had better thrust-per-weight ratio. Further, the new product had a long "pot life," allowing longer casting time needed in large grains. When used, it did not have fibers, as did the older nitrocellulose, but took a form more like a powder or talc. Using nitrocellulose nitrated at 12.6 percent, processed with nitromethane and ethyl centralite, a solution was made. Then it was emulsified in water, the nitromethane dissolved out, and then centrifuged. The plant incorporated a solvent recovery system for the nitromethane. By 1958 production got up to four hundred pound capacity per eight-hour shift, but could be expanded six-fold.[15]

Plastisol nitrocellulose was only one of a group of products processed through Indian Head when Browning took over. The Propellant Plant purchased several products which it had previously produced locally, including nitrocellulose, sulfuric acid, and nitric acid. Furthermore, it bought the oxidizers and fuels, as well as the Radford-manufactured carpet roll sheet stock. In 1960 Indian Head produced eight "product lines": plastisol nitrocellulose, smokeless powder, double-base casting powder, cast propellants loaded into motors, nitroglycerin, composite JATO (Jet Assist Takeoff) propellants loaded into motors, 2-2-dinitropropanol (DNPOH), and high-bulk density nitroguanidine (HBNQ). Indian Head used the purchased nitrocellulose in the production of plastisol nitrocellulose and in making smokeless powder. With the locally produced nitroglycerin, Indian Head made double-base casting powder, cast the propellants, and then loaded them into motors.[16]

THE PROPELLANT PLANT IN 1962

A tour of the facility in 1962, when Browning took over as the technical director, would have revealed a host of interrelated research, development, testing, production, and quality assurance activities, spread over the two thousand acres of the peninsula between the Potomac and Mattowoman Creek. Across the creek, on Stump Neck, the Explosive Ordnance Disposal School and technical centers—successors to the Explosives Investigation Laboratory of the war years—continued as independent operations under the same naval officer commanding the Naval Propellant Plant.

Starting at the Research and Development (R and D) building, building 600 housed James Wilson, serving as head of R and D, and Browning as technical director. The R and D Department had three groups: "Re"—Design, Evaluation, and Planning; "Ro"—Pilot Plant; and "Rr"—Research. Each of these groups was further divided into divisions and, at the next layer, into branches. "Rr" had four divisions: Polymer, Applied Chemistry which served as a consulting branch for the pilot plant, Analytical Chemistry, and Chemical Physics Divisions. "Re" had Administrative, Applied Math and Statistics, Ballistics, and Advanced Planning and Design Divisions.[17]

Behind the R and D building was a sprawl of shops, where the Public Works Department did electrical, sheet metal, boiler, and pipe work. An instrumentation shop, a carpenter shop, and a paint shop assisted in the maintenance of station structures. A machine shop had been transferred to work under the jurisdiction of the Production Department.[18]

Moving to the plants, the visitor would find quite an array of production facilities. The extrusion plant was responsible for motor loading. Building 580 handled two JATOs (2.2KS11,000 and 5-KS4500) and the testing of the Antisubmarine ASROC and RAPEC (Rocket-Assisted Pilot Ejection Catapult). Building 588 loaded air-launched Bullpups. The extrusion presses turned out rockets varying from small gas generators to 12-inch ASROC grains. Among units produced were the Hawk and Sidewinder gas generators, a bomb ejector, the RAPEC pilot ejection grain, the Zuni and Sidewinder, and the grain for the ASROC submarine torpedo launcher.[19]

Along the river, the Ballistics Division of Research and Development provided the testing for rocket grains. There were three sections: Ballistic Studies Section; Experimental Section; and Type-Life Section. Here the type-life programs were maintained, and missiles were withdrawn from storage at intervals for examination and tests. Quality Assurance operated a bay in the ballistics area for static firing of small production grains and tested the larger grains in the JATO area.[20]

The Polaris Base Grain Line began operation in 1962. About eighteen acres were devoted to the production of 120,000 pounds of base grain a month. The base grain, containing nitrocellulose, nitroguanidine, aluminum, and oxidizers, was produced as small right-circular cylinders about .035-inch in diameter and .035-inch long. The base grain was made by extrusion of the mix through small dies, then cut to length using a power cutter containing many blades on a spinning disc. The cut grains would be dried, coated with graphite powder to prevent build-up of static electricity, and then screened for correct size and blended to reduce variation between lots. These grains would be placed in a mold or motor case and then saturated with casting solvent, which consisted of nitroglycerin desensitized with triacetin, which would cause the nitrocellulose to swell and jell to form the cast grain. The material would set up during the cure.[21]

The Biazzi Plant continued to produce nitroglycerin in the early 1960s, with a capacity for 2,200 pounds per hour. All the nitroglycerin used on the station was produced at the plant, where a total crew of eight prepared raw nitroglycerin and desensitized the product.[22]

The Research and Development Department's "Ro" group operated the Pilot Plant. The purpose of this facility was to investigate new propellant ingredients, new propellant systems, and processes for the manufacture of the propellants and their ingredients. As a step between the laboratory and the full-scale production facility, the Pilot Plant was constantly called upon to work with scaling up formulations which had been experimented with at Indian Head and at other laboratories. Small workable quantities of new ingredients were produced and evaluated. If the new substance proved worthwhile, the variables in the process would be studied to enable scale-up. The Pilot Plant would produce half-pound lots to study the compatibility of the mix with the developed formulation. Then five-pound mixes would be made to study propellant burning rates and properties. Finally, large mixes in the range of fifty to two thousand pounds would be made to load motors which could be static fire tested to get specific impulse and burning rate data.[23]

The Cast Plant produced all the larger grains made at Indian Head. Most of the larger rockets at that time were made from cast grains, utilizing nitrocellulose, nitroglycerin, aluminum, and oxidizer. Terrier and Talos grains were produced at the plant by this method. Plastic sheets would be heated and sealed into a hollow cylinder or "beaker" and surrounded by a fiberglass or steel cover to hold the cylinder upright. To form holes in the eventual grain for burning, metal cores of the proper shape and size would be mounted vertically on a metal plate and then the beaker was lowered over the cores and bolted in place. The beaker would be filled, by remote control, with casting powder, usually small granules of nitrocellulose. Casting solvent, still at this time nitroglycerin desensitized with triacetin, would be forced into the mold under air pressure, and the mixture would then be cured or cooked for as long as nine days. The metal cores would be removed and the ends of the grain sawed off to proper length for motor loading. Motor cases would often be coated with rubber or asbestos before the grain would be inserted to provide insulation which would prevent burn-through of the motor case. An igniter would be attached and the motor crated for shipment to an ammunition depot or to the fleet.[24]

The "JATO" area was used for testing of large-size motors. In September 1960 an RCA 501 computer was added for rapid evaluation of the test results and was in full operation in 1961. Twelve different types of large motors were regularly tested in the early 1960s.[25]

In January 1962 the Propellant Plant developed a list of facility requirements, establishing priorities for construction through 1968. Forty-eight projects were given high priority and planned for fiscal year 1963. In addition to maintenance and utility work, the plans called for ambitious expansion in several technical areas, especially in the construction of more facilities for the Polaris plant. A cold conditioning building, a new x-ray facility, new mixing facilities, new equipment for the presses in the pilot plant, and development of a toxic and radiological laboratory were planned for 1963 and 1964. The following year's plan called for temperature and humidity controls in the production buildings. In justifying these expenditures, the proposal pointed to safety factors, efficiencies, improved ability to meet specifications, and economies in time and money.[26]

MANAGEMENT CONCEPTS

The changes brought about in the period 1958–62 greatly increased the size and volume of the Propellant Plant's operation. In those years, the facility went from about 1,500 employees to over 3,200, and a $60 million investment changed the nature of the Plant. Among other changes, production of sulfuric acid and nitric acid was dropped since adequate commercial sources for the acids had developed. By early 1963 the opening of twenty-three buildings associated with the Polaris program alone represented a major realignment of facilities and workload.[27]

The growth had caused problems in management with which Browning struggled. Some of his associates also recognized the problems and attempted to face them squarely. Staff member Jim Wilson believed that the rapid changes and growth of the organization made coordination of all the personnel in production and in research quite complex and inefficient. In a careful study prepared as a master's thesis in Engineering Administration for George Washington University, Wilson suggested a set of reforms designed to utilize project management techniques without fundamentally altering the functional management design of the plant. Using a research system based on communications theory and operations research, he conducted a study of management's information requirements in project direction, focusing on the flow of communications. Based on interviews with ten staff members, he evaluated preference for, and actual time spent, on eleven types of communication. Wilson found, among other things, that long-range planning, although a "highly preferred activity" of management, received only a small portion of effort. Top management utilized verbal communications in scheduled conferences and in personnel work to a greater degree than did the department or division managers. Department managers spent more of their time in current, rather than long-range, planning. Written communications tended to be sent to too many addressees. Wilson recommended several changes, and as director of R and D, he was able to implement some of them himself.

Specifically, Wilson recommended

a. linkage of work described in each monthly progress report to a goal or milestone
b. preparation of weekly abstracts of trip reports, with fewer full copies of the full trip report
c. more limited distribution lists on memoranda

d. preparation of one-line-per-subject minutes for scheduled meetings
e. preparation of single-format financial reports
f. further consolidation of department files at division level
g. consideration by Industrial Management of recommendations a through f for the support and staff departments

Wilson himself took up recommendations a, d, and f. He submitted his other reforms for consideration by management and by the commanding officer.[28]

WORKLOAD DEVELOPMENT: NASA EFFORTS AND INERT DILUENT

In the search for new "customers," Browning and his associates turned several times to the National Aeronautics and Space Administration (NASA) in hopes of expanding the Propellant Plant's role in the space program. In early 1964 the Plant submitted a proposal to NASA to provide support services to Langley for $50,000, for "scientific services." In making its presentation, the Propellant Plant listed a diversified capacity reflecting a "small-scale cross-section of the propellant industry." Despite the fact that the proposal was hastily patched together out of printed public relations materials and resumes, the Propellant Plant did get a consulting contract to do hazards evaluation for NASA. On a more significant level, the Plant became involved in the production of several of the larger rockets employed in space exploration over the early 1960s, particularly the X-248 Scout, used in the third stage of the Delta rocket, and the X-259 Athena motor produced for the Air Force.[29]

Browning initiated a continuous flow method of producing propellants which was both safer and more capable of being standardized than the older batch process, which relied on large batches produced in bowls with rotating beaters similar to oversized cake mixers. The batch process was fraught with danger to both operators and equipment. In 1963 the Propellant Plant, under the leadership of Browning and Commanding Officer Captain O. F. Dreyer, prepared a two-year plan with the cooperation of Rocketdyne to develop and put in operation an inert-diluent production facility for solid propellant. The concept would allow for continuous, rather than batch, production of propellant by suspending the ingredients in an inert liquid, mixing them in liquid form, and then removing and reprocessing the inert carrier for reuse. The carrier would transport the ingredients from separate, remotely controlled dispersion areas to an area for mixing and casting, blending them in a specially designed rotating jet blender. The plan had several virtues beyond its continuous production nature. It would allow for safe, remote control operation at all stages. The inert carrier itself would reduce the hazards involved, since the amount of energetic material at any location would be reduced and the energetic materials were both diluted and wetted. The lines could be sized and laid out to prevent propagation of any detonation.[30]

Rocketdyne, a division of North American Rockwell, Inc., had experience with the concept in a process they called "Quickmix," used in the production of several composite propellant formulations. Rocketdyne cooperated in developing a statement of work as part of a Design Criteria Guideline for bidding by architects and engineers. According to preliminary plans, the process would be used at first to produce Mark 12 Mod 1 Terrier Booster double-base formulations. It could later be adapted to produce high-energy propellants and other composite-modified propellants, and be versatile in the production of a wide range of motors. It was hoped that the process would reduce the unit costs by taking advantage of the economies inherent in the process. Furthermore, the process held out hope of a more uniform and controlled product.[31]

The complete plan called for twenty-four casting pits in groups of six at each of four "pit farms," but the beginning plan would be based on an expandable fraction of that system using two pit farms, each with one pit. The completed twenty-four-pit design would be a system of disperser systems and magazines, together with supporting utilities, roadways, conveyor systems, and control rooms which could work on a twenty-four-hour, seven-days-a-week schedule. The initial, partial phase would be built so that it could be expanded to the full scale if funding later became available.[32]

In order to demonstrate the advantages of the inert diluent system, the Propellant Plant constructed and operated an Engineering Demonstration Plant, employing the process on a continuous scale. Construction began in January 1963, and the Demonstration Plant shakedown was completed in May 1963. In July plant operation began with single-base propellant, and in November the plant produced nitroglycerin plasticized propellants. By early 1964 the plant produced cast propellant which was installed in a full-scale Bullpup rocket motor, and the plant produced several motors in a single processing run.[33]

Early in the demonstration phase, however, the plant revealed problems. Heptane was used as the carrier, and it worked fine with existing double-base and composite-modified double-base propellants. However, chemical engineers anticipated difficulties using heptane as a carrier with more complex composite propellant formulations which would contain metals, and began planning the use of a pneumatic mix system. Like the inert diluent, the pneumatic mix would rely on a fluid carrier, but it would utilize a gas instead of a liquid to carry solid and liquid droplets into a mixing unit.[34]

FROM RESEARCH TO PRODUCTION: OTTO II

During the early 1960s Indian Head became involved in the research, development, and production of a new product. Such a full involvement, from research all the way through production, was unique at Indian Head for this period, and the product became one of the major success stories of the middle Browning years. Called "Otto Fuel," the product was an adaptation of a locally produced solvent used in the casting process and was to be used as a fuel for high-speed torpedoes.[35]

In the early 1950s, the principal liquid fuel used in "wobble plate" or cam-driven engines for torpedoes was normal propyl nitrate, a monopropellant. At the torpedo research facility at Newport, Dr. Otto Reitlinger warned against using this chemical because it was too dangerous and unstable. His judgment was vindicated when there was an ignition of vapor, a deflagration, and a fire in a storage tank, which killed five people.[36]

Reitlinger moved from Newport to a management position at the Bureau of Weapons (BuWeps). He visited Indian Head to discuss the problem of torpedo propellants with the researchers there. Even though the facility was not "tasked" with torpedo work and its evolving mission focused on rocket and gun propellant, he sought advice from the Indian Head chemists. In spelling out the torpedo fuel requirements, Reitlinger made it clear that a torpedo monopropellant had to run clean and not produce a lot of carbon to foul the torpedo engine. Unlike a missile, a torpedo is run many times; in peacetime, ship captains run and recover torpedoes over and over in practice drills. Reitlinger sought a liquid monopropellant that would be an improvement over existing fuels.[37]

At Indian Head, Reitlinger talked to John Murrin about the advantages of using a double-base casting solvent, a nitroglycerin-based composition, as a highly energetic, clean-burning propellant; but it had a severe psychological disadvantage. No naval officer would allow a nitroglycerin-containing formulation in a torpedo propellant. A substitution in the casting solvent with 1-2-dinitroxypropane for nitroglycerin was selected. The monopropellant worked out was 76 percent propylene glycol dinitrate (PGDN), 22.5 percent triacetin, and 1.5 percent 2-nitro-diphenylamine. This formulation was designated "Otto Fuel I."[38]

Indian Head performed tests on the new fuel. At first, Otto Fuel I looked good, and it could have served as a torpedo propellant; however, it was never used in a torpedo. The problem was that the triacetin was slightly soluble with water, and if some of the monopropellant were spilled and then washed with water, the residue could lead to a high concentration of the sensitive, energetic PGDN. Thus, shipboard spills could soon become hazards. Reitlinger had Murrin look at replacements for the triacetin, and a substitute was found which was less water soluble, di-n-butyl-sebacate.[39]

The resulting fuel, "Otto Fuel II," worked fine and avoided the hazards of both the normal propyl nitrate and the earlier experimental fuel. Indian Head received credit for developing the substance and soon went into production; Reitlinger not only had the fuel named after him but took out the patent in his name. Otto Fuel was used in the Mark 46 air-dropped anti-submarine torpedo and in the ASROC rocket-launched shipboard torpedo in the 1960s without any accidental deflagrations. Murrin did the calculations at Indian Head and later at BuWeps, working out the optimum formula and its characteristics.[40]

Otto Fuel was not only a success story for Indian Head. The new fuel brought some further renown to Dr. Reitlinger, and some personal advancement to Murrin, the primary Indian Head researcher who had worked on the improvement of the fuel.[41]

In developing specifications and warning of hazards associated with Otto Fuel, a later Indian Head report pointed out that the diluent, or desensitizer (the triacetin in Otto I and the di-n-butyl-sebacate in Otto II), had to be water insoluble in the event of a spillage and a water rinse. There were five other safety criteria: low vapor pressure; the diluent and the nitrate ester had to have very close vapor pressures to prevent fractional distillation in storage; the fuel had to have high flash and flaming points to prevent bulk ignition in air; the fuel should decompose when not under pressure; and it should have high thermal stability. Experimenters subjected Otto Fuel II to a whole range of safety tests with these criteria in mind. They tested for fire hazards, explosion hazards, impact of shells or fragments, dropping of drums, and compatibility with a wide variety of shipboard chemicals and materials, from wax and vinyl tile through sulfuric acid, bleach, and nitric acid. In general, strong acids, strong bases, and oxidizing agents reacted vigorously with the fuel. Otto Fuel was toxic, and extrapolating from animal tests to humans was difficult. The problem was handled thus: "It has been estimated that ingestion of 2 ounces of Otto Fuel II would kill a man-sized rabbit." The tests resulted in a series of guidelines and recommendations for handling the material which were incorporated in the Safety and Handling Procedures "OP 3368."[42]

By 1967 production of Otto Fuel II through the Biazzi Plant had become a regular procedure. PGDN itself was produced with little modification of the plant; later, some changes were made to the Biazzi Plant to assist in mixing the PGDN with the other ingredients to produce the full Otto Fuel II. Eventually the Biazzi Plant was equipped with separate additions, including a PGDN storage and transfer system, separate spent acid systems, and a storage and transfer system for the desensitizer di-n-butyl-sebacate. Further expansion included a system for mixing the desensitizer and diphenylamine into "premix," a system for mixing PGDN and premix, and a system for drying the final product, Otto II.[43]

Operators gave particular attention to maintaining temperature and avoiding build-up of PGDN. Years later, Lu Grainger, who had been in charge of the process, remembered the problems with the red fumes which sometimes popped the lids on the Biazzi stainless steel containers. On one occasion he noticed a maintenance worker near the building surrounded by a cloud of the red fumes which had seeped out, almost totally engulfing the man as he casually cut weeds. Grainger drove him off and then drowned the facility in cold water, luckily preventing a major deflagration. Eventually the problem was traced to the formation of a small build-up of droplets that were not fully washed through the system, and a minor redesign of equipment corrected the difficulty. Experience gained in handling and manufacturing Otto Fuel II became incorporated in revisions of the safety manual for handling the fuel, loading torpedoes, handling spillage, and extinguishing accidental fires.[44]

The fuel was used as the internal combustion fuel in the Mark 46 torpedo. The engine was started, not with an electric starter system, but with a small propellant grain in an igniter-booster assembly. In 1967 Indian Head was involved in the final procurement of the starter assembly, evaluating one produced by Thiokol, the other by Hercules. Tests involved control firings, environmental tests (including temperature, vibration, acceleration, and shock), and visual inspections. A problem of rubber bonding of a spacer in both grains was discovered and a minor change recommended. Otherwise, Indian Head recommended both the Thiokol and the Hercules grains as qualified.[45]

THE BOOTSTRAP PROBLEM MASTERED

Through the early 1960s the Propellant Plant's Advisory Board tried to redefine the "official mission" several times, and in 1964 and 1965 recommended a change of name to Naval Weapons Propulsion Plant. The board's ruminations on such issues had little effect; Browning simply continued to get new pieces of work, using recent success to argue for more and concentrating on his pragmatic marketing style. By 1965 the Propellant Plant's programs included a wide range, which demonstrated how successful Browning and his young researchers had been in getting new business and developing further some of the old standbys. In promoting the Propellant Plant's ability to manage contracts on Polaris, Browning could point to several related projects: production of a gas generator for Polaris, production of two large space rockets, surveillance on Polaris motors, work on Sidewinder, and contract management in other areas.[46]

The Polaris Mark 13 Mod 0 Gas Generator, produced locally, involved management techniques common throughout the Polaris project. The Propellant Plant provided technical drawings, specifications, tooling, and subassemblies, and the program was closely monitored through both the tracking system "PERT" and "Line of Balance" control charts. Quality Assurance spent 34,000 man-hours and Production spent 30,000 man-hours in developing this gas generator. Each unit produced had documents such as ordnance drawings, ordnance specifications, weapons specifications, and any waivers or deviations, as well as full results of all tests. The strict control was necessary because the gas generator had to "conform to a limited ballistic envelope."[47]

The two rockets with space applications, X-248 and X-259, were being produced in the Pilot Plant. Both the rockets were Hercules-ABL developed and met both NASA and Navy-Air Force specifications. After June 1962 the Propellant Plant received orders for 117 of the X-248 motors, and by June 1965 Indian Head had delivered 69 motors, with the rest scheduled for 1966. The X-248 space motor work was run by a Project Group under a Project Team Concept. The X-259 was designed for the Air Force Athena program, as the rocket for the second stage. Some seventy engineers and technicians from all departments worked on a Project Group management system and were proud of the fact that they could demonstrate quick response time. Both motors utilized fiberglass case construction and had high-energy propellant.[48]

The Polaris Surveillance Program had been incorporated into the existing "type-life" program. Over 2,500 motors were tested each year. There were over forty types in the program, ranging from the 2.25-inch Mousetrap, through Tartar, Terrier, Talos, and Polaris A2 and A3 second-stage motors. Polaris surveillance involved one hundred movements of the motor per year, all accomplished without damage.[49]

To demonstrate the ability of the Propellant Plant to manage external contracts, Browning could point to the fabrication of prototype quantities and "preproduction evaluation quantities" of the Sidewinder 1C propulsion unit. Furthermore, Indian Head took over technical administration of the procurement of motor hardware and the loaded motor, managing the Navy's contract with Rocketdyne.[50]

The Propellant Plant also handled procurement management on the Shrike/Sparrow missile, and Browning claimed that the responsibility came because of earlier success on the Sidewinder propulsion unit. For Shrike/Sparrow, Indian Head wrote the request for proposals and the statement of work, evaluated bids, and worked closely with the program management desk at BuWeps. Indian Head specialists planned on-site inspections to ensure that the winning bidder conformed to minimum standards in quality control and facilities.[51]

After about two years of building the research and production facilities under Browning, an independent evaluation of the technical capability of the Naval Propellant Plant at Indian Head by Booz, Allen, and Hamilton, performed for the Special Projects Office (SPO), showed how far the Plant had come since its doldrum years in the mid- and late 1950s. The SPO had ordered the independent evaluation to determine which naval facilities should take over the technical direction of various parts of the Polaris missile system procurement. On this evaluation, Indian Head ranked "excellent" in its ability to support private contracts in the rocket propulsion area, based on the prior Polaris, Sidewinder, and "X" space rocket work, and the Sparrow/Shrike management.[52]

Booz Allen, in evaluating the Propellant Plant, distinguished between "capability" and "capacity." Capability referred to technical competence, while capacity would include personnel limits and required facilities. The question addressed by the outside evaluation team was, who would provide management on first-stage missile motors produced by Aerojet, on second-stage missile motors produced by Hercules, and on the gas generators produced by Hercules? Booz Allen found that Indian Head was an "excellent candidate" for those management tasks, for several very solid reasons. The Propellant Plant had qualified people. It planned to phase out programs which would free up personnel just as they were needed. In addition, the "educational profile" of the engineering staff was excellent, and the Plant had extensive prior experience working on aspects of the Polaris program. The report took into consideration the basic problem of how to staff a new program with current people and found that 77 percent of required staff could be assigned to the proposed project. The study had held that only a nucleus of 25 percent needed to be immediately available for a facility to be qualified. By 1965 Browning had developed the entrepreneurial ability to match capacity and need for work with opportunities and funding sources.[53]

SHOTS IN THE ARM: 2.75 REVIVED, ZUNI, LINE-THROWER, C-3 PLASTIC

Contract management and increasingly automated production of chemicals and propellants did not spell out a need for more personnel on a large scale. Total civilian manpower at Indian Head fell from about three thousand in 1963 to about twenty-five hundred in April 1964. In August 1964 the Gulf of Tonkin Incident and the response of Congress to it marked the beginning of Vietnam escalation, which continued over the next three and a half years until 1968. During this period, the ability of Indian Head to respond rapidly to wartime needs, as it had in both the World Wars and during the Korean War, once again brought a burst of new tasks and successes, and further personnel growth.[54]

The 2.75-inch folding-fin aircraft rocket had been out of production since 1957. Indian Head, in mid-1965, took on production of the propellant grain as well as contract management responsibility for the full motor and tube design. Called a "suppression" weapon, the rocket was the major armament of the helicopter gunship as employed by all services in Vietnam. Full assembly of the weapon was handled at McAlester, Oklahoma. Little had changed about the 2.75-inch rocket since World War II; its length was about forty-eight inches, and with its warhead it weighed about eighteen pounds. Originally designed as an air-to-air weapon, it had evolved into an air-to-ground standard. Several different types of warheads could be attached in the field. Because of involvement in production of extruded grains for other rockets, Indian Head had continued to improve its ability to meet specifications, and the rejection rate of the 2.75 was much improved over 1957 levels.[55]

Since the grain was to be produced by the millions, Indian Head worked on a number of projects to reduce the cost of the product, if only by pennies per unit. Propellant grain M43 Mod 1 for the 2.75-inch was inhibited with ethyl cellulose washers on the ends and ethyl cellulose tape around the outside. Three layers of tape would be applied, with a bonding agent—ethyl lactate-butyl acetate, or ELBA—which would be poured over the grains during the bonding. Excess ELBA would be gathered and sold. In an effort to reduce the cost of the 2.75-inch rocket, a process for recovering the excess ELBA solvent, purifying it, and recycling it was studied. The method was reported and implemented in the production of the grains. After putting the system in place in November 1967, Indian Head estimated that savings amounted to $362,000, but inspectors from Naval Ordnance Systems Command put the estimate at $203,000 on the production of some 31,000 2.75 grains and about 2,000 Zuni grains.[56]

During the Vietnam War, the 5-inch artillery rocket, the Zuni, operated using an old propellant, N-4. The existing Zuni was Mark 16, Mod 2; the grain was Mark 49, Mod 1. Indian Head prepared a proposal to upgrade the propellant with a high-energy propellant having a better specific impulse [225 to 235 pound-foot-seconds to a pound mass]. Indian Head had two formulations it wanted to try out, both containing copper and lead salts and aluminum. The proposal—to modify nozzles, load one hundred grains, and provide them for testing—ran about $40,000–$50,000 and would take about eighteen to twenty-two months, depending on whether the testing was to be done at Indian Head or not. Over the next several years, program managers continued to fund Zuni improvements at Indian Head.[57]

Through these years, Indian Head continued to be the center of information for a variety of weapons and propellants. The Naval Ordnance Systems Command ordered that any records of shots of a still-experimental line-throwing rocket should be sent to Indian Head. The rocket itself was a standard 2.75-inch, but the grain was half-length, the space filled with styrofoam; the rocket case was standard length, with the folding fins removed. Launched from a slotted pipe, the range was about six hundred to seven hundred yards. Indian Head collected performance data for potential improvements.[58]

Among the projects later remembered as successfully and hastily put together was the production of plastic explosive "C-3." Browning had originally spotted a young chemical engineer, Jay Thornburg, during his tenure on the Assistant Management Board. Browning assigned the C-3 job to him. Working through the

Christmas holiday, 1964, Thornburg took a truck convoy to Ohio to pick up equipment and had it installed and running in a total of six weeks. While a small project, the success in this work epitomized several of Browning's management styles: a responsiveness to market needs and his reliance on young and energetic specialists. He was especially proud of the rapid scale-up and production capability of the facility.[59]

GUNS: OLD WEAPONS REDESIGNED

Although Indian Head had played a role in developing an improved gun propellant, "Navy Cool," or "NACO," some gun specialists argued that research into classic problems of interior ballistics and focused improvement on the gun and the shell had been neglected in the 1950s and 1960s. The gun engineering section at Indian Head was committed to redressing the imbalance. Funding for several gun projects picked up in 1964. Browning got a commitment of money to evaluate the state of the art in gun technology, which led to studies of a patented chemical for reducing bore erosion called "Swedish Additive," plans to improve the 20mm gun, and work on a combustible cartridge case.[60]

One project which caught the interest of Browning and his associates was "HARP" (High Altitude Research Projectile), a Canadian-originated plan to launch a satellite utilizing a gun. In 1964 an employee of NASA had written an article suggesting the application of several of the HARP concepts, and Browning agreed to hire the author to help in recruiting funds for the project. Initially, Browning could not get external funding for the concept, and the effort was funded from internal surpluses which were generated out of fixed-fee work during the period, in the amount of $150,000 in 1966. The original group worked on projectile design and evaluation and on propellant development, compiling a four-hundred-page report submitted to NASA.[61]

One line of research was improving the concept of fin-stabilized subcaliber rounds supported by a "sabot," or spacer-ring, together with a "front rider" which helped support the shell in the barrel. The sabot which ringed the shell and the front riders would allow space between the shell and the barrel for the fins. These designs had been researched a bit by the Germans in World War II and in the United States in the 1950s, then more or less shelved during rocket development.[62]

In 1967 the initial conceptual design of the "Gunfighter" weapon system was worked out using a subcaliber fin-stabilized projectile, a 5-inch shell fired from an 8-inch barrel. The shell would be ringed with a sabot, approximately three inches thick, which would not only allow space between the shell body and the barrel for the stabilizing fins, but would ride in the rifling lands of the barrel. The sabot would also serve as an "obturator" or gas-retainer, much like a piston ring in an automobile engine. The front riders, supporting sabot, and obturator would all be shed on firing, as had the rotating band which almost hit Wilson's yacht in 1913. With an aerodynamic shape, stabilizing fins, and high muzzle velocity, the shell would achieve extremely long range. The Long Range Bombardment

Ammunition, or LRBA, when first produced in 1968, achieved ranges up to 72,000 yards or about 36 nautical miles.[63]

In 1970 an operational evaluation was carried out with thirty shells against a selected target complex in South Vietnam. USS *St. Paul* (CA73) was the only cruiser in active service with the appropriate 8-inch gun. After a test firing at Okinawa of three inert-loaded shells, *St. Paul* proceeded to a point off Vietnam and conducted, on two days, a shelling exercise against Viet Cong targets, over 35 miles away. Aircraft spotters called in the positions of impact, and corrections were made. On both exercises, after a few bracketing shots, rounds were placed within 500 yards of the target. The shells were loaded with PBX-w-106, a castable explosive developed at White Oak and scaled up by NOS in the Pilot Plant. In each exercise, a few of the rounds apparently landed within feet of the designated target, proving the accuracy and value of the ammunition for area target saturation. No fuze malfunctions were observed. Air Force pilots in the vicinity questioned the presence of Navy spotter planes for artillery; when told they were spotting for *St. Paul*, some 60,000 yards away offshore, they were convinced that the Navy pilots were deranged. The LRBA more than doubled the range of the 8-inch gun; the fire control and spotting system, although experimental, proved satisfactory; at least six structures were destroyed.[64]

The gun work was consolidated under Dominic Monetta into a "gun group," formally named the Gun Systems Engineering Branch. One study worked on the optimum form of the sabot and obturator in order to reduce the number of shed parts, which represented a hazard to friendly ships and personnel. A sabot design with several "petals" which would flake away was worked out in 1971. The designers used a "constructive iterative" approach, or design tool, which had other applications. This method would employ one simple, basic design; the design's flaws would be addressed and corrected, and the resulting "iteration" would be studied for new problems. In this fashion, improvement stages, sometimes entirely on paper, would lead to a solution.[65]

A CRISIS IN WORKLOAD—PESSIMISTS' VIEW

By 1970 and 1971 the mood had changed at Indian Head. With the cutbacks in troops in Vietnam in 1969 and the decline in NASA work with the scaling down of Apollo, the Station began to face once again the problems of diminished workload and possibly being replaced by private sector facilities.

While Indian Head staff members saw the cutback as a local problem, the general reduction in resources affected all Department of Defense, Atomic Energy Commission, and NASA laboratories. According to one group of analysts of technological research management, the early 1960s had been a "golden age of Federally-sponsored, mission oriented technology development," but the end of the decade saw the beginnings of a long-range decline. In 1968, for example, the Department of Defense closed the Naval Radiological Defense Laboratory, combined the David Taylor Model Basin in Carderock, Maryland, with the Marine

Engineering Laboratory in Annapolis, and carried out other closings and consolidations around the country.[66]

Thus it was with good cause for trepidation that Indian Head staff viewed their situation and prospects in 1971. An extensive self-study described past achievements which fell in four areas. One area was improved products, such as the 2.75-inch rocket and the Zuni. Indian Head had produced a million rounds of the 2.75, before Sunflower, Kansas, picked up the production. A second area involved solutions to technical problems and helping contractors in qualifying for ordnance production, as in the work on the Sidewinder and Chaparral documentation packages. A third area of accomplishment was the reduction of contractor prices through competitive or dual capability. The Station could boast that it had reduced the price on the Terrier Sustainer from $10,000 per unit to less than $4,000; it had reduced the price of the rocket catapult from $1,300 to $800 between 1968 and 1971. In this regard, the Station played a role very similar to the Powder Factory in 1912, when it offered a competitive model for du Pont.[67]

A fourth area of accomplishment was the performance of vital work "unattractive to the private sector" for health, safety, or profitability reasons. In this regard, the Station pointed to small-scale extruded products, the emergency production of the plastic explosive C-3, scale-up and production of Otto Fuel II, production of a pyronol torch for ordnance disposal, and evaluation of fluorine compounds for the Atomic Energy Commission.[68]

Despite such achievements, the tone of the Station's advocacy in 1971 had changed from the confidence in its quality and role as displayed in 1964 when the Booz Allen report had endorsed the station, and when Thornburg had raced to Ohio to put together the plastic explosive plant. The implication of the 1971 presentation was that all of the areas of achievement were ancillary to the private sector. Summaries of projects gave details of tasks later picked up by GOCOs, assistance to private companies in handling a contract, comparative work designed to limit the excess profits of corporate giants, and jobs too risky or unprofitable for businessmen to take on.

Later, Browning himself would remember the period as one of crisis and attribute the decline in morale and confidence to personal friction between himself and Commanding Officer Bernard W. Frese. From a broader perspective, outside factors contributed as much or more to the problem, however. The reduction in work assignments and the general cutbacks in defense expenditures induced a rather desperate search for funding and a self-contradictory search for supporting arguments. The general decline in research funding coupled with a political environment which favored private sector over public sector work were clearly national, not personal, causes of the problems. Falling back on an argument used in the 1920s with no success, advocates of the Station pointed out how vital the Indian Head facility was to the nearby economy, with about twenty-five hundred employees and $30 million in payroll in February 1971. That hat-in-hand line of argument had never been Browning's style. Washington did not owe them a living, he would point out.[69]

Some of the arguments presented in 1971 seemed based on the line of reasoning that the millions of dollars required to keep Indian Head afloat were a drop in the bucket when compared to the amounts flowing to the private sector.

"The corporation giants that Indian Head must compete with for funded work load in the propulsion and explosive areas are impressive. It should be obvious that the amount of business required to keep the Naval Ordnance Station solvent would not affect competitors." Even when using stronger arguments, a lack of precision marred the presentation, as when it was claimed that "incalculable millions of dollars" were saved to the taxpayer by holding down prices. More convincing would have been a considered calculation of those millions.[70]

The authors of the rather ineffectual complaint that more money was needed had good reason to be concerned in 1971. The problem was that a projected drop in funded workload to nineteen hundred civilian employees by the end of calendar year 1971 would make the Station "completely noncompetitive and a significant liability to the Navy." To avoid a "disaster area" in the county, a range of rather desperate solutions were suggested; but in each case, the authors presented not only the arguments in favor of the solution but the reasons why the solution could not be implemented. Each suggested procurement to be placed with Indian Head was listed with the reasons why a competing Army facility, GOCO, or separate naval facility had a better claim to the work.[71]

THE OPTIMISTS' VIEW

Such ambivalence, self-doubt, and helpless frustration at the structure of military procurement had not stood in the way of Browning's earlier successful efforts to round up new business. Although some of the personnel in the early 1970s were affected by the downturn in morale and funding, there were several bright spots. The difference between a pessimist and an optimist is often described as the difference between those who see the glass as half empty and those who see it as half full. Looking at the same picture, the "optimist" faction had an entirely different vision of Indian Head in 1971.

The Engineering Department took a positive tone, listing a group of "Major Technical Accomplishments" and proceeding to propose further work based on those accomplishments. "One is inclined to dwell disproportionately," the Engineering Department chided others, "on the general concern for reduced budgets and curtailed or cancelled programs . . ." Fiscal Year 1970, Engineering claimed, was a year of technological achievements "as well as the beginning of the transition from support of a war effort to peacetime retrenchments."[72]

The Engineering Department pointed with pride to a series of successes in guns, rockets, chemical production, and recently acquired budget lines. The list was extensive. It included the success of the sabotted subcaliber LRBA in Vietnam and work on Rocket-Assisted Projectiles for 5-inch, 38-caliber guns. The department also noted the planned substitution of a new solventless propellant—NOSOL—produced at Indian Head, for the early "Navy Cool," or NACO, gun propellants developed at Indian Head in the mid-1950s. By 1971 NOSOL had been successfully tested aboard USS *Norton Sound*, and further production would be expected.[73]

In the area of rocket engineering, the "optimist" faction pointed to these works:

◇ Mark 66, an improvement on the 2.75-inch
◇ Work on the 5-inch wrap around fin aircraft rocket
◇ Work on small ships Chaffroc

On torpedoes and gas generators, the optimists in Engineering pointed to other recent successes, including the use of Otto Fuel in the Mark 48 Torpedo and work on the torpedo starter-grain. Another successful product was shown in the use of a low-migration cellulose acetate in the inhibitor for a mud-agitator, a device for freeing mines from the bottom of rivers and harbors.[74]

A host of other products showed the vitality of Indian Head: continuous processing of plastisol nitrocellulose; pilot loading of PBXN (Nitranol) in torpedoes; use of low-migration acetate in beakers for cast propellants; reduction in nitroglycerin absorption in inhibitors; and development of a flammable adhesive for a 152mm Army consumable cartridge case.[75]

Budget figures for 1971 for the Engineering Department included some $39 million for the Rocket-Assisted Projectile and over $3 million for other gun work. The improved Zuni proposal was funded, and such projects as small-caliber ammunition, the line-throwing rocket, and production of the solventless propellant NOSOL each had six-figure funding.[76]

In particular, the "gun group" within Engineering took a positive tone. The successes of the gun group engineers had been accompanied by an experimental approach to management itself. In 1967 Browning had appointed Lucien Rositol as director of planning. Rositol developed a planning system with the claim that it "integrated behavioral science with systems analysis." He called the planning system "Socio-Structural Strategic Planning." In effect, the system called for the creation of "task forces" made up of internal personnel to develop long-range plans in three areas: fleet support, electronics, and gun systems. The "behavioral science" side of such an approach apparently meant that the task forces would develop a group commitment and sense of shared involvement which would facilitate the matrix or project management.[77]

The task force for guns was headed by Jay Carsey on a consulting basis. Carsey, who had worked at Indian Head in the early 1960s and had served as the first chairman of the all-station Assistant Management Board, had accepted an appointment as president of Charles County Community College. Carsey remained a consultant to Indian Head in the area of management through these years and into the 1970s. With the gun planning group he served as "facilitator." The task force which worked on gun systems had the unlikely acronym "LORAP TAFT FOGS," or Long-Range Planning Task Force Team for Gun Systems. TAFT FOGS studied both the history of the gun effort at Indian Head and current opportunities, working out several long-term objectives.[78]

The objectives, set in 1969 for accomplishment by 1976, included these:

◇ tripled funding for gun work
◇ doubled direct man-hours in gun work
◇ initiation of a projectile development effort
◇ getting 10 percent of gun work funded in applied or basic research
◇ becoming the cognizant field activity for all gun propulsion
◇ developing a new gun propellant

◇ setting up production of combustible cartridge cases
◇ getting a contract for a new ordnance concept
◇ developing six subcaliber long-range projectiles
◇ integrating all the Station's gun work
◇ training six technicians in gun work
◇ becoming the liquid propellant R and D agency for the Navy
◇ getting production up for 250,000 pounds/month of 50,000 yard range propellant

The TAFT FOGS objectives did prove useful; many of them were accomplished well within the seven-year time frame.[79]

In addition, the group produced three white papers, or "WHIPS," regarding different aspects of gun development: one on liquid gun propellants; one on projectile design and guidance system capability at Indian Head; and the third on combustible cartridge cases. The "WHIPS" pointed to problems as well as capabilities in developing the new programs. For example, the white paper on projectile design and guidance system capability at Indian Head noted that for projectiles planned with ranges over one hundred miles, some system of projectile guidance would be required. Such a system would require capability in electronics beyond that which Indian Head could mount. The white paper suggested that the way the laboratories were structured worked against such an innovation—to put inertial guidance systems on gun shells. This paper and the others took a long-range historical view and reflected a realization that the institutional evolution of the weapons business provided a limiting matrix on lines of future development.[80]

The beginning of work in guns boded well. In 1971 NOS was assigned as design agent for Navy ammunition and projectiles (one of the TAFT FOGS objectives), and several projects for gun improvements began to flow to the Station in the early 1970s. Special projects included SCAMP, which grew out of the destruction of the Israeli *Eliat* by an Egyptian-launched, Russian-made Styx missile in October 1967. SCAMP utilized hundreds of thirty-grain flechettes, or nail-sized darts, which could clear decks like grapeshot. Because of their high velocity the flechettes could intercept and destroy incoming missiles.[81]

Other engineering successes included the Rocket-Assisted Projectile, which grew out of the "General Operations Requirement" that Russian guns should not out-range American guns of the same caliber. Another TAFT FOGS plan that bore fruit was the liquid propellant gun program which started in 1973.[82]

One of the developments in chemistry springing from the new emphasis on gun work was the work in NOSOL gun propellants. From 1970 to 1975 Stephen Mitchell and others worked on such propellants, evaluating several high-energy types. Since these propellants did not have the problem of volatiles, they offered increased long-term chemical and ballistic stability. Several of the NOSOLs used triethylene glycol dinitrate with ethyl centralite and metriol trinitrate instead of nitroglycerin and/or nitroguanidine.[83]

When compared to a standard army propellant, a NOSOL produced an equivalent impetus at a reduced flame temperature and a reduced muzzle flash. NOSOLs would produce less wear and tear on the weapon from heat and cooling and show better stability in storage.[84]

FACING THE SEVENTIES

In 1970 a reorganization of naval research required that Indian Head's Research and Development Department, insofar as it did not directly support the production facilities, be attached to one of the existing naval laboratories. Accordingly, most of the R and D personnel were detached and administratively controlled by the Naval Surface Weapons Center, the laboratory which represented the merger of two of the Indian Head "daughters"—the facility at Dahlgren and the Naval Ordnance Laboratory at White Oak. While some of the Indian Head Naval Ordnance Station staff members remained in place in the old location of the R and D Department, building 600, several productive researchers were transferred to White Oak and not replaced, beginning a long-term decline of the group. No longer would Indian Head be able to claim officially as part of its mission the full range of research and development in propellants, even though much of the industrial development work and testing of both propellants and chemicals continued to involve various aspects of scientific research and engineering development.[85]

In mid-1972 Browning created under Dr. James Wiegand a new department for Special Programs. This department had four divisions: Electronics headed by Vince Hungerford, Cartridge Actuated Devices headed by Jay Thornburg, Gun Systems headed by Dominic Monetta, and Documentation headed by Wayne Dennison. The department would become a "wellspring" for many of the new major efforts at Indian Head over the next years.[86]

For several reasons, Browning took a somewhat diminished role in leadership at Indian Head in the period from 1972 to 1974. For a long period, he absented himself from the Station for training in management techniques. In addition to the frictions which developed between himself and Commanding Officer Frese, Browning grew discouraged over several other aspects of the position. In 1974 he announced his retirement and took an extended sick leave. Thus, through much of 1974 and early 1975, the Station operated under acting technical directors. Captain Frese's successor, Captain Stanley Gary, helped fill the gap in leadership with a strong commitment to good management and to Station upkeep and safety. But for a full two years, the Station suffered a hiatus in civilian leadership.[87]

The decade from the early sixties to the early seventies had been one of innovation as the Station responded to the winds of change under charismatic leadership. New management styles brought participation of newly recruited professionals and rewarded those who showed initiative and creativity with greater responsibilities and rapid career advancement. New ideas from the private sector and from graduate schools of management brought invigorated planning and participation, much of it with demonstrated success. The Vietnam War provided both a bulk of new work, in the area of extruded rocket production, and an opportunity to prove out the sabot fin-stabilized concept with long-range ammunition, as well as a variety of other new products. But the rapid decline of war production in 1969 and 1970 and the increasing emphasis on private sector and GOCO responsibilities in defense and space procurement did not bode well for Indian Head at the end of Browning's tenure.

Browning had left a strong legacy: a cohort of youthful, ambitious scientists and engineers, a host of adaptions and new programs, and a commitment to excellence. Success had depended a great deal on the ability to win in "workload development," the kind of marketing in which Browning excelled. In a period of growth, optimism was natural; similarly, in a period of constriction, it was much harder to exude confidence in the future. Browning's forceful personality, his obvious dedication to excellence, and his ability to communicate his dedication to both employees and potential "customers" seemed to his admirers to have produced a period rich with achievements.

The station had thrived in the mid-1960s, and in the eyes of many of the younger staff members, that growth and success could be attributed to Browning personally. Yet in retrospect it was clear that the Station had grown and thrived when there was an ample naval budget for new programs. In the period of constriction of the early 1970s, success was far more elusive. From the close-up perspective of the engineers and chemists at Indian Head, it seemed as if the good times and the bad could be attributed to one or two personalities, and that problems as well as successes derived from the local leadership. However, much larger forces were at work, both for expansion and contraction, and by the decade of the 1970s, the business of naval procurement had changed. No longer could a single individual knock on doors and round up program managers in a one-on-one sales effort. The new leadership at the Station would need to adapt to the more sophisticated and more restricted world of procurement in the 1970s. Would the Browning style be adaptable, or would it have to be entirely replaced with a different way of doing business? The next decades would reflect continued efforts to sustain the growth and to recover the sense of optimism of the middle years of Browning's administration.

CHAPTER 10

The Post-Vietnam Years

Goddard Power Plant
Building 873
1957

THE DECADE AND a half from the mid-1970s through the late 1980s was another period of adjustment to peacetime and restructuring of administration at Indian Head. While the years following the end of the Vietnam War have not conventionally been called the "postwar years" by the American press or by scholars, many of the problems and adjustments in the world of weapons procurement in that period bore a close resemblance to the years following the end of the Second World War. As in the post-World War II period, the changes begun in the war years carried forward; the diversification and expansion brought on by the war once again required a redefinition of identity and mission. In the 1970s, as in the late 1940s, funding declined and the Station seemed threatened with closing. The Station once again sought support and help from its political and community friends in Charles County and Maryland.

On the national scene, other wider developments again influenced the shaping and direction of the Station. The crisis of Watergate, the resignation of President Richard Nixon, and the collapse of the pro-Western regime in South Vietnam combined to bring a period of depressed public morale and a loss of respect for all government employees.

In the Navy a further reorganization of the systems commands in 1974 required that programs formerly administered separately by ordnance and ship specialists be combined. The attempts to coordinate the two styles of ordnance and ship design in the areas of paperwork, computer systems, maintenance, procurement arrangements, and engineering funding resulted in a proliferation of studies, analyses, instructions, and reporting schedules. From 1977 to 1981 the administration of Jimmy Carter led to a de-emphasis in weapons development and further restriction of the budget.

The early 1980s would see a turnaround in naval budgets and an upturn in morale throughout the armed services, as President Ronald Reagan brought a new commitment to military readiness and higher budgets. Early in the first Reagan administration, the decision to build to a "600-ship" Navy, and the need to arm those ships, brought a stimulus to propellant manufacture and weapon innovation. Indian Head staff members, however, tempered their enthusiasm over these trends in the face of renewed emphasis on private, rather than government, work. By the end of the 1980s, international developments hinted at an era of improved United States-Soviet relations, and renewed constraints on the budget once again suggested that defense expenditures would be subject to tight controls, cuts, and resulting cutbacks at institutions like Indian Head.

David Lee took the position as technical director effective October 1975 and served through November 1985. Lee held a B.S. degree in Mechanical Engineering and had pursued a technical and executive career in the private sector, with experience and increasing responsibilities at Aerojet in Sacramento, at LTV Aerospace Corporation, and at Shilstone Engineering Testing Laboratory. For a short time he had served as president of a firm which manufactured mobile homes. His practical business background provided the source for a number of the managerial techniques which he brought to bear at Indian Head.[1]

When Lee arrived at the Naval Ordnance Station (NOS), he encountered a range of problems, some which seemed endemic to the locale, others which grew out of the hiatus in leadership, both local and national, and still others which arose from limitations imposed by legislation and policy. A few of the issues he dealt with echoed those from the earliest days of the Station. The relationship between the corporate and the government sector in weapons development and manufacture continued to present a set of issues, as it had in the time of Captain Joseph Strauss and Secretary of the Navy Josephus Daniels. Concerns continued over the impact of the Station on the surrounding community, both in employment and in environmental effects. For Ensign Robert Dashiell, hiring local farm boys had not entirely offset the negative impact of falling shot and thundering guns; in the 1970s, the issues of Charles County's dependence on federal payroll and concerns over chemicals grew far more complex, but they sprang from similar causes. In 1904 George Patterson had complained of the difficulty in recruiting talent due to Indian Head's isolation. Lee would continue to face the question of motivating young experts to come and to stay. No longer a matter of geographic isolation, the challenge would be to compete in the national market with higher private pay scales. As always, the managers of the facility attempted to clarify the mission, now expanded with a long list of "functions."

Some new concerns required changes in old ways of doing things. National legislation in the areas of civil rights and environmental controls brought new responsibilities to the administrator of a major chemical and industrial plant, especially one operated by the federal government. Low percentages of women and minorities in the professional ranks pointed to the need for more aggressive recruiting to correct the appearance of discrimination. Furthermore, the protection of employee rights which grew out of equal opportunity law and policy made it far more difficult in the 1970s to reassign, relocate, or force resignation than it had been when Joe Browning took over. Concerns over potentially toxic waste from the chemical plants, which for decades had been spilled in Mattawoman Creek, required new procedures and conformity with state as well as federal rules.

The rapidly evolving naval procurement bureaucracy developed its own language, as managers hoped to harness the ever-growing defense establishment through the use of systems approaches, clarification of tasks, and increased documentation and computerization of information. Unfortunately, many of those efforts appeared to make matters worse, producing ever more red tape. Survival in that world required that one learn the new definitions, use the proper "buzz words," file the correct forms and reports, keep abreast of bureaucratic as well as technological changes, and, as before, look for opportunities to maintain or increase the workload.

Browning had brought a distinctive style to marketing which had worked particularly well in the early to mid-1960s, when Program Managers at the Bureau of Weapons had looked for places to purchase research and development. Convincing one or a few people in a chain of command required personal presence, talented staff, and an ability to match capability and need. But with the decline in funding, increased technical competence in military purchasing, and the growth of structured multistage review processes, new ways of marketing the Station's services would be required.

Lee installed a more systematic method of getting business and offering services. He kept in mind that he had "customers" in Washington, and he used the term explicitly. But the nature of ordnance procurement in the Navy had become complex and was "layered" with many reviews by knowledgeable officers and civilian experts. No longer could one simply drop into an office and conversationally ask what needed to be done. The Station had to have a much more sophisticated means of determining how and where its services would be needed, and Lee set out to create just such a structure. When he reorganized Station management, he oriented the new offices toward particular customers and toward particular product lines, much as one might in the private sector. Where Browning had acted as the chief marketer, Lee began to divide up that responsibility and delegate it to specialists in each area.

Even before Lee arrived, the flush times of the sixties had ended. At the Naval Ordnance Station, the hiatus in local leadership before he arrived coincided with broader national developments that made a clear definition of the Station's mission difficult. At the same time that the presidency was immobilized over the

Watergate scandals, Indian Head drifted afloat without a rudder due to Browning's extended leave and the time-consuming search for his replacement.

During the months before Lee came aboard, records of meetings held by the support departments with the commanding officer and by the operating departments with the acting technical director clearly showed this decline in morale. Rather than successes and upbeat plans, the major questions discussed included personnel ceilings, set at about nineteen hundred people in 1975, refusals by headquarters to provide more help, and the always worrisome problem of personnel safety. The discussions generated very little in the way of new approaches to these continuing issues. Managers even discussed the futility of the meetings themselves. In 1975 the managers made an attempt to install a system of management by objectives as an internal improvement. The operating department heads often cancelled their meeting with the acting technical director, although the support departments, responsible to the executive officer and the commanding officer rather than the technical director, faithfully held to their meeting schedule. Commanding Officer Captain Stanley Gary provided leadership and support during the difficult transition, but the absence of a strong civilian leader in the period before Lee's appointment was sorely felt.[2]

LEE ABOARD—REORGANIZATION AND PUBLIC RELATIONS

Dave Lee took the post of technical director at Indian Head effective 16 October 1975, and immediately planned the reorganization of the Station along "program" lines. Lee announced his plans for the new structure a week after his appointment. Under a single program management officer, separate individuals would now focus on groups of programs. From the start, it was clear that Lee would resolutely face the issues and act to get the Station moving in a new direction.[3]

Lee expected each Program Manager to respond to different purchasing units at headquarters and to concentrate on the marketing and administration of the programs, funneling money to the various functional departments which would do the work. Over the next years, Lee rotated personnel from one post to another and made other changes, but he maintained the dual structure of program management staff and functional line departments throughout his ten-year tenure.[4]

Lee clearly expressed the purpose of the reorganization. "Program management," he said, "will more actively address and be responsive to our customer markets." Program Managers would "oversee scheduling, marketing, and finance," leaving the functional or technical departments more free to "pursue excellence." As a means of relieving technical people of paperwork, delegating marketing, and focusing on products and buyers, the new arrangement made eminent sense.[5]

In addition to his concept of customer-oriented organizational restructuring, Lee brought from the private sector a concern with marketing and managing through communications and public relations. Lee looked for opportunities to give speeches and over his stay at Indian Head developed a thick speech file. He even collected speeches given by other technical directors and naval officers at other facilities, together with materials from published articles that he could use to provide background for his own talks. In this regard, the difference between his approach and that of Browning immediately struck the staff. Browning had excelled in the small, face-to-face meeting, where he had been noted for listening rather than talking. By contrast, Lee seemed to thrive on the lectern, view-graph, and large audience. Where Browning had relied on the telephone and personal commitments, Lee resorted to memoranda, position papers, and public statements. Lee became rather proficient in the speech-writing art. At Indian Head he gave at least two "State of the Station" addresses to the whole staff, in 1978 and 1981.[6]

Again, a change in the times was perceived locally as a change in personal leadership styles. Many of the methods and procedures installed by Lee represented his attempt to bring Indian Head into conformity with the changed styles of the 1970s. Where Browning had operated personally, Lee operated organizationally. Yet in another sense, Lee continued one of Browning's innovations: the development of younger engineers into a domestic source of management talent.

Lee worked on public relations in a number of ways. He arranged for exhibits displayed at the Pentagon, Crystal City, Dulles Airport, and elsewhere. The locations were well chosen to bring the Indian Head message to specific customers in the defense establishment. He wrote a monthly column published in *Profile*, the Station's own monthly newspaper. The files of the Station's public relations office soon bulged with his directives, suggestions, drafts, deadline schedules, and ideas. And he did not neglect local community feelings.[7]

Jay Carsey, then serving as president of Charles County Community College in La Plata, advised Lee to make contact with the local people, particularly those in the up-county wealthy social class who felt little identity with Indian Head. Carsey himself helped in the effort by providing hospitality from time to time at his large traditional home, "Green's Inheritance," which could easily host a hundred guests at a cocktail party. In addition to attending the "by-invitation" gatherings there, Lee made a practice of stopping in to the informal but gracious "at-homes" which characterized Southern-style weekend social life in Charles County.[8]

Early in his stay, perhaps at Carsey's suggestion, Lee gave a speech which he called his "First Annual Report," which in fact consisted of a slide presentation and talk to the local people of Charles County. In the "Annual Report," Lee described the work of the Station, its recent achievements, and its impact on local employment. He promised to be a "good neighbor" by keeping the local people informed, by controlling pollution, and by working to provide a steady payroll. He vowed to improve the looks of Indian Head with maintenance, and he would "run the business to stay around." He echoed Josephus Daniels when he called the Station the "Navy's insurance for preventing industrial monopolies." But he spoke of his sensitivity for contemporary concerns with environment and fair employment practices as well. Lee's activist role and his hard work at the public relations side of his assignment soon paid off with evidence of increased pride and sense of direction.[9]

Lee and his associates more explicitly described and addressed the complex world of naval procurement than had preceding generations. He and his Program Managers became quite expert in recognizing opportunities for conducting work for a wide variety of customers in the Naval Sea Systems Command (NAVSEA), the combined systems command which incorporated the former Naval Ordnance Systems Command (NAVORD) and Naval Ship Systems Command (NAVSHIPS). They also marketed their services in the Naval Air Systems Command (NAVAIR) and to a lesser extent in the Army and Air Force. In his speeches and correspondence Lee constantly used the new, acronym-riddled vocabulary which described the process of planning, development, procurement, program review, and consultation which the systems commands had generated. Moving from the private sector to the intricate world of naval program funding required that one immediately become a team player, master the unfamiliar language, and find one's way through the labyrinth of bureaucracy. Lee appeared to adjust to that new world quite rapidly.

The "Report From the TD—Comments on Trends and Developments at NOS" communicated the news of programs, activities, and funding from department to department, as well as tidbits of information picked up at headquarters regarding future needs and prospects for business. These publications, which Lee issued as an in-house publication on a weekly or biweekly basis faithfully for over five years, served several purposes, including managerial communication and still another form of public relations. The reports also had the obvious function of morale-building, with a constant upbeat tone reflecting achievements, new funding, and recognition of various kinds. While meetings or the "grapevine" would often spread the news before Lee could get word into his weekly public report, appearance of a staff member's name in news of an achievement, over the technical director's signature, would constitute a form of personal and institutional recognition. As a motivational device, it could assure wide recognition of an individual's accomplishments. The style was different, but the larger scale of the operation and the multitude of small achievements made the new methods appropriate.[10]

ADDING TASKS THROUGH THE SEVENTIES

Through 1976 and 1977 Lee's efforts began to bear fruit. At first, many of the achievements and funding came as outgrowths of work already at the Station, such as the specialization in cartridge actuated devices (CADs), which had been brought to the Station in 1966 under the leadership of Bob Spotz, the manufacture of a variety of propellant grains, and the continuing production of Otto Fuel and other chemical products.

ORDNANCE DEVICES

The further consolidation of the responsibility for cartridge actuated devices and propellant actuated devices (PADs) for all three services in the CAD/PAD department at Indian Head stood out as one of the major successes of the era. The work on CADs had supplemented the earlier PAD work when Spotz had led the effort to fill in a gap in private supply in mid-1965 with the closing of the Naval Ordnance Plant in Macon, Georgia. The process of officially consolidating the responsibilities had started with a decision of the Joint Logistic Commanders in 1973; by 1977 a series of "Working Agreements" with officials in NAVAIR, in the Air Force, and in the Army gave Indian Head increasing tasks in the areas of engineering and production of the devices. The total market consisted of over eight hundred different cartridge actuated units, with about five hundred used in the Air Force alone.[11]

CADs not only propelled air crew escape systems with their ejection seats but worked in "stores separation," that is, freeing bomb racks, fuel tanks, and missile launchers from aircraft. Other small cartridge devices provided the energy for engine starters, for fire extinguisher systems, for fin deployment on missiles, and for other systems requiring one-time application of intense force. The department's work included not only production but product improvement, surveillance programs, design engineering, some exploratory research and development, contractor surveillance, and maintenance and service on older units. Funding for fiscal year 1977 had climbed to over $18 million. By that time the CAD/PAD department employed 150 people. In other departments 76 employees worked in direct support of CAD/PAD activities.[12]

Indian Head continued to make the Mark 1 Mod 1 Rocket Catapult for ejection seats used on the McDonnell Douglas A4 A, A4 B, A4 C, and A4 E jet fighters. The grains lasted forty-eight to sixty months, depending on the storage temperatures, giving Indian Head a steady market for replacement grains. The whole rocket catapult measured 45.6 inches long, and about 3 inches in outside diameter. The catapult fired in a launch sequence which went from primer to igniter, booster, auxiliary igniter, and sustainer, all of which took only a few seconds, to launch the seat and pilot from the aircraft to a safe distance, separate the pilot and seat, and then deploy the parachute. Any failure of the Mark 1 Mod 1 catapult or others provided by private manufactures would be reported to the Indian Head commanding officer for investigation by the Explosives Ordnance Disposal group.[13]

In August 1973, during Browning's administration, the Navy CAD engineering support and test jobs had been transferred to NOS. In October 1973 the Joint Logistics Command issued a directive for CAD/PAD tri-service consolidation at NOS; in December 1975 the Army and Air Force completed the transfer of most of their CAD/PAD activities to Indian Head. Under Jay Thornburg, as head of the CAD/PAD Department during 1975–80, the Station's work in ordnance devices expanded into new areas. CAD continued to thrive through 1976, and a new CAD Test Building opened on 12 April. Frankford Arsenal in Philadelphia, which had previously worked in this area, shipped its CAD material and records to Indian Head. The local CAD Department inspected and tested the Frankford materials and used some of the equipment. In mid-1976 the CAD group began to participate in the design of the B-1 bomber crew escape system. Gradually, the Air Force transferred its units of CAD work, including, in June 1977, responsibility for a delay cutter development program.[14]

CAD work involved close liaison with private contractors and with foreign governments. A CAD representative from Indian Head served on the "interface working group" with General Dynamics working on the Tomahawk cruise missile. Martin-Baker funded Indian Head for $355,000 to qualify their design of an ejection seat for Navy use. CAD designers also worked on an escape system for a West German jet aircraft. A purchase order came through for an ejection system to be used on the F-18 developed by McDonnell Aircraft. The F-18 seat involved people from CAD Engineering, Ballistic Test, and Public Works, and Lee boasted that it showed the Station's "can-do attitude."[15]

CAD experts became involved in accident investigations either from unexpected ejections, failures to eject, or "man-seat separation failures," frequently travelling around the United States to Navy and Air Force bases. Indian Head investigators also travelled to the Philippines to investigate a nonfatal accident there.[16]

As pilot ejection conditions became more demanding due to higher speeds and altitudes, researchers looked for new methods of sequencing the various stages of ejection. A division of Martin Marietta, on contract to the CAD/PAD Department at Indian Head, investigated a system of fluidic sequencing, using a pressurized gas bottle to set off the sequence of events. The authors of the report found that inaccurate timing and unreliability of the technology required further work before the system would hold much promise for replacing the CAD approach. The chief of Naval Development included the proposal for further work on the fluidic approach in his "6.3 budget" for advanced development in 1977.[17]

STANDARD ANTI-RADAR MISSILE

Other projects in Lee's first years included getting ready for the Standard Anti-Radar Missile (ARM) and doing work as Procurement Agent, In-Service Engineering Agent, and Design Agent on this project. These tasks involved preproduction engineering for chamber lining, booster mixing, and booster casting of the Standard Missile. The Pilot Plant turned out a booster batch with excellent mechanical properties that gave promise of a good burning rate. By May 1977 Ordnance had made seven consecutive lots of the Standard ARM motors and had shipped 49 units. By mid-July, of 140 units put together only 6 had been rejected.[18]

SIMULATORS

In 1969, through the efforts of Lu Grainger and Vince Hungerford, Indian Head had moved into a new area, the production of mock-up models of missiles for training purposes. Utilizing electronics specialists, some transferred from the Station's own Public Works shops, a new unit set up simulators that would give crews experience in handling and firing of expensive, larger missiles without the expense of actually firing and losing the missile. In 1977 the Harpoon Program yielded a contract with NAVAIR to construct twenty-two training shapes. By the end of the year, Lee announced that the Harpoon work had been so successful that he expected it to lead to follow-on work with Tomahawk.[19]

JATOS

Representatives from Indian Head spoke with Program Managers from NAVAIR at a meeting hosted in 1977 by the Solid Propellant Information Center in Mechanicsburg to discuss the question of "Jet Assist Takeoff Units" (JATOs) production. All parties agreed that the 2.2 JATO did not lend itself to commercial procurement "because of antiquated design," little changed since the late 1940s. Therefore, the 2.2 JATO would remain at NOS for the next two years, during which time an improved design for JATOs would be worked on.[20]

Later, Indian Head received $300,000 for improvement of the 2.2 JATO from NAVAIR. By June Ordnance had made three batches of the new JATOs using a new resin (P-10) and began producing up to two batches a week. By December 1977 Indian Head shipped reloaded JATOs, especially designed for cold weather performance, for testing in Antarctica in Operation Deep Freeze.[21]

WORK FOR OTHERS

Projects for units outside the Navy began to come in, with successive program accomplishments breeding new markets. The Army wanted to draw on Indian Head's experience in conducting surveillance on the liquid-fueled Bullpup and set up a LANCE Life Cycle Coordinated Test Program under the Weapons Quality Engineering Center (WQEC) Department, which had taken over the long-standing surveillance work. Both Bullpup and LANCE used a similar liquid propulsion system. NAVAIR agreed to put another missile, the new Harpoon, into WQEC's type-life program. This marked the first time, Lee remarked, that Indian Head had experience with surveillance over a turbojet or turbofan engine.[22]

Another Army project came through the Harry Diamond Laboratory, which provided funding for a Pyrotechnic Thermal Simulator in the amount of $38,000. The Army laboratory would use the pyrotechnic material from Indian Head to fabricate test arrays for burning tests, which would simulate the heat of a nuclear explosion against armored vehicles.[23]

The move to find customers beyond the Navy through these years included international cooperative efforts with North Atlantic Treaty Organization (NATO) partners. Indian Head obtained three-year funding for $885,800 for the Sea Gnat Motor Development program, a "decoy" rocket which deployed chaff to mislead incoming radar-guided missiles. In June Indian Head staff static tested some of the motors to verify design changes which would increase the efficiency of the motor while decreasing its size.[24]

GUNS

Work in the gun systems engineering division, under Dominic J. Monetta, had been a bright spot in an otherwise largely gloomy picture in the early 1970s. Gun work continued to provide some breakthroughs. Through 1977, work picked up on the liquid propellant gun program. One project provided a mathematical description and computer program for predicting the electrical and thermal behavior of an ignition system for liquid propellants, to help in the design and prove the feasibility of a new electric ignition system. A young ensign assigned from the Naval Academy, Stephen A. Wynne, assisted in the development work.[25]

Indian Head's record in type-life surveillance and in gun systems engineering led to a Conventional Ammunition Surveillance Program. Admiral Michaels, the chief of Naval Materials, and Rear Admiral White, deputy chief for Operations and Logistics, approved the program for about $3.4 million in 1977.[26]

REORGANIZATIONS

Through his tenure, Lee continued to experiment with the local organizational structure. By mid-1977 he had refined his original arrangement. In a staff position reporting directly to him, he had Lu Grainger heading the Program Management Office, with six administrators each focusing on one area of marketing and management: Air Launched Weapons, Surface Launched Weapons, Gun Systems, Underwater Weapons, NAVSEA-Manufacturing, and Electronics. Al Camp, who had headed Research and Development (R and D) before its shift to White Oak control, took a special staff position as "Chief Technologist." Camp, who was well known for earlier work in perfecting and stabilizing the 2.75-inch rocket while at the Naval Ordnance Test Station-China Lake, was widely recognized in rocket and propellant circles. As the holder of patents and a man of considerable charm and repute, Camp's position as Lee's representative at meetings added to the stature of the Station.

The six technical departments, more or less following the older functional division, included Ordnance (which had inherited the work of Production), Pilot Operations (which had inherited the pilot process work from the R and D Department), WQEC (which had inherited the ballistic test and some of the R and D units devoted to manufacturing support as well as type-life surveillance work), Engineering (which had separate internal divisions for air, surface, guns, and electronics), CAD/PAD, and Engineering Support. In addition, nine "Support Departments" not responsible to Lee reported through the executive officer to the commanding officer. These included the Comptroller, Supply, Public Works, Personnel, and other administrative departments.[27]

Besides the growth of the CAD/PAD Department, other areas had grown to include new responsibilities and services to various funding agencies in the Navy and in the other services, requiring some slight fiddling with the functional organizational line-up. The Gun Systems Program Management Office performed several roles to assist NAVSEA in maintaining and upgrading naval guns; the new electronics engineering division, under Vince Hungerford, began to man-ufacture and service the missile simulators. For such reasons, Lee added or ex-panded divisions within the existing departments and within the program management offices from time to time. For personnel at Indian Head, multiple promotions and lateral shifts in career reflected the need to adapt to broad changes in the ordnance business.[28]

PROBLEMS AND SOLUTIONS

As an administrator, Lee demonstrated a patience and acceptance of the complex and frustrating world of the burgeoning military bureaucracy. Never-theless, he admitted that the paperwork blizzard threatened at times to overwhelm him. In one of his in-house speeches pointing to achievements, he outlined a host of difficulties characteristic of civilian service to the military in the era: public disdain for federal employees, hiring freezes and ceilings, and "administrivia beyond belief." "We relegate," he said, "the conduct of our daily affairs to in-struction writers, rather than letting logic and common sense prevail on how we do our jobs."[29]

Lee found some ironies and humor in the flood of instructions and counter-instructions. The Office of Management and Budget urged that work be contracted out, while the Department of Defense urged work not be contracted out if it displaced a federal worker. Among legislative harassments, Lee listed both the Privacy Act, "which says you shall not tell anybody anything," and the Freedom of Information Act, "which says we have to tell everybody everything." Efforts to conform to new rules regarding saving energy, protecting the environment, and preventing employment discrimination made a tough job even tougher.[30]

In spite of the headaches, by 1978 Lee could claim that his Program Managers had done a good job "seeing to it that the successes of the station's efforts are properly publicized at the proper levels. We are now getting due credit at the right levels of management in headquarters." He pointed to gains in the CAD area, in the electronics area, in the quality engineering work, and in engineering in general. For the first time in twenty-six years, Lee pointed out, Indian Head had been removed from the base closure list. That list of military installations slated for review and possible closing was dreaded by all civilian employees of the military across the country; removal from the list was a concrete measure of his personal success and that of the commanding officers and employees at the Station. Public relations and an internal structure that closely mirrored the funding agencies and their needs had begun to work, bringing the total funding from $123 million in 1975 to $180 million in 1978.[31]

Although Lee's public addresses tended to emphasize the positive, he was often quite frank about shortcomings and unresolved problems. In several cases, he pointed out, he had to step in personally to correct difficulties. In order to make the standard missile effort successful, he had to establish a "standard missile project office," suggesting that he thought the regular functional department should have been able to handle the new task without such supervision from his

195

own staff. In the case of the Shrike warhead pilot production program, "severe trouble" of a technical and financial nature also required that "management address" the issues. With regard to the NATO Sea Gnat program, he needed to place "special emphasis" on the "interface between NOS and NRL" (the Naval Research Laboratory). "Our Sidewinder production," he noted, "ended in a program disaster in that we could not fulfull [sic] the extent of the contract and requested that it be terminated."[32]

While he couched such critiques in bureaucratic language that in retrospect may seem impersonal and analytic, for those involved it was clear that Lee was quite ready to identify problems and to point out shortcomings when they affected the Station's performance or reputation. Although Lee had set up program management staff to get the new business, he did not expect his Program Managers to oversee the details of internal compliance and continued liaison, leaving those tasks to the department heads in the functional departments. When there was a shortcoming, he would admit it and deal with it.[33]

Often the difficulties arose from factors beyond the control of the Station. Lee's notes revealed his awareness of external causes such as the failure of contractors to comply with schedules or specifications. Nevertheless, some managers found the public spotlight and the high standards of performance very demanding. Through the late seventies, the tempo of production increased and the volume of business expanded. With larger dollar volume came a necessary price: living in the spotlight. Any discontinued program, shortfall in production, or liaison problem would be very noticeable.[34]

PRODUCTION AND ENGINEERING: STRIKING THE BALANCE

Lee paid particular attention to the "workload balance" between production and engineering, achieving a dollar ratio by the late seventies of about 60–66 percent engineering and about 40–33 percent production. He explained his reasoning carefully. If the production work shrank to less than a third of the budget, the Station would be easily closed and the various engineering tasks, as "paper functions," could then be shifted to another location at the whim of the program people in Washington. However, too great a reliance on production would be dangerous as well. Congressional pressure, he believed, would continue to drive the most attractive or profitable products from Indian Head to the private sector. To find a way through this dilemma he tried to balance the two, while keeping an eye on changes.[35]

The two-thirds engineering, one-third production ratio represented Lee's practical compromise on two long-standing issues: the mission of the Station and its relationship to the private sector. In order to maintain this proportion of work in the face of changing products and opportunities, Lee set up some long-range planning groups. He intended to work on five-year goals, carefully stabilizing the two-to-one ratio of engineering to production at about the same level. In this fashion, he believed, NOS could be available to upgrade for higher quantity

TABLE 14

Wage and Professional Employees, NOS, 1973–1978

Year	Total Employees	Percent Blue-collar	Percent White-collar
1973	2026	51	49
1974	1920	47	53
1975	1816	40	60
1976	1851	41	59
1977	1858	40	60
1978	1977	40	60

Source: David Lee, "State of the Station-1978," 28 September 1978, State of the Station-1978 by D. E. Lee, Box 13, Indian Head Technical Library, Naval Ordnance Station, Indian Head, Md.

production should an unexpected national emergency call for such participation, as it had in the past. Keeping the plants running while relying on new and continuing engineering work for the majority of the budget would be the core strategy for planning.[36]

Although total personnel numbers had declined slightly from 1973 to 1978, hovering just under the two thousand mark, the proportion of professionals as compared to blue-collar workers had increased, reflecting the shift away from production to engineering and contract management. Table 14 details these changes.

The long-standing question of the relationship with the private sector required a slight redefinition. In 1981 Lee described the "interface" between NOS and the private sector in a view-graph supported speech that he delivered jointly with Lu Grainger to business audiences. He spelled out a far more complex relationship than simply watchdog to prevent monopoly. First of all, NOS had taken on the technical management role, in which it would demonstrate the feasibility of a new product, provide documentation to contracting officers, review the technical aspects of production, and provide technical knowledge to contracting officers. In addition, Indian Head remained ready to take a larger role in producing tactical propellants during a military emergency or mobilization. Furthermore, the Station would provide a "backup source" as an alternative supplier for the Navy. In case of accidents or failure in quality that would prevent a company from providing a product, the Navy might "find itself hanging from a rope." Indian Head could "come on line" if needed.[37]

In their presentation to private companies, Lee and Grainger explicitly stated that NOS did not compete with the private sector. "NOS does not bid on contracts. Our purpose is to help private industry in any way we possibly can." They went on to spell out how production of the Standard ARM formerly made by Aerojet, and ammonium nitrate for the Mark 6 Gas Generator, formerly produced by AMOCO corporation, both showed how Indian Head moved in to pick up a product when the private sector stopped production.[38]

The exact position of Indian Head in the ordnance procurement picture had grown complicated, and it could not be neatly summed up in a catch phrase such as "bureau of standards for ordnance," or "yardstick for private enterprise." Lee recognized the complexity and articulated it in a vocabulary which business leaders could find acceptable. As an engineering center with knowledge of industrial-scale production, Indian Head provided the unique and complex interface between development and private production. In scaling up a new process, in checking quality, and in providing impartial documentation to competitive bidders, Indian Head had found a series of roles that it was uniquely qualified to fill.

By the early 1980s, with the national mood turning in a more pro-business direction, the Navy-owned facilities at Indian Head had found several niches in the propellant market. In addition to providing engineering services to translate a laboratory process into a production process for the private sector and providing a national "back-up" resource in the case of emergency, full-scale production at Indian Head would concentrate on several materials too unprofitable, too dangerous, or too difficult for the private sector.

The phenomenal expansion of private sector weapons manufacture, while changing the opportunities for the Station, created some difficulties as well. Personnel recruiting and retention in the professional ranks remained a sore point for Indian Head as it did for many government and academic institutions through this period. Recently hired professionals left in great proportions through these years, making "attrition" a serious concern. Lee blamed the losses on several factors, including salary differences between government and private sector work, government constraints on promotions to higher grades, and excessive bureaucratic delays in hiring replacements for higher-grade professionals who retired or resigned. People "bump up against the GS 12 rank," he pointed out. After "stagnation" at that level, they would seek employment elsewhere. As a solution, he relied on more active hiring at the junior ranks, and he sent Hugo Lopez, Al Camp, and others on recruiting trips through this period to help make up the professional shortfall.[39]

Even with the shift in emphasis to engineering rather than production, manufacture of several old products and movement into new areas continued apace through the late seventies. Otto Fuel continued to be a major product, with production hitting 250,000 pounds for the month of March 1978 and a total production of 1.5 million pounds by mid-August. Standard ARM production began to increase, and in March 1978 Indian Head shipped 48 of the missiles in one week, including some for the Air Force. By August 460 of the units had been manufactured and shipped. With the resolution of quality assurance on the Shrike warhead, Production turned out several hundred of these in early 1978, but the plant shut down for four months until August 31.[40]

Some of the engineering tasks really involved management more than engineering, as shown in nature of the jobs which Indian Head performed in connection with the fleet gun maintenance programs. Through the late seventies, the Gun Weapon System Replacement Program continued to engage Indian Head personnel, as the surface weapons program there managed a contract for NAVSEA to study and make recommendations about the Navy's whole gun overhaul and maintenance system. A series of studies by a private firm, Aeronautical Radio, Inc., examined the history and evolution of gun maintenance programs and recommended several changes.[41]

The studies sprang from the fact that earlier gun overhaul systems had relied on a repair pool of guns and mounts, a group of duplicates from which spares could be selected and to which rebuilt or reconditioned elements of ordnance could be returned. But from the mid-1950s through the mid-1960s, the Navy acquired a more complex and sophisticated "armaments profile" which included the 5-inch/54 Mark 42 Gun Mount and the Mark 68 Fire Control system. The Navy had bought these components without a large back-up supply, apparently out of a concern to economize. Without the pool of back-up equipment, a weapon would simply be out of service until a shipyard became available to perform the repair work.[42]

Reorganization further complicated the maintenance picture when the Bureau of Weapons became Naval Ordnance Systems Command, and then when the Bureau of Ordnance merged with Naval Ship Systems to form the Naval Sea Systems Command in 1974. These changes required that various maintenance schedules and procedures, for ships on the one hand and for weapons on the other, be coordinated and merged. It would make good sense to do periodic repair and upkeep work on the ship's engines and structure at the same time the ship came into port for gun and fire-control overhaul. If the schedules remained uncoordinated, as the fleet expanded, more and more ships would be laid up for yard work.[43]

Predicted requirements for about thirty overhauls a year would require that the whole process be sped up as much as possible, if the "necessary armaments posture is to be maintained." Indian Head managers worked with the consulting firm in recommending improvements in information flow, timetables, scheduling, and systems that would reduce the shop-time for the various types of upkeep, care, and remodelling. In addition to recommending changes in the paper flow, the report recommended buying a limited supply of depot spare parts and subassemblies to help speed up some of the processes.[44]

The details of the Indian Head work on this project are rather typical of many of the nontechnical, more managerial and systems-oriented tasks taken on by Indian Head staff through the 1970s. The work involved making management recommendations rather than focusing on hardware, processes, and development. Under the plans for improvements in information flow, Indian Head computers would be used to gather and transmit data on maintenance, keeping the number of transmissions to a minimum.[45]

As Lee had recognized, by calling the new jobs "paper functions," more and more of the tasks at Indian Head became administrative or managerial work devoted to information handling for the Navy, moving away from the more old-fashioned, hands-on work that one usually associated with production and engineering. In the late 1970s, such work included studies of computer programs useful for working with interior ballistics data and studies of fluid-pressure sequencing devices for pilot ejection systems. Another report focused on the discrepancy in results between two methods of testing the pressure generated by a propellant—the strand-burning test method and the closed-bomb method. The comparative study contributed to a larger Indian Head effort to evaluate testing methods and devices and to correct contradictory

TABLE 15
Program Funding, NOS, Fiscal Year 1979

Program	Program Funding $ in Millions
CAD/PAD	25.8
WQEC (type-life surveillance)	2.2
Other	13.5
Program Management Areas	
PM-1 Air-launched	7.0
PM-2 Surface-launched	24.5
PM-3 Gun Systems	9.0
PM-4 Underwater Weapons	8.0
PM-5 Chemicals/Products	2.5
PM-6 Electronic Systems	12.0
Total	104.5

Source: Naval Ordnance Station, "A Briefing for Capt. R. D. Buchwald," July 1979, Box 12, Indian Head Technical Library, Naval Ordnance Station, Indian Head, Md.

data in a 1964 manual published by the Chemical Propulsion Information Agency.[46]

Another investigation along the same lines focused on a comparative investigation of various theoretical models used in studying interior ballistics. The study examined the assumptions about exactly how the burning process worked in propellants, considering such questions as the porosity of the propellant, the compaction of the propellant bed, the flow of hot gases, and the motion of the projectile. The report summarized work in various laboratories toward developing a computer model of burning rates, which could become a valuable design tool for the gun community. The report pointed out that at least seven governmental and private laboratories engaged in research in the area, and each used different measuring devices. The authors recommended a single piece of hardware for such testing, to serve as a "referee" to standardize laboratory-to-laboratory variation in testing, and developed a preliminary conceptual design for the gadget. The question of producing standard information from different testing methods was a recurrent one in technological advance. Two decades before, similar issues had led to the "round-robin" programs set up by Browning.[47]

By the end of the 1970s, Lee's system of program management had worked well. Of the $104 million in the fiscal year 1979 technical budget, $25.8 million came from CAD work and about $15 million from a variety of minor programs, including the ongoing type-life surveillance work. The Program Managers could take credit for about 60 percent of the project-related money which came in, as shown in Table 15.

NAVSEA accounted for about 44 percent of the budget, NAVAIR for about 22 percent, and the Army and Air Force for 19 percent. The rest came from a variety of sources within the Navy, with about 1 percent from the private sector.[48]

NEW ASSIGNMENTS

The continual addition of specific, ongoing engineering tasks and assignments gave proof of success at defining a place in the changing world of naval procurement. Through the 1970s and 1980s, the commanding officers, technical directors, and various Program Managers constantly pushed to get Indian Head designated as the locale for specific ordnance-related jobs.

Pollution abatement work at Indian Head had begun in earnest in 1970, with an executive order from President Richard Nixon that charged all federal facilities to take the lead in reducing their own polluting activities. Over the early 1970s chemical engineers worked with staff members from the Public Works Department to develop several systems for removing organic and heavy metal materials from effluents, to construct a nonpolluting destruction facility for ordnance, and to clean up the sewerage system of the Station. With the addition of an environmental engineer, Indian Head's competence in this area led to an unusual assignment. The Navy designated Indian Head, jointly with a group at Dahlgren, as the Ordnance Environmental Support Office in 1975, with some thirty engineers responsible for working on environmental inquiries and resolution of particular pollution crises in the naval ordnance establishment.[49]

In regard to a wide variety of weapons and missiles, Indian Head got NAVSEA or NAVAIR to designate the Station as an "agent," which carried a specific management role. By 1980 most of the types of designated agent roles which Indian Head had acquired fell into three areas—design, in-service support, and procurement support—for a variety of specific weapons, devices, and systems.

> Design Support
> > Engineering Design Agent
> > Production Support Agent
> > Technical Support Agent
> In-Service Support
> > In-Service Engineering Agent
> > Maintenance Engineering Agent
> > Fleet Support Engineering Agent
> > Designated Overhaul Point
> > Inventory Control Point
> > Documentation Repository
> Procurement Support
> > Procurement Agent

Regarding the CAD/PAD work, Indian Head had earlier been designated as the "Receive, Segregate, Store, and Issue Authority" for all rocket catapults used in air crew escape systems. NAVAIR had designated Indian Head as the "cognizant field activity" for JATOs and for rocket systems and rocket motors.[50]

Each of these designated agent, point, repository, field activity, or authority roles provided assured work and payroll, and the legal authority for each could be traced to a specific, dated systems command order or instruction. In planning for the workload, these "agencies" represented at once a current mark of achievement and a future worry. When a weapon or a system became obsolete, the

TABLE 16

Evolution of Sea Systems Management Information Activity

Year	Acronym	Name
1964	NUMIS	Navy Uniform Management Information System, Crane, Indiana
1967	NOMIS	Naval Ordnance Management Information System, Indian Head, Maryland
1969	CENO	Central Naval Ordnance Management Information System Office, Indian Head (successor to NOMIS, also called "Central NOMIS Office")
1965	CASDO	Computer Applications Support and Development Office, Portsmouth, New Hampshire (formerly CDA, or Central Design Activity, of NAVSHIPS)
1977	SDA	Proposed name of CASDO/CENO merger, Sea Systems Design Activity during site study
1979	SEAADSO	Sea Automatic Data Support Office, Indian Head (combined CASDO and CENO, incorporating Portsmouth activity)
1982	SEAADSA	Sea Automated Data Systems Activity, Indian Head (successor to SEAADSO)

Source: Review of Acc. 181-85-75, Record Group 181, Washington National Records Center, Suitland, Md., and information provided to the author by Bill Jenkins.

"agent" role would vanish; when a new system developed that matched the evolving mission of the Station, Indian Head had to be ready to get new assignments. As in the past, the rapid changes in technology required flexibility and constant readjustment in planning.

New weapons opened new opportunities. The "Phalanx" Close-in Weapon System used a Gatling-gun design of extremely high speed to shoot down incoming antiship missiles. The Surface Weapons System maintenance branch at Indian Head contracted for a study of how the maintenance depot for Phalanx should be established. The resultant study called for Indian Head personnel to play a part in the team that would evaluate different sites for the final "depot overhaul activity." In this case, while Indian Head would participate in the site validation team, the Station had no guarantee that it would gain any significant final role in this advanced gun system.[51]

Indian Head obtained the location, in the late 1960s, of a central computer systems facility for the central design of management information systems that reported directly to the Naval Ordnance Systems Command. This office, which evolved to a staff of about two hundred, grew out of the attempt to coordinate information systems, first within Ordnance and then within the combined systems command, NAVSEA, after 1979. The growth of this independent tenant activity also resulted from the merger of ordnance and ships desks at headquarters into the new systems command, NAVSEA. The evolution of the acronym designations for this computer facility are found in Table 16.

The crucial step in the evolution of "NUMIS" into "SEAADSA" came with the selection of Indian Head in 1977 as the location for the combined office to operate jointly the ordnance information system and the ship information system.

Each of these computer systems grew out of different paper management systems, that could trace their ancestry to different bureaus in the days of World War II and earlier. The decisive factor in locating the combined system at Indian Head cannot be viewed as a victory of the "ordnance" outlook over the "ships" outlook within the merged systems command. Indeed, at SEAADSA, the two management information systems continued with their different characteristics well into the 1980s. Rather, the site selection decision came about because of Indian Head's closeness to Washington, and because Indian Head could demonstrate a greater negative employment impact if the CENO personnel had shifted to Portsmouth than if the Portsmouth CASDO personnel shifted to Indian Head. This office, like the designated agencies and activities, represented a solid role for Indian Head.[52]

PLANNING CONFERENCES, REVIEW PANELS, TRAINING

As weapons budgets picked up through the early 1980s, Lee engaged senior and middle management in a number of planning exercises. Some of these met over a two-day stretch, held at the Charles County Community College. Using the tried-and-true briefing method of projecting an outline of the presentation from transparencies, each Program Manager would give a talk describing the present role of the Station in a particular activity and the desired future roles on the basis of technological and programmatic changes. Planners would define the approach, strategy, or liaisons that would be required. The planning cycle, using a flow-chart system, indicated that required and existing manpower, equipment, and facilities should be compared and decisions made by senior management on how to fill the needs. The planners set tasks, assigned them to particular staff members, and reviewed them the following year.[53]

By the mid-1980s, the Weapons Systems Simulation program had grown, and its "strategic thrusts" for acquiring new business followed the approved pattern at the planning sessions and offered an example of the kinds of information the presentations covered. Vince Hungerford, as Program Manager, discussed the specifics of exactly how many trainers, simulators, or dummy rounds of each type of missile had been delivered, on the Standard Missile, on the Tomahawk, and on the Harpoon. He then spelled out the "balance of programs" which would need simulators and the requirements to take on a specified number each year. He also indicated the sponsoring "customer" at the systems command, by name, for each of the programs he would attempt to gain over the next year.[54]

Making the needs specific, targeting the customers by name and address, planning a strategy for gaining new business, and specifying which individual would take on the task all represented a systematic approach to workload development in an era of complexity and vast growth. Where Browning had originally made the contacts with customers himself and then used a representative, Carsey, to assist him in knocking on doors, Lee developed a more elaborate system which took into consideration the increase in offices, sponsors, and potential sources of funding.

The planning sessions produced an important by-product—that of wider involvement of Station personnel in the issues. At the 1984 planning conference, several of the support departments, including Security, Public Works, and Supply, participated in the discussions, making presentations about their future needs and modernization plans. The planners called the 1985 meeting a "corporate" rather than a "strategic" planning session, to reflect the employment of the corporate planning concept at NAVSEA headquarters, where a "Corporate Management Plan" had been put in place.[55]

In 1985 Lee indicated that he had seen improvement in the accuracy of forecasting needs for manpower. He estimated that the Station had stabilized at about fourteen hundred man-years on direct labor and about twelve hundred on indirect labor, which he estimated would continue through the rest of the decade. While no new large-scale production programs had "surfaced," a few large ones had "slipped."[56]

All the Program Managers presented a review at the planning sessions, following a set pattern of discussing objectives, current workload, current assets and projected requirements, projects under way, planned projects, issues requiring management decisions, and a tentative milestone chart reflecting the estimated years when future tasks would be acquired. While the managers showed a degree of individuality in the reports, all tried to develop sensitivity to future weapons requirements, looking forward as much as five to ten years to such developments as the long-range Aegis missile, insensitive explosives, and new applications for the CAD/PAD and trainer-simulator technologies.[57]

In addition to working on means of improving the planning, Lee focused several administrative reforms on the continuing serious issues of safety and quality control. An earlier generation had identified collectively as process control the various measures involving adherence to procedures, use of safety precautions, in-process measurement and testing, and product testing. In Lee's opinion, process control in some areas had eroded to the point that both the safety and quality of the products and processes were endangered.[58]

Lee dealt with these issues in a set of "Guidance and Policy Papers," modelled in their style and format on such papers issued from NAVSEA by Admiral W. E. Meyer. One policy required what he called "Critical Program Reviews" in order to ensure better attention to detail and production of quality products. A second policy required "Test Readiness Reviews," a thorough review before a test of a new device would be permitted.[59]

Lee put the first of these policies in place early in 1983, when he noted that some of the "products cause some major problems in service use and inventory." He viewed the "situations and results with great alarm." Lee attributed the failures to a lax attitude toward detail and required a set of close reviews of design, documentation, and process as they pertained to all programs, products, and services. The reviews consisted of hearings conducted formally and under disciplined procedures. Lee did not convene the full schedule of reviews which he originally outlined, but several of the hearings exposed weak points in safety and quality which led to corrective measures.[60]

The Critical Program Reviews sometimes singled out individuals for criticism, and panel reports submitted to Lee frequently resulted in personnel transfers and changed procedures. Lee himself attended many of the reviews, personally questioning staff members. Difficult as the sessions may have been for some of the individuals, most administrators at the Station found them a useful means of identifying responsibility, holding people to a high standard, and insuring a consciousness of both quality and safety standards.[61]

Lee thought of the Test Readiness Review sessions, which he initiated in January 1985, slightly differently. Here he intended that prior to a major article or device test a panel of specialists would determine that the test objective could be achieved, that the test article quality could be traced, that the methodology of the test would support the objectives, and that all personnel involved understood and were capable of the test. The reviews would be held whenever design called for a major change or when a major test series began, such as Design Verification, Qualification, Preproduction testing, or Lot-Acceptance testing. Again, in retrospect, administrators agreed that the Test Readiness Reviews accomplished their purposes.[62]

Lee implemented several concepts to focus on the younger professionals. In his first year he tried to institute a plan which would rotate young engineers through production work and get them experience working with wage-grade personnel. In 1976 Lee reconstituted Browning's Assistant Management Board as the Professional Development Council (PDC), with a focus on training. He dissolved the PDC in 1977 and reestablished it in late 1981, with concentration on training and on producing case studies.[63]

In the early 1980s Lee increased his focus on the perennial question of training and orientation of younger professionals. He issued a Guidance and Policy Paper asking his department heads to work up training plans for engineers, spelling out familiarization required as well as formal training. Another attempt at training went back to his first concept of rotational assignments. Through a policy paper issued in March 1985, he ordered a plan by which selected professionals with high motivation and potential would be rotated to other assignments. He intended that the training develop those general skills which would make for a good project engineer or manager.[64]

After Lee's resignation, senior management convened in early 1986 for discussion of a variety of management reforms, under the direction of the executive officer. The group discussed the planning process, which, the managers agreed, was valuable and needed further study and improvement. The planning process was retained, and it continued under Lee's successor. Particular issues that needed reconsideration included who should plan and how and when it should be done. Although some steps had been achieved in breaking down the wall between the operating departments, under the technical director, and the support departments, which reported to the executive officer, the managers felt more should be done to bring Supply and Public Works "into the mainstream." The Test Readiness Review sessions were continued after Lee's departure as a means of insuring accuracy, quality, and safety.[65]

SUCCESSES AND ADJUSTMENTS

While some of the administrative issues dealt with by Dave Lee came out of the complexities of the burgeoning defense establishment, others had been with the Station in one form or another for years. Perennial issues such as how to balance the innovative spirit and the requirements for safety and quality, how to attract and hold young talent, and how to adjust a functionally organized structure to constant changes in technology were typical of the administrative issues faced by technical directors throughout hundreds of defense and private research and development institutions in America. Lee's experimentation with reorganization, public relations, communication, training, and various ways of holding people responsible for quality and detail did not exist in a vacuum. His methods serve as illustrations of how one facility attempted to deal with these major, widespread issues. The measure of success in this case was the adaptation to new needs and the solid establishment of Indian Head as the center for certain specific ordnance tasks.

Technologically, the major successes of the era included an expanded, interservice role in the CAD/PAD work, development of simulators and trainers for missiles, and manufacture of a new generation of missiles. In case after case, the Station acquired a new task or role as design agent or other consultant position in support of the rapidly expanding NAVSEA and Defense Department procurement systems.

To adapt to the changing market, Lee set up concerted planning sessions, involving wider circles of middle management. With the growth of the procurement system in ordnance through the post-Vietnam years, the Station had adjusted—and adjusted well—from the individual approach and personal style of the 1960s to a more organized, institutional approach. Planners, managers, and commodity specialists now had an intimate knowledge of the needs of clients, and a set of offices had been established to deal with the more complex procurement structure. On Lee's resignation in 1985 to return to the private sector, the Station would require someone to fill the post of technical director who could continue the process of adaptation and continue to address the ongoing issues.

A NEW TECHNICAL DIRECTOR

From November 1985 until the fall of 1986, the Station conducted the search for a new technical director. The interview teams eventually settled on Dr. Dominic J. Monetta, who served as technical director from November 1986 to June 1989.

If background and experience in management could suit one to take on such a list of complex issues, Monetta was certainly a good choice. Monetta had earned a B.S. in Chemical Engineering and had spent the years 1963 through 1968 working in the Production Department at Indian Head, followed in the late sixties and early seventies by work in the expanding gun engineering programs. Monetta himself was a product of the Assistant Management Board established by Browning, and he served as chair of that board in 1967 and 1968. He had gained more experience in the theory and practice of management in jobs which he had held after leaving the Station in 1975.

Monetta left Indian Head during the hiatus between the administrations of Browning and Lee and took a position with the Energy Research and Development Administration (ERDA), the organizational successor to the Atomic Energy Commission. In 1978 he took over as Director of Planning at the Gas Research Institute (GRI) headquartered in Chicago. GRI was a national organization that had been established to help ERDA manage funding in gas research and development. In addition to his post at GRI, Monetta worked on and received a doctorate in Public Administration, concentrating in R and D management, and established his own consulting firm. Since leaving Indian Head, his positions had dealt with the corporate and strategic questions of allocating resources and managing technical work.[66]

To encourage each department, division, and branch head to see himself as part of a "corporation," Monetta organized "affinity groups" which would bring together managers at each level to discuss plans and problems. As a means of communication, he preferred the meeting to the memo, because, as he pointed out, "The written word has two defects: it is not interactive, and it is not real time." From the historian's point of view, this approach itself has a defect in that it leaves a rather lean record from which to document the changes, programs, and decisions.

Nevertheless, from a series of oral history interviews, it was possible to reconstruct the ambitious program which Monetta set for himself and the Station during his tenure. Like Browning and Lee before him, Monetta concentrated on the continuing problems of communication, planning, motivation, and marketing. Some of his methods drew from his own training and background, and others drew from contemporary or recent works in management. He recommended that all his middle managers read W. E. Deming's *Out of Crisis* and apply its points regarding communication and participative management, and he provided copies for the staff to read.

Monetta attempted to address management issues with several simultaneous measures:

◇ executive training
◇ extensive participatory management
◇ a "Joe Browning Award" for excellence in management
◇ conscious professional role-playing
◇ affinity group meetings

Monetta utilized a kind of working vocabulary for the roles that he believed middle management should play, using the slightly ironic terms current in management thinking about roles. The terms were made popular in the "multiple advocacy" theories of Roger Porter, author and advisor to Presidents Ford and Bush. One type of role Monetta called "Robber Barons," reflecting the natural

207

tendency of managers to acquire and use resources for their own programs. At Indian Head, Monetta viewed this as the proper role for the Program Managers; the concept was that they would try to get funds and control as much as possible.[67]

A second category was the "Honest Brokers," or facilitators. At Indian Head, that role would be played by the department managers, who would try to get as much done as possible with the resources that came their way from the Robber Barons and would help allocate resources among their divisions. In order to prevent the natural tendency of department heads to build their own units at the expense of others, and to keep them to the role of Honest Brokers, several reforms would be required.

A third category he called the "Praetorian Guard." Like the Executive Office of the President, this group of managers would be devoted to ensuring the interest of the institution as a whole and would report directly to the technical director. At Indian Head, the group which played that role was the Corporate Planning Division.[68]

The "operative paradigm," Monetta stressed, would be "win-win," not "win-lose." In an attempt to set up a "win-win" situation, he reorganized the program management concept. First, he increased the number of Program Managers from five to nineteen, making each responsible to one particular client or to a natural cluster of clients. Secondly, Monetta moved program management from his own staff office, where it had been located for a decade, into the operating departments. Within the departments, a division head was given the responsibility for program management. The particular divisions within departments tended to be those that already played a management or support role in the department, rather than a more pure engineering or developmental role. Program Manager-division heads would have thirty or forty people who would report directly to them, but because they were limited as to space and facilities, they would have to go beyond their own division to achieve the work. Thus, Program Managers would have local engineering skills they could directly call upon, particularly in the planning and proposal stage, but would be required to "purchase" further services in the institution.

Department heads would continue to play the role of Honest Broker, rather than empire-building Robber Baron, partly because they would all meet biweekly in the Senior Management Board. That affinity group of all the department heads would tend to keep each other "honest" and to ensure that resources not be duplicated and that work be properly distributed. Another of Monetta's strategies to keep the department heads from building empires at each other's expense was to regularize the process of rotation in office, formalizing a tool which Lee had used extensively. He made it explicit that department heads would serve no more than three to five years in a position before moving to another department. Thus, the cadre of department heads would tend to develop an all-Station interest rather than a more parochial, department-based identity. It might appear that such transfers would remove individuals from the technical fields they knew best, but Browning, Lee, and Monetta shared the concept that good engineering management talent needed to be nurtured and advanced by opening opportunities.

Monetta consciously worked to build the institution so that it would be less dependent on what he called the "cult of personality," that is, less subject to the style of the individual technical director or commanding officer, and would have a stronger collective sense of purpose and identity. Thus, he tried to build structures in the areas of training, management, planning, and "affinity groups" which would last long after he left, and which would continue to strive for the goals of quality and participative management along the lines which he had initiated. By working to share power and to define responsibility, he hoped to get away from the concept that the institution would revolve around a single individual. Yet Monetta's leadership was so emphatic and the role of technical director so central to the institution that it was difficult to achieve the stated objective of reducing the role of the individual.

The planning system continued to evolve. The annual planning conferences now focused on two-, seven-, and twenty-year objectives, revising the plan each year to take into account unpredictable developments and changes. The Navy's procurement budget followed a two-year cycle, and the military construction plans of the Navy for new facilities was planned on a seven-year basis, so the Indian Head plans matched those requirements. The twenty-year cycle and the work of the commodity managers kept the Station alert to indications of future developments which might open new opportunities.

The commanding officers at Indian Head, Captain James D. Tadlock through 31 July 1987, and Captain George (Fritz) Wendt through 21 July 1989, cooperated in restructuring the departmental organization of the Station. As in earlier periods, the exact relationship between the commanding officers and the civilian technical directors continued to depend upon the personalities of the specific individuals involved. Monetta successfully worked with both Captains Tadlock and Wendt in bringing the administration of the facility into closer touch with the administration of the technical programs. They arranged that several of the "support" departments be shifted directly to the jurisdiction of the civilian technical director, including Public Works, Supply, and Civilian Personnel.

Monetta also created a rank of "associate technical director," which comprised three staff positions reporting directly to the technical director. Here he appointed individuals who had long service to the institution. He drew one from Product Support which was responsible for the engineering and development facilities. A second represented Industrial Operations which worked with the departments involved in all explosive operations, including production. The third position, Resource Management, dealt with resources, planning, technical information, civilian personnel, and the Station's own computer activities. The three individuals holding these associate technical director positions together represented nearly a century of experience at Indian Head, including Jay Thornburg, A. J. Perk, and Vince Hungerford. All three had come to the Station during the Browning era. Monetta found it appropriate that despite their comparative youth (all in their forties or early fifties) they were known on the Station as "the three wise men."

While all of the administrative changes of the period 1986–89 are not detailed here, this partial description may convey a sense of the intense concentration on management issues in the period. Utilizing ideas from the Browning era, and adding concepts from the 1980s drawn from the work of Deming and Porter, the Station concentrated on the same issues which Lee had regarded as crucial: planning, marketing, and management for quality manufacturing and engineering services.

With all the concentration on new management ideas, technology and engineering were not neglected, however. Continuing a trend which began during the 1970s, the Station increased the proportion of engineering and administrative roles over production. NAVSEA recognized the achievement of the Station by reconstituting several of the previously acquired agency roles as official "centers of excellence" for the Navy. This designation meant that the Navy would not duplicate the effort elsewhere and would treat the facility as the lead collection of experts in a particular area. The Station had acquired the "center of excellence" designation by mid-1989 for six technologies:

◇ Guns, Rockets, and Missile Propulsion
◇ Energetic Chemicals
◇ Ordnance Devices (CAD/PAD)
◇ Missile Weapon Simulators
◇ Explosive Process Development Engineering
◇ Explosives Safety, Occupational Safety and Health, and Environmental Protection

Each of the centers of excellence had grown out of an established track record. Some, like the CAD/PAD work, could be traced back only two decades or so to the mid- and late 1960s, and some, such as gun propulsion or energetic chemicals, had their roots in the days of Dashiell and Patterson. Management at the Station had reason to take pride in the designation of the six centers of excellence, and they stressed this role in public presentations of the Station's capabilities.

As Captain Fritz Wendt and his successor, Captain Ed Nicholson, issued their monthly narrative reports, they would structure reports of achievements largely around the accomplishments of the six centers of excellence. The center for guns, rockets, and missile propulsion continued to report on production, to investigate accidents, to provide crews for on-site support in rocket installations and safety inspections, and to conduct a wide variety of development tasks. The center continued to produce old standbys such as Otto Fuel and the 2.75-inch rocket, as well as new products such as LOVA, or Low-Vulnerability Ammunition. The latter propellant used an inert binder with RDX, or cyclonite, and gave equal or better ballistic performance at reduced barrel wear rates than the older smokeless powders.

The center for energetic chemicals continued to produce large orders of chemicals such as high-bulk density nitroguanidine (HBNQ) and liquid gun propellant, to assist in developing new explosives, and to work on new chemical plant design. The center for ordnance devices expanded into new devices, including cutters and retrorockets, and continued to provide services in manufacture, survey, evaluation, and design. The center for missile weapons simulators constantly reported more installations of their training shapes, new missile designs simulated, and feedback from trainers around the country. The center for explosive process development engineering worked on design of new explosive warheads, worked with Dahlgren and other testing facilities on experimental explosives, provided technical assistance to commercial and single-service managed production facilities, and worked closely with Air Force and Army as well as Navy researchers. The center for environment and safety, which was officially designated as a center of excellence on 1 October 1988, worked with OSHA—the federal Occupational Safety and Health Administration, with the Environmental Protection Agency, and with other federal and state agencies. The center worked on training and on developing rules and procedures for limiting the risks of ordnance materials to individuals and to the environment.

GLASSWARE TO STAINLESS STEEL; LABORATORY TO PRODUCTION

As Monetta left the Station in 1989 to take a position in the Department of Energy, he explicitly stated his views about the role that the Station had come to play since the middle 1970s. In the areas of weapons development and energetic materials, the competitive model of purchase from the lowest bidder could not apply. Instead, the government had to assure that it was obtaining the safest and most reliable product, and preproduction engineering had to be performed "in-house." After the process was understood and specified, it was appropriate in many cases for the private sector to compete. Fair competition required that potential bidders knew that the process and specifications were accurately described and that the method of getting the product "from glassware containers to stainless steel containers" had been achieved by an independent, objective, sophisticated, and reliable facility. Getting the process from glass to steel—from the laboratory to the production line—would be the role for Indian Head. Further, if the Station could build a national and international reputation as the U.S. Center for Energetic Materials, it could soon acquire responsibilities not only for Defense Department propellant and explosives work but for other agencies as well.

Both Dave Lee and Dominic Monetta recognized that growth into new areas dealing with the complex and sophisticated world of procurement required a considerable expansion of administration. Following a trend initiated by Browning in developing younger management talent, they moved away from relying on the technical director alone. Both directors used a series of boards, planning groups, and middle-management positions to develop a structure that could handle the complexities and be responsive to the Navy's multiple needs in the area of ordnance. While production was still at the heart of much of the work at Indian Head in chemicals, propellants, and devices, it was clear that earlier recognition by Browning and Lee of the managerial and engineering tasks as a high and growing percentage of the work at Indian Head had been accurate.

One consequence of the concentration on management tasks was that the total money spent on indirect, management activities had increased. Over the years it became apparent through the Navy's cost accounting system that there was a dollar price attached to the increased quality. While clients ultimately would find themselves served with a higher degree of responsiveness, better quality, and better engineering, they would also find that they had to pay more for what they got.

As world tensions declined during 1988 and 1989, with the Soviet withdrawal from Afghanistan, with democratic reforms in Eastern Europe, and with *glasnost* and *perestroika* suggesting that the relations with the Soviet Union would improve, all defense installations found themselves facing new budgetary conservatism. Base-closing lists, reduced military expenditures, and a search for ways to cut the federal budget in the defense sector would once again, as they had in the 1920s and the 1940s, place a premium on the mean and lean facility rather than those heavy with administrative overhead. As new leadership came aboard in 1989, the Station once again faced the question of how to maintain its excellent reputation and its claim to fame in several areas at a reasonable cost.

CHAPTER 11

The Second Century Begins

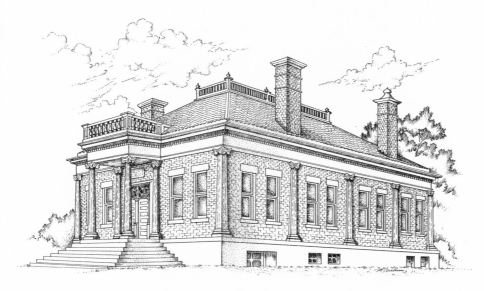

Chemical Laboratory
Building 101
1899

SHROUDED IN DAMP fogs that drift up from the Potomac River, the aging brick buildings of the naval compound seemed to slumber in complacent disdain for the changes that swept the nation, the defense establishment, and the world's economy. After a century, the enclave at Indian Head still felt remote from Washington, D.C., despite the four-lane State Highway 210 that tied it to the Capital Beltway.

Behind the chain link fence that ran across the neck of the Indian Head peninsula, the Ordnance Station stretched into the woods like a quiet and secluded private park. Past the guarded gate, beneath the dark southern pines and towering oaks, neatly raked lawns surrounded the scattered accumulation of a century of building styles. The structures were an eclectic bunch, ranging from the original brick homes in which Dr. George Patterson and Lieutenant Commander Joseph Strauss had once resided, to factory plants converted to offices, to brick supply sheds hunkering beside long-quiet railroad sidings. The very names of some of the structures echoed almost-forgotten technologies—a powder drying shed here, a mixing plant there. Isolated picnic tables and a small boat dock nestled below the Valley

213

Storehouse, Building 54, built in 1908, in the valley where Ensign Dashiell's first guns had thundered next to the river front in 1890. One might suspect the ordnance station was a thing of the past, quietly wrapped in its shroud of history.

However, appearances were deceiving, for behind the forested façade, the living pulse of innovation and production still throbbed, and the Indian Head activity constantly adapted to the whirlpool of forces that swept the nation, the Navy, and the ordnance community.

A DECADE OF TURBULENCE

In 1989, 3,000 people labored at the research, development, test, evaluation, and manufacturing technology of energetic materials at the Indian Head Naval Ordnance Station. Workers compounded powder and propellants for the fleet here; they loaded shells and rockets; others carefully crafted the hundreds of different types of cartridge- and propellant-actuated devices crucial to the safe operation of all modern military aircraft. Increasingly, electronic simulation replaced physical testing, reducing risk and saving money, while offering many of the same valid test results. The station remained the largest employer in Charles County.

But deeper changes and potentially destructive forces were afoot. The period from the end of the Cold War through the end of the century, slightly more than a decade (1989-2000), would become a period of wrenching transition for the U.S. defense establishment. This "decade-plus" would be characterized by defense spending cutbacks and government restructuring, by a vastly reduced ordnance budget, and by increasing concern with environment and safety. As the tensions of the Cold War diminished and then, in a short two-year period, seemed to evaporate entirely, the Department of Defense ordnance budget would be cut by three-fourths, from over a billion dollars to less than 300 million. Although historically the late 1940s had been identified as the "post-war years," the ensuing 40-year Cold War had structured the American economy around the defense budget and the maintenance of a massive military-industrial infrastructure. But by the 1990s, the United States moved into a true "post-war" period, even more clearly than the post-Vietnam years during which the Cold War had kept the defense establishment and especially the Navy at high alert. To realize the benefits of the "peace dividend," military procurement would shrink, defense civilian and military personnel totals would be cut back, and government real estate would be shut down and sold off.[1]

Congress approved the Base Realignment and Closure procedures of 1991, 1993, and 1995. Under these "BRACs," a commission would decide on a list of military base closures that Congress could only approve or disapprove as a package, to prevent state delegations from voting in protection of local constituencies. In a wave of economy measures, the BRACs closed and consolidated one military facility after another, and sheer historical longevity was no protection against the cutbacks. Nearby, three venerable facilities fell victim to closure under BRAC 95. The Navy's White Oak Laboratory in Silver Spring, Maryland, that had opened in 1923, the Annapolis branch of the David Taylor Research Center that could trace its origins to 1906, and the Philadelphia Navy Yard that had built and repaired naval ships since 1801, were all closed within two years. Bottom-up reviews and "data calls" required every military installation to demonstrate why it should remain open, and more painfully, how its closure could be cost-effective. To survive a BRAC required extensive analysis, intelligent strategic understanding, and patient production of rationales, documents, and reports to be reviewed by the Navy and Defense Department. Some community support could also help.[2]

With declines in the budget for the Department of Defense through the 1990s, Reductions-in-Force (RIFs) swept through the Defense establishment. Although a reduction in personnel might in theory allow management to eliminate the least productive or valuable employees, in practice, RIFs seemed to cut into clusters of talent that managers would rather retain. For Navy shore facility personnel, RIF and BRAC became the most dreaded acronyms of the decade.

Other external changes would confront all military installations. In a frenzy of technological innovation, the nation and the world became wired together in the Internet, an information network that held great promise for revolutionizing communication and the conduct of business. Although Desert Storm in 1991 led to new definitions of military need, the overall decline in the purchase of munitions led to shrinking budgets. Through most of the 1990s, ordnance procurement dried to a small but steady trickle of training rounds and replacement of aging stock.

Several other technical "drivers" in the form of changed requirements, some already apparent in the 1980s, became more urgent: the toxic chemicals of explosives and propellants and their production processes were not environmentally-friendly, and they had to be changed. Munitions aboard ship posed danger to those who sailed with them as well as to potential enemies, and rendering them insensitive to shock, fire, electrostatic and radio emissions led to many projects and developments. Miniaturization of electronics, the impact of weapons testing on the marine environment and a few specific "lessons-learned" from Desert Storm added further new technical challenges and opportunities. At the same time, the Navy continued to seek ordnance that was more effective, lighter, and less hazardous to handle.

Some administrative factors that had already shown up in the 1980s, such as increasing reliance on "outsourcing" or purchase from the private sector came to fruition in the 1990s. The Goldwater-Nichols Act of 1985 had led to a different and competitive environment in which naval procurement officers sought the best prices and lowest overhead from suppliers within and outside the Navy. Wherever possible, commercial-off-the-shelf items had to be substituted for unique, government-only designs. Collectively known as "acquisition reform," the changed ways required that managers adapt to a new culture, more akin to the practices of the business community, than to the long-established traditions of government and Navy.

Through the Clinton-Gore administration (1993-2001), the federal administration gave great emphasis to "re-engineering" or "re-inventing" government. These re-engineering efforts took the form of looking more thoroughly at efficiency of service, at defining "customers," emulating participatory management styles of business, and reducing waste. "Customer" as a word, did not exist in government lexicons in 1981; two decades later, government administrators had to identify their customers and work to satisfy them. Despite widespread skepticism about the effectiveness of such governmental re-invention, by the end of the decade, the reforms began to produce results.

Within the Navy, the shifting of research and development budgets from the systems commands to independent weapons program officers, underway in the 1980s, put the laboratories and R&D centers in a weakened position. The change meant that no longer did those designing and ordering ordnance have a vested interest in protecting a particular in-house facility if it were less efficient than other government facilities or private competitors. If a program officer shifted development or production work away from Indian Head to another facility, the administrative and overhead costs for maintenance and for management at Indian Head would have to shrink as well. If they did not, the increased cost of product, just as in private business, would reduce the competitive position of the station, in a downward spiral.

In 1992, the Indian Head Naval Ordnance Station was administratively reassigned and renamed as the Indian Head Division, Naval Surface Warfare Center (IHDiv-NSWC). At the IH Division, managers struggled with the barriers to customer satisfaction imposed by old internal structural divisions, colloquially known as the "stovepipe" structure of separate departments, each pursuing support independently and not finding inter-disciplinary opportunities or efficiencies. Several different Indian Head departments might approach a single customer in a project office, each seeking a different piece of a single development project. Building integrated product teams (IPTs) that cut across departmental boundaries would appear on paper to address that issue, but making teams work effectively depended on the ability of people to adapt to change. A new drive to actively market the Division's capabilities, rather than simply waiting for the assignment of work, had to be developed.

Indian Head had to convert its culture that had been a century in the making, to fit into a suddenly different world. Everything outside the fence changed in the 1990s: a new relationship with headquarters, smaller budgets, different procurement rules, revolutionized communications, a truly post-war world, new technical demands, government reinvention, and the looming threat of closure. Inside the fence, could the pulse keep beating?

Fundamental cultural traits stood in the way. All institutions are saddled with a degree of conservatism, weighed down by their history, making resilient reactions to change difficult. At Indian Head that conservatism was exacerbated by the fact that a century of manufacturing, storing, testing, packing, and shipping energetic materials produced a culture that was "risk-averse." When handling explosives, zero tolerance for risk was not only wise, but also a matter of life and death. A single mistake could damage facilities, or worse, injure or kill staff members. Yet when faced with changes of the magnitude encountered in the 1990s, organizations had to adapt or die. A risk-averse, conservative mind-set, a long-bred survival trait, could prove fatal when times changed so rapidly.

Yet there were features of the Indian Head culture that set it apart from other Navy laboratories and R&D centers in the region and that worked to keep it adaptable, adjusting, and alive. Joe Browning, Dave Lee, and Dominic Monetta provided their successors, Roger Smith and Mary Lacey with specific management tools and devices that could help navigate Indian Head into the 21st century. Browning's emphasis on grooming younger researchers for management positions had produced more than one generation of internal leaders, experts in the technical side and experienced in administration. Lee's Program Management initiative got people to work around the stovepiped, vertical administrative structures. Monetta's stress on strategic planning and analysis strengthened the ability of the staff to examine where they were and where they were going. In the period 1986-1989, Monetta had helped move the Indian Head self-image from that of a Navy production facility to an RDT&E laboratory for the entire federal government in the field of energetic materials. The crew was wise and experienced, but the winds of change blew strong in the first decade of Indian Head's second century.

NAVIGATING INTO A NEW CENTURY

The heritage that Joe Browning, Dave Lee, and Dominic Monetta left, of growing and developing internal managers among the younger technical experts, was one of several strengths that helped explain the survival of Indian Head through all of the changes and external pressures of the 1990s. The Assistant Management Board (later restructured under Dave Lee as the Professional Development Council), originally influenced by Browning's devotion to Charles P. McCormick's prophetic *The Power of People,* was firmly embedded in the station's management tradition. The AMB/PDC provided a constant cadre of in-house managers. Branch managers and Department heads all had years of experience on the technical side, but at the same time, training, understanding and a taste for administrative issues, strategic analysis, and planning.

Roger Smith, from his arrival in October 1989, through his sudden death in January 1999, brought a management style that was flexible and well suited to changing times. From his first days at Indian Head, Roger Smith impressed others as a caring, even compassionate administrator, with an ironic sense of humor that soon endeared him to many of the staff. Soon after his arrival, he began to use the base monthly newspaper as an outlet for his views and ideas, at first through interviews, and then, more frequently, through a column he wrote. Smith held the BS in Physics from Dickinson College, Pennsylvania, and the MS in Physics (Acoustics) from American University in Washington, D.C. Smith's background with the Navy went back to 1962, when he began in the Bureau of Ships, as a GS-5 in Mine Countermeasures Research. He spent the next 17 years in mine warfare, with an assignment that took him to Vietnam at the end of the war. He served as a technical advisor in Haiphong in the clearing of mines, as part of the peace settlement. In 1979, Smith left mine warfare and entered anti-submarine warfare at NAVSEA. In 1982, he took a Senior Executive Service (SES) position as technical director for undersea warfare. In 1988, he left the position due to a ruling by Secretary of Navy John Lehman that SES personnel rotate out of their jobs every five years. He went to Keyport Washington, where he served as Technical Director of the Naval Undersea Warfare Engineering Station.[3]

Smith described himself as being "anti-bureaucrat," despite his more than two decades in government service. On his arrival at Indian Head, Smith explained through the base newspaper that he hoped to serve as a "coach, not the star player." He said that he sought to breed initiative through the delegation of authority. He pointed out, "Many people delegate work—and there are two bad things about that.

If you delegate work, you get it back again and people don't think. On the other hand, if you delegate responsibility, you take full advantage of everyone's brainpower. So what it boils down to is that managers who delegate responsibility become more of a coach than a player." During his two years at Keyport, he developed his coaching style and regarded it as part of the Total Quality Management (TQM) method that had become popular through the late 1980s. He hoped to spread TQM ideas at Indian Head. In his first interview with the base newspaper, Smith revealed his personal qualities, as well as his management concerns—his admiration for "Far Side" cartoons, his hobbies of squash, racquetball, mountain climbing, wood-working, and eating. He showed a light-hearted side. "I need to have fun and everybody should have fun, too. It doesn't mean we aren't serious about what we are doing."[4]

As Smith settled into the position, it became clear that his first remarks about coaching, TQM, and having fun were all genuine. TQM training began November 19, 1990. Smith appointed a group of 26 employees to serve as Total Quality Facilitators. Smith explained that in his view, the core of TQM was to get people to think for themselves. "Within DOD, there are a lot of changes going on, and many organizations will not survive. We have to strive for 100 percent customer satisfaction. A pilot will not be satisfied with an ejection seat that only works 75 percent of the time."[5]

After a surge of demand created by Operation Desert Storm in early 1991, it became clear that the facility was in for a period of decline. Smith warned his managers, using TQM methods, to adapt to the anticipated changes. In July 1991, both he and Captain Edwin P. Nicholson urged the attendees at a Program Management review to seek to expand the customer base, even as downsizing occurred within the defense establishment. Smith believed managers should avoid complacency and recognize that the 1990s would be a period of greater challenge for the ordnance station. The "strategic thrusts" outlined at the July 1991 meeting showed that Smith was well aware of many of the local consequences of broad national and international developments. To enhance the station's role, he said, managers had to seek expansion of the customer base, obtain recognition as a leader in environmental technology, achieve a full spectrum of technical knowledge, modernize facilities, develop personnel, and achieve performance improvement in cost, schedule, and quality of products, goods, and services. These management goals were not simply empty phrases, for over the next few years, the facility took on specific measures to get new customers, to win recognition in the environmental area, to modernize facilities and to work at performance improvements.[6]

Smith's monthly columns in the base newspaper, *Flash Point*, revealed the independent mind and sometimes quirky sense of humor that endeared him to those who worked closely with him. Soon many of the more than 2000 workers at Indian Head looked forward to opening the paper to page three to see what Smith had to say this time around. He took on serious and light subjects alike. In September 1991, he noted that the number of people with "Smith" as a family name at Indian Head lagged behind those named Johnson. The Smiths would have to work harder at it, he said. In a more serious vein, in October 1991, he wrote an editorial about the annual performance ratings given by supervisors to those working under them. He found the system flawed, and based on false assumptions. He suggested several ways to "minimize the pain."

The performance appraisals did not recognize that poor performance was most frequently due to systemic problems, such as the work system, poor training, or being in the wrong job. He especially suggested that supervisors ask how they could make it easier for staff to perform, not to focus on ratings, and to concentrate on rewarding teamwork. This editorial, among others, was often remembered as evidence that Smith himself was willing to think "outside of the box," just as he wanted others to do. At the same time, some managers who believed that performance ratings were one of the few disciplinary measures available to supervisors felt that his position could weaken their authority.[7]

Other columns ranged over a variety of topics, serious and amusing. One urged employees to treat customers like royalty while others humorously compared a list of "What's in" with "What's out." In December 1991, for example, pizza was in and beef was out; environmental protection was in and pollution was out; e-mail was in and memos were out; fun at work was in and management by fear was out; and so forth. A later, well-remembered column dealt with the issue that engineers were regarded as "nerds." In a tongue-in-cheek essay, Smith defended "nerdliness," complete with a photograph of himself with taped eyeglasses, plastic shirt pocket protector, and ill-matched clothes. His ability to relax, have fun, and to engage in self-deprecating humor, made for a dramatic and refreshing contrast with most naval civilian administrators.[8]

When the Naval Ordnance Station was officially re-designated as IH Division, Naval Surface Warfare Center (IHDiv-NSWC) in January 1992, Smith gave a balanced appraisal of the new structure. The Indian Head Division was one of five divisions within NSWC—the other four being Dahlgren, Virginia; Crane Indiana; Pt. Hueneme, California; and Carderock, Maryland. Altogether under the new arrangement there were thirteen field activities in the NSWC, with eight of the smaller facilities ranked as remote field activities of the larger divisions. Smith had argued that the Indian Head division should be called the "Energetics Division," but all of the divisions and field activities had been designated geographically. Smith pointed out that the outcome was favorable to Indian Head, in that it was not reduced to a subsidiary field activity of another major laboratory, but stood alone as a Division, organizationally equivalent to Carderock, Dahlgren, Crane, and Pt. Hueneme.[9]

The many changes and shifts in the defense picture began to seriously affect morale in the early 1990s, despite Smith's best efforts to stress the positive and to provide elements of good humor and TQM recognition of individual responsibility. He continued to try to counter the negativity in a number of ways. One was to stress the central mission: work in Energetics. He noted that Indian Head was an "essential element of the ordnance industrial base and will become increasingly important as ordnance companies go out of business." He looked on the bright side in other ways. The mission statement recognized Indian Head as having a full range of functions in the area of energetic systems. The Army was asking for IH production of the 2.75-inch rocket and Low Vulnerability Ammunition (LOVA) gun propellant. Later in 1992, Indian Head received designation as the Tri-Service Center for the 2.75-inch rocket. The construction contract for the new cast plant was underway. Indian Head was gaining recognition in the environmental area. Despite the "gloom and doom" sentiment regarding defense cutbacks, Smith made every effort to view the glass as half-full, rather than take the pessimistic view that it was half-empty.[10]

BUFFETED BY BRAC

Even with his optimism, however, it was a trying time. As BRAC 93 approached, the Indian Head Corporate Operations Office gathered from each department massive quantities of information to provide the Navy with reasons why Indian Head should keep its mission, identifying its unique strengths and stressing those features of its location that could not be replicated elsewhere. Even two apparently contradictory aspects of Indian Head could both be presented as positive features. The IH Division was close to Washington, so that personnel from NAVSEA and the Pentagon did not have to go on overnight travel to visit, and similarly, staff from Indian Head could attend meetings in Crystal City or Washington on a day trip. On the other hand, Indian Head was isolated, surrounded as a peninsula by the Potomac and Mattawoman Creek, limiting the degree of suburban encroachment. The facility had production capabilities, allowing for "surge" production in time of emergency, keeping its similarity to an industrial production facility in the private sector.

The Corporate Operations Office compiled further justifications for continuing to operate Indian Head, amassing details and statistics for the data calls. The body of experience in dealing with energetic materials, together with the unique production facilities all added up to further lines of argument. Specialists could help the Navy be a smart buyer of technical products. They served as experts on investigation teams when there were accidents or storage anomalies. In one specialized discipline after another, Indian Head had been designated as a "Center of Excellence" during the Monetta years. Closing or moving the facility would result in loss of talent as personnel took other positions and as teams were dispersed; some equipment simply could not be dismantled and relocated.

All of the reasons for continued operation were true, and in the end, they were well-received. Nevertheless, the defensive exercise was nerve-wracking to those developing the documentation and following the efforts of NAVAIR and the China Lake Naval Air Weapons Center to consolidate ordnance work in the California laboratory. Support from local political and community leaders and friends in NAVSEA all could help.

Re-districting the Congressional district as a result of the 1990 census combined Charles County with Prince Georges County, resulting in Indian Head being represented in Congress by Democrat Steny Hoyer. Considering the generally warmer support that the military received from Republicans, many at Indian Head viewed this change with trepidation, but Hoyer soon turned out to be a strong advocate for his new constituents. Smith and the commanding officers who succeeded Nicholson all welcomed Hoyer on many visits: Capt. David G. Maxwell (1992-1994), Captain Wayne J. Newton (1994-1997), and Capt. John J. Walsh (1997-2000). Whenever a new facility was opened or a major award presented, Hoyer received an invitation, and he responded often. Several times a year through the decade, *Flash Point* featured pictures of the current CO together with Roger Smith and Steny Hoyer, or with Senator Paul Sarbanes or Barbara Milkulski. Although the IH Division could not "lobby" for its position as a government agency, there was no rule against showing hospitality to a member of Congress nor any rule against a member of Congress visiting a federal facility in his or her district.

Despite the fact that the BRAC procedure left many feeling defensive, Indian Head benefited in specific and concrete ways from the downsizing of institutions through the decade. Although many administrators remembered the BRACs as periods of crisis, in fact the Indian Head Division actually came out ahead as a consequence of the closure and realignment decisions. In two separate moves, hundreds of new personnel came to the division from units that it absorbed.

An explosive incident at White Oak, Maryland, on 28 June 1992, put that laboratory in the news. Encroaching suburbs north of the Washington Beltway had pushed the metro area right up next to the once countryside location of that laboratory over the decades. Despite the fact that the government promptly paid claims for a few broken or cracked windows in neighboring housing developments, the detonation gave strength to arguments for closure. White Oak fell victim to BRAC 93. Personnel were shifted in fiscal year 1994, some to Dahlgren, some to the Carderock Division (previously known as the David Taylor Research Center), and some to Indian Head. Over 250 researchers from White Oak's Explosive Development and Underwater Warheads branch, part of the Dahlgren Division's detachment at White Oak, were added to the Indian Head roster. This addition helped to enhance Indian Head's research capability, and brought it closer to the "full-spectrum" model that included research, development, test and evaluation, production, and technical service to the fleet.[11] With the incorporation of the new people into Indian Head, Dr. Kurt F. Mueller was appointed head of the Energetic Materials Research and Technology Department, Code 90, and Bill Wassman was chosen as Head of the Underwater Warheads Technology and Development Department.[12]

The second accretion of personnel did not result directly from a BRAC, but from the fact that various ordnance facilities were "orphaned" as a result of restructuring. John Trick, working under Roger Smith, negotiated the transfer of four such detachments on 27 September 1998. The Naval Ordnance Center, a short-lived administrative unit established by NAVSEA in the wake of Desert Storm to improve logistics to the warfighter, wanted to "offload" the detachments. Trick developed the rationale for their shift to Indian Head with the NSWC command. The logic was that Indian Head could provide just the sort of technical expertise in the chain of command above them that they needed.

The four detachments all added separate strengths. All four were part of the Quality Evaluation staff of the Naval Ordnance Center. At Concord, California, a unit brought expertise in non-destructive evaluation, while another, at Seal Beach, California, specialized in Strategic Systems. Earle, New Jersey housed engineers designing and evaluating systems for packaging, handling, storage and transport. Another unit at Yorktown specializing in airborne weapons was added. The NSWC Crane Division, in Indiana, got one of the five Naval Ordnance Center detachments—one at Fall Brook, but Indian Head succeeded in adding the other four.[13]

John Trick later credited Roger Smith, both for his understanding of how these additions would enhance the full-spectrum capability of Indian Head, and for his many contacts developed through his long career in naval technical administration that helped facilitate the transfers. Altogether, nearly 240 new personnel came from the four technical detachments.[14] Together with two other existing remote detachments, a strategic weapons group at the McAlester Army Ammunition Plant in Oklahoma, and the Explosives Engineering group at Yorktown, by 1998, Indian Head now had detachments nationwide. The division could demonstrate that it was substantively involved in almost all functions related to energetic materials, supporting its claim to be the National Center for Energetics.

Detachments	
Location	Specialty
Earle, NJ	Packaging, Handling, Storage, Transport
Concord, CA	Quality Evaluation Technology
Seal Beach, CA	Strategic Weapons
Yorktown, VA	Airborne Weapons
McAlester, OK	Special (nuclear) Weapons

[Table Source: 1999 Command History]

Although buffeted by the BRAC process, in the end, Indian Head not only survived the structural reorganizations, but it emerged in a stronger position. After the 1992 reorganization, Indian Head was a co-equal NSWC division; it had a nationwide presence with its remote detachments; it had established its role as the National Center for Energetics. In keeping with the Navy's emphasis on full-spectrum Research, Development, Testing and Evaluation (RDT&E) centers, Indian Head had moved in two directions along the scale of RDT&E. Personnel from White Oak enhanced capabilities on R and D side, while the quality evaluation personnel at the detachments strengthened the "E" end of the RDT&E spectrum.

GROWING ROLE IN A SHRINKING BUSINESS

Perhaps even more fundamental than the management issues and personal management styles were several technical and economic factors that contributed to the survival of Indian Head while other nearby naval facilities closed under the rounds of BRAC. The very fact that the ordnance budget shrank to a third of its former size over the decade meant that the private sector turned away from ordnance production, as Smith had recognized as early as 1992. Weapons, propellants, explosives, rocket and torpedo fuel, and cartridge-actuated devices for defense no longer represented a huge market. Company after company went out of business or converted to other products. In some commodities, Indian Head produced a specialty product in a smaller market, and the Navy simply had fewer and fewer places to turn for the products, the services, and the expertise. Program officers at the Naval Sea Systems Command and at the Office of Naval Research could find the research and development and production facilities at very few places, and the Navy's own facility at Indian Head was prominent among them.

The consequence for staff could be nerve-wracking. The shrinking private sector participation in ordnance meant that for the technicians, engineers, chemists, and "rocket scientists" of Indian Head, there were fewer and fewer places competing for their talents. Roger Smith and Mary Lacey both recognized that there was not likely to be a "brain drain" of the key people with technical expertise from Indian Head to the private sector or to other government facilities during a time of ordnance retrenchment. While limiting the job opportunities for ordnance specialists, the changing economic environment could help the institution as it became somewhat easier to retain talent. So in these ways, Indian Head benefited from the hard times in ordnance. The half-empty glass could be seen as half-full.[15]

Beginning in 1991 staff at Indian Head was reduced by about one third, or 1,000 people. Smith could see the RIFs coming, and by starting early, he was able to avoid massive layoffs of less senior personnel. By attrition and early retirements, Smith was forced to lay off only six of the 1,000 employees separated from service. Coupled with reductions in overhead and improved speed of hand-off from development to production in rocket motors, Smith's management style began to gain wider recognition. The base won the U.S. Senate Productivity Award for the State of Maryland in 1994, and the Hammer Award, presented by Vice President Al Gore on 7 October 1997.[16]

Senior managers at Indian Head recognized the broader ordnance-economy changes and worked to use them in their marketing and in their strategic planning. In 1993, Dennis Chappell, of the CAD/PAD department at Indian Head, initiated a request that the U.S. Department of Commerce study the national CAD/PAD industry. In response, the Commerce Department's Office of Strategic Industries and Economic Security, in the Bureau of Export Administration, developed a survey over 1994 and produced in 1995 an industry study. The report found that while non-military applications of CADs and PADs had increased rapidly in the period 1991-1995, military uses had declined. Defense shipments fell from $210 million in 1991 (representing 73 percent of the overall CAD/PAD market), to $178 million in 1995, representing only 42 percent of the total market.[17]

The data in the survey revealed that several firms had retired from the business, and those that remained tended to concentrate on the growing market in automobiles and other private sector applications. With new models of cars equipped with airbags, deployed by CADs, a steady and growing market completely unrelated to defense absorbed the profit-driven private sector. Airbag CADs offered several attractions to manufacturers, most notably the huge orders for single designs. The number of companies in the business had shrunk from about 60 to about 35, owning about 44 establishments or production facilities.

The CAD/PAD office at Indian Head asked the Department of Commerce for a follow-up study in the year 2000. Some of the same trends noted in 1995 continued through the end of the decade. The number of establishments producing CADs and PADs further declined, from 44 to 34. The surviving companies struggled with defense downsizing, larger and more demanding customers, increasingly restrictive environmental policies, and a slow-moving export licensing process. According to the report, "stronger, more aggressive firms gobbled up weaker firms." Increasingly, the private sector tended to abandon the manufacture of the specialized, relatively small numbers of CADs and PADs used in military applications. Other Indian Head Division offices followed up with similar Department of Commerce studies to evaluate changing market conditions.[18]

The work of the Indian Head Division in CADs and PADs, with most applications in aircraft, served not only the Navy, but increasingly, the Air Force as well. In recognition of this joint service role, on 16 April 1998, the CAD/PAD Joint Program Office was established at Indian Head. The Joint Program Office had an official "stand-up" ceremony 4 September 1998, attended by Senator Paul Sarbanes who served as keynote speaker, and by Rear Admiral Jeff Cook, who was the Program Executive Officer in charge of the CAD/PAD joint program.[19]

Dennis Chappell was appointed Deputy Program Manager in PMA-201 Conventional Strike Weapons Program Office, the acquisition authority. At the same time he served as head of the Joint Program Office and as Program Manager for CAD/PAD at Indian Head. The direct channel through the Program Executive Office allowed for more efficient operation; with its increased visibility, the Indian Head CAD/PAD program could get resources to execute programs and follow through effectively. In effect, Indian Head now housed both customer and supplier.

The joint program served to consolidate separate Air Force and Navy programs for sustaining CAD/PAD production and playing a role through the whole life cycle of the commodity. Sustaining production meant more than simply keeping up production, and implied a much broader set of benefits. The Joint Program eliminated unnecessary duplication in engineering, acquisition and testing; it consolidated contracting, and it yielded further savings through combined surveillance testing. Economies of scale, elimination of competition between the services for the increasingly shrinking industrial base, and eventually less proliferation of multiple CAD/PAD types were envisioned as future benefits. As a new application of "jointness" between the services, the assumption by the Navy for Air Force readiness in this area represented a great leap forward in trust between the services.

In the new world of naval acquisition, the sense of marketing services that had developed under Browning and that Dave Lee had advanced became ever more essential under the new procurement environment of the 1990s. Lee had established a Program Management office, with program managers who would work across division lines to sell products and services to customers among the NAVSEA project officers and to special new funding sources at the Office of Naval Research. Other new clients under Monetta's leadership had included offices in the Defense Intelligence Agency and the Central Intelligence Agency. The logic of the Program Management office was simple enough. The stovepiped departments at Indian Head each had a separate focus: Production, Testing, Weapons Engineering, or Weapons Simulation Technology. But customers developing a new weapon would need to be sold on the whole range of capabilities as a package during the process that took a new weapon or new chemical composition through engineering, production, testing, and computer simulations. Hence, Lee's development of a Program Management Office with different areas of concern flourished through the 1980s and 1990s. By 1999, there were six separate Program Managers (PMs): Conventional and Air Weapons, Surface Weapons and Ammunition, Undersea Weapons and Explosives, Environmental and Chemicals, CAD/PAD, and Surveillance and Predictive Technology.[20]

Roger Smith created and supported a number of devices to enhance customer relations and to measure customer satisfaction. One was a Customer Advisory Board, set up in 1992. Another method, developed by the Corporate Operations Office, was to survey customers with a form, established in 1994, that would measure customer satisfaction on a number of vectors, such as cost, quality of work, meeting objectives, and keeping the customer informed. Such measures were subjective, but by comparing results to prior years, a "metric" of performance was soon established. The different PMs could compare their success with customers from year to year and from one PM area to another. After a slight drop in year-to-year overall scores from 1994 to 1995, the scores generally climbed from 1995 through 1998.[21]

An initiative of the Program Management Office or "PM shop" was to establish "cornerstone teams" that would supplement and overlap with marketing teams from the various departments. The concept was that the cornerstone teams would cut across several technologies and departments at the division, with focal points on such newly developing areas as unmanned vehicles, micro-electronic mechanical systems, environmental technology, or weapons simulation technology. Three of the cornerstone teams were additionally staffed by outside consultants.

By the admission of participants, the effectiveness of cornerstone marketing teams and departmental marketing teams varied widely. One objective was to get technical people engaged in understanding the marketing side of the activity. The meetings, proposal-writing, discussion of prioritizing strategies, developing tactical objectives and making recommendations for investment of money, all served to expose the scientists, engineers, technicians, and computer programmers to the business side of the enterprise.[22]

The method, although highly structured on paper, had varied success, partially dependent on the personnel involved, and partly dependent on policy changes at the program offices, at NAVSEA, and at ONR. The teams would conduct analysis of the defense environment, including a close look at the DOD budget. Potential business would be identified with customer requirements specified, and new sponsors and markets sought out. After a risk assessment and cost/benefit analysis, select opportunities would be pursued. The teams would develop a Plan of Action and Milestone, and the PM shop would review proposals. The PM Office would track the success and win/loss statistics on the proposals, developing "lessons learned" that would presumably be used in the following year to improve the bid and proposal process. The PM Office would use the evaluation of customer satisfaction and marketing success to shape the next year's marketing plan.[23]

Although the system could engender new business, all the internal work could not alter the fact that decisions beyond the control of the PM shop and the earnest efforts of the team members would shape the expansion and contraction of work at Indian Head. On the one hand, the shrinking defense industry left some project management offices at NAVSEA or ONR with very few alternatives from which to choose. Thus Indian Head could count on steady and even growing business in a few areas, because private sector competition had atrophied, and because the number of government facilities with capabilities in ordnance had been reduced. On the other hand, the overall decline of workload at Indian Head from 1989 to 1999 with overall production decline, led to a reduction in staff, from about 3,000 personnel to about 2,000, with further cuts indicated over the next two years. That shrinkage itself had a deleterious effect on Indian Head's ability to keep costs down, because the necessary minimum of overhead had to be spread across an ever-smaller direct labor base. The Customer Satisfaction survey revealed that the lowest "score" was in the area of "Cost." Internal measures of workload, budget, and current and projected funding all painted a picture of declining size, increasing proportion of overhead, and continuing fiscal crunch. At some point, the IH cost of producing even a unique specialty item would make it attractive for the private sector or another government facility to compete and undercut price.

Through the late 1990s, the effect of the decrease in direct workload was felt by those outside of management in numerous small ways, as the CO and TD cooperated to find ways to cut back on overhead. Staff members emptied their own trash; some maintenance of sidewalks, roads and landscaping was deferred. But there was no easy solution, and law mandated essential parts of overhead. The very buildings and equipment that made Indian Head unique became a burden. Some apparently abandoned buildings contained stored items that could not be disposed of; it was expensive to keep buildings in use, but more expensive or impossible to demolish them. Even a locked and closed structure had to be inspected and checked for fire-safety, maintenance, and repair.[24]

Welcoming outside tenants to come in to share the burden was only a partial solution. Although the 3,200 acres at Indian Head seemed full of available space, much of the compound was marshland, while other wide areas were precluded from use by the "explosive arcs" that surrounded buildings processing or storing energetic materials. In 1999, the EOD School departed, and a few months later a Marine unit, the Chemical Biological Incident Response Force, took up quarters as a new tenant. The Marines brought some advantages such as improvements in "Quality of Life" expenditures, with sports and recreational facilities and increased payroll expended in the surrounding community, but the basic overhead of the base, such as infrastructure maintenance, received no relief.[25]

When Mary Lacey suddenly found herself taking up the reins of leadership after Roger Smith's death in 1999, she confronted a series of challenges, some continuing, some new. With background as an engineer at White Oak, and head of the Science and Technology Division at Dahlgren, she brought precisely the kind of mix of technical and administrative experience that NAVSEA sought for facility Technical Directors, now designated as Executive Directors. Lacey had been a member of the Senior Executive Service since 1996. She had served as program manager for advanced technology in the Office of the Chief of Naval Operations. A graduate of the University of Maryland, she had won the Navy Meritorious Civilian Service Award, and the Navy Women in Science and Engineering Achievement Award. Of course, an immediate issue was establishing her authority, just as Monetta and Smith had been required to do before her. But almost immediately, other issues presented themselves.

Although Indian Head had survived both BRAC 93 and 95, Admiral G. P. Nanos, Commander of NAVSEA, asked Rear Admiral M. Mathis of the Naval Surface Warfare Centers to work jointly with staff from NAVAIR to examine consolidation of facilities and staff between Indian Head and China Lake. Working through channels, Indian Head staff sought to demonstrate that Indian Head would be better situated continuing in its NAVSEA setting, rather than under NAVAIR and consolidated with China Lake. Lacey and other Indian Head staff worked to develop a chain of advocacy to defend their own position, while at the same time, they tried to visualize how to make the best of the consolidation should it occur. Within a year, NAVSEA and the Naval Surface Warfare Centers agreed that they preferred to retain the ordnance research and production capability of Indian Head, rather than to allow it to shift to NAVAIR, and consolidation was deferred, at least for the moment. Meanwhile, Lacey dealt with several other major issues.[26]

In May 1997, the local union had held a quiet election, with 45 members of the 800 represented personnel at the Indian Head division present. Local 1660 of the AFL union, the American Federation of Government Employees (AFGE), was voted out and replaced by AFGE Local 1923. The new local representing the collective bargaining unit employees took a far more active stand than the prior leadership, filing grievances, invoking arbitration, and asking for participation in worker safety violation inspections. Captain John Walsh, whose own background included a stint in the Navy as an enlisted man before finishing college and being commissioned as an officer, found himself engaged in more than one tough controversy with the union. Even the exact number of arbitration cases resolved favorably by each side became a matter of public debate.[27]

When Lacey came aboard, she inherited a friction-filled relationship between management and labor that did not make dealing with other issues, such as survival, downsizing, and overhead expenses, any easier. Thirty months of negotiation had resulted on 18 December 1998, in a negotiated contract between the union and the division. In early January 1999, Walsh and Roger Smith had signed a Settlement Agreement regarding overtime pay between the IH division and AFGE Local 1923. Despite these settlements, prior arbitration cases and the rancorous and long-drawn out negotiations left an uncomfortable legacy that Lacey inherited a few weeks later.[28]

As she worked through the issues of potential consolidation with NAVAIR and with internal labor relations, Mary Lacey let it be known that one of her main administrative goals was mentoring. She defined mentoring as providing career guidance and support to junior staff people, particularly for women and minorities. She believed her emphasis was not so different from the tradition begun under Browning, that of growing internal talent and building from within. Indeed, "mentoring" as she defined it, and "coaching" which Smith had stressed, both represented the type of management that nurtured and relied on the intellectual independence and the talents of subordinates.[29]

Lee, Monetta, and Smith had all made contacts with local educational institutions, and the liaisons continued to bear fruit in Lacey's first years. On 14 December 1998, Indian Head and the University of Maryland, College Park, signed a cooperative agreement establishing the Center for Energetics Concepts Development. The goal of this CECD agreement was to develop a new generation of energetics experts, as well as to share expertise and facilities. Students in the University of Maryland Masters of Engineering program could work with experts in specialized fields at Indian Head, such as fiber optics, explosives manufacture, and a variety of energetics-related R and D areas.[30]

Lacey, the program managers, and department heads still confronted the fundamental issue of finding an adjusted role for the division in the post-war defense environment. The mission of the division and its relationship to the private sector, a recurring question over the prior century, continued to require constant attention and renewed definition. Browning, Lee, Monetta, and Smith had all struggled to navigate the same difficult waters.

RESPONDING TO REQUIREMENTS

As in the prior 100 years of Indian Head's history, technological change and military requirements shaped the technical work schedule at the division. Literally hundreds of specific projects, studies, technical developments, and innovations continued to flow from the division through the decade, but only a few representative items can be detailed here. Although on the whole, the decade of the 1990s was one of retrenchment in ordnance, the short battles of Desert Storm in 1991 did produce a few lessons-learned that had an impact on the ordnance community. Other "drivers" and requirements that shaped the research, development, test, and evaluation agendas of the division sprang from concerns with the environment, and continuing efforts to make munitions safer and less dangerous to the health of those who made and handled them. The advancing front of technology more broadly, especially in the realm of electronics, micro-electronics, computer simulation, and the development of the Internet, all spurred other innovations.

Some issues remained on the ordnance agenda since the 1970s, such as the concern to develop munitions that would not detonate from external heat, an issue brought home by the devastating fire aboard aircraft carrier *Forrestal* in 1967 that killed 134 personnel. Making insensitive munitions (IM) that could withstand fire, and also stand up under the electro-magnetic and static electrical environment aboard aircraft carriers and other ships became an even higher priority item in 1992, when the Navy issued IM as a requirement. Reformulating the rocket propellants for such standby weapons as the 2.75-inch ("Mighty Mouse") and the 5-inch ("Zuni") became a high priority through the 1990s. Further modification to the 2.75-inch required that its adverse, polluting effects be reduced, with methods to eliminate lead from the composition and even to change the formulation of the paint on the casing.[31]

In 1999 Plastic Bonded Explosives (PBX) of several types, developed under the Navy's Insensitive Munitions Advanced Development—High Explosives (IMAD-HE) Program were announced as recently qualified. For each application of PBX, extensive studies evaluated the safety, vulnerability and performance of the explosives. For example, a set of studies evaluated potential explosive fills to replace the existing warhead explosive (70/30 Octol) used in the M72A6 Light Anti-Armor Weapon. The explosives tested were PBXN-9, PBXN-110 and PBXC-129, and two booster explosive materials were also evaluated as part of the study. PBXN-9 was selected as the main charge explosive and PBXN-5 as the booster. In the next phase of the program, 81 PBXN-9 explosive billets were fabricated and the best quality billets were then used in 63 M72A6 warheads. The warheads were tested for flight characteristics and in performance tests against a variety of target arrays. As a result of the test programs, PBXN-9 met or exceeded all performance requirements, and showed better vulnerability characteristics than the previously used 70/30 Octol warheads.[32]

Under the IMAD-HE program, workers at China Lake, Dahlgren, and Indian Head cooperated in finding a number of applications for newly developed insensitive munitions. IH served as the lead laboratory in developing an insensitive explosive, PBXW-17, as a grenade fill to be used in anti-personnel obstacle breaching systems (APOBS). Altogether by the year 2001, more than a dozen new explosives with insensitive qualities had been introduced and deployed in over 43 new weapons using the insensitive munitions developed at Indian Head. The APOBS, under development in 2001, represented a case of cooperation with industry, in which Indian Head not only provided the new formulation originally used in the 5-inch 54 caliber projectile, but worked with developing a new weapon system, using braided nylon line to contain the grenades for the breaching system. PBXW-17 contained 94 percent RDX and served as a low-cost alternative to other explosives.[33]

In recognition for the insensitive munitions work, Anh Duong at Indian Head received the ONR Arthur Bisson Prize for Naval Technology Achievement for the transition of insensitive PBX into a total of 16 weapons, representing the teams at Indian Head who had worked on the PBX compositions. The prize was named for the late Dr. Arthur E. Bisson (1940-1996) who had established a reputation in transitioning science and technology to naval applications. The award was granted annually to individuals in the Department of the Navy whose program best exemplified the qualities of Dr. Bisson's achievements, through having a major impact widely recognized in the Navy and the nationwide technical community. ONR presented two Bisson prizes on 16 May 2000, one to Indian Head for the insensitive munitions work, and the other to the team that developed the Advanced Enclosed Mast/Sensor System, an innovative radio and radar mast design for future ships.[34]

During Desert Storm, the Mark 57 Destructor, an anti-mine device, helped clear minefields. The Indian Head Pilot Plant and the Cast Plant stepped in as the sole producer of the Mark 14 explosive component of the Destructor and the MK 47 acoustic safe-arm device, when the industry producer failed to produce.[35]

However, Marines found Iraqi deployment of thousands of shallow water mines impaired their chances of making an amphibious landing, and a new requirement for shallow water mine breaching devices drove technology through the early 1990s. Under ONR funding and then under anti-mine warfare funding, Indian Head researchers developed "Distributed Explosive Technology" systems. A device that would fire rockets carrying a deployed explosive line charge in the form of a net was developed and successfully demonstrated by 1996. Although not made a major acquisition by the Navy or Marines, the pioneering work in this technology helped establish the IH division with a place in the niche of line charges used in mine-clearing.[36]

In addition to testing inert versions of the system, which used nylon cord instead of the detonating cord, researchers developed and worked with a computer simulation model. Automatic Dynamic Analysis of Mechanical Systems (ADAMS) was a deployment model developed to represent the surf zone array, based on an earlier rocket-deployed array computer model. The model included equations, based on empirical and physics principles, representing the aerodynamic forces, the structural members' mechanical properties, line tensions, and features of the launch platform. ADAMS allowed the researchers to determine the stresses on the lines that would break the harness, bridle, and other elements of the array. As a consequence, without the expense of running a set of physical tests, researchers could recommend 13 modifications and improvements to the array system, such as adding fins to the rocket motors, modifying the harness, adding parachutes, and so forth. Such virtual experimentation, using known factors and subjecting the materials to a variety of conditions, very clearly illustrated how far simulation had developed and how it could save time, money, and effort in the weapons development process.[37]

As in prior eras, defense technologies could often find peaceful applications. One of the computer codes developed to explore the effect of the surf zone arrays was called Line of Bombs Effectiveness, or LOBE code. The LOBE code predicted the formation of craters and channels through the use of empirical scaling, a technique based on the principle that the behavior of one system could be predicted from the behavior of another similar system by multiplying some empirically derived scale factor. Of course, just such principles had been in use in hydrodynamic modeling since the days of William Froude in the 19th century, and aerodynamic scaling since the work of Samuel Langley and the Wright brothers. LOBE applied a scaling technique to cratering phenomena, using a library of empirically derived data. The code included such variables as the size of the charge, the depth of the charge in the soil, the amount of water overburden and the spacing of the charges, applying scaling principles to predict the depth of the resulting channel. Soil types were only differentiated in general categories as hard, medium, or soft, noted as an area for future improvement of the code. Possible commercial applications of the use of lines of charges for canals suggested that the LOBE program and its results might be used outside the military.[38]

At the Indian Head Pilot Plant, experimental lots of new propellants and explosives had been made in two batch processing systems. The plastic-like mixes of propellants would have consistencies running from pancake batter through very thick bread dough. The more fluid mixes would be processed through a vertical mixer, while the more viscous, thicker mixes were processed through a horizontal mixer. The fluid mixes would be appropriate for cast propellants and explosives and the more viscous materials would be extruded in spaghetti-like strands and then chopped into grains for gun propellants. The knowledge-base for batch manufacturing was based on long experience and the accumulation of shop know-how; very little science had informed the process. The plant had acquired new equipment in 1985, a twin-screw mixer/extruder to demonstrate continuous processing, rather than batch processing.

In this context, the terms "continuous" and "batch" were slightly deceptive, as an outsider might at first believe that the continuous process would mean large-scale production. However, the twin-screw mixer/extruder made possible flexible, custom production of small experimental lots of material, exactly suited to the experimental engineering technology role of Indian Head. Researchers at Indian Head sought ways to expand the use of the continuous processing system for research into manufacturing technologies.

The end of the Cold War and the decrease in ordnance acquisition reduced even further the incentive in the private sector for scientific research into the mixing and manufacturing technologies of energetic materials. In 1994, Roger Smith obtained funding to establish an Energetics Manufacturing Technology Center of Excellence at Indian Head, to pursue research into the science and technology that lay behind the mixing and production of energetics material. This "ManTech" Center, established under the leadership of John Brough, would bring together the decades of shop experience and know-how with scientific work in the field of rheology, the study of the deformation and flow of matter. Working through contracts in academia, particularly at Stevens Institute in Hoboken, N.J., the ManTech Center began to build an understanding of how the shear imparted to the energetic materials in the processing stages affects the product producibility and performance. [39]

John Brough, as head of ManTech, often pointed out that for the continuous processing technology to return even greater benefits, it would have to be scaled up. Through the 1990s, the department developed the scale-up plans, hoping to move from the 40mm twin-screw extruder to larger equipment. In February 1998, an explosion in the gun propellant batch facility at Indian Head set back the Navy's ability to meet requirements for high performance gun propellants, but the crisis "catalyzed the need to move forward quickly" on the continuous processing plant, according to Brough.

In 1999, the Indian Head Division received an emergency MILCON allocation of $6.5 million to begin construction of a continuous processing scale-up facility, with an 88mm machine. This scale of continuous processing would represent a long-sought departure from the batch processing that required a series of operations involving multiple facilities, wasted material, and safety hazards as the energetic materials were moved from one processing site to another. The new continuous processing plant held promise for reduced labor costs, minimizing waste, enhancing safety, improving quality control, and improving flexibility for processing new materials. In addition, the scaled-up plant would reduce the facility "footprint" and require smaller amounts of solvent. Construction began in May 1999, with dedication of the facility expected in Spring, 2002. [40]

INTEGRATED CIRCUITS AND THE INTERNET

Developments in electronics brought several new areas of work to the division through the 1990s. Microelectromechanical systems (MEMS), which used miniature integrated circuits and related mechanical devices to move switches, taking the printed circuit from the realm of information handling to physical motion, found early application in fuzing and safing devices. Although the MEMS concept went back to the origin of the transistor, actual MEMS devices were first introduced in the mid-1980s. Although Integrated Circuits were basically two-dimensional or planar, MEMS added the third dimension to the technology that allowed the creation of miniature mechanical and miniature electrical components, such as gears, levers, beams, capacitors, coils, and relays. These new components could be used in the design of electromechanical sensors and actuators. As the technology matured, whole new electromechanical systems could be created that would sense a change in the environment and initiate a physical motion. Whole machines could be constructed on the micrometer scale. During 1994, Indian Head personnel began an investigation into adapting MEMS technology for use in the next generation torpedo. Funded by the Office of Naval Research, the seven-year project allowed the evaluation of the performance of individual sensors and actuators, and the identification of critical design issues. The work found application in the small-diameter "anti-torpedo torpedo" being developed at the turn of the century. By the end of the decade, new ordnance applications were being found, as MEMS could be used to implement ordnance supply and inventory conditions and to monitor munitions conditions.[41]

The coming of the digital age provided other revolutionary opportunities. In the CAD/PAD department, a "re-engineering" group brainstormed the question of who exactly was the customer. After deciding that the customer was the individual replacing or installing a piece of equipment, not the procurement officer who placed the order, the group analyzed the 18 steps through which a request from a user found its way through the complex bureaucracy of the Navy. Some 50,000 CAD/PADs were changed out annually. As inventories declined, there were shortages at the wholesale level, and an increase in backorders. Out of the project, the office developed an on-line catalog, a system of verifying the credentials and identity of the ordering individual, and a toll-free telephone dial-in system through which orders could be filled and the necessary notifications and budget adjustments could be made. The original order period had been over four months between request and delivery; with the dial-up system in place, customers received their parts by FedEx within eight days. The achievement was recognized by the presentation of the David Packard Excellence in Acquisition Award to the CAD/PAD Reengineering Team, announced in April 2001, and presented at the Pentagon in September 2001.[42]

Although similar systems had been put in place for inert items, this was the first time a call-up system had been established for re-supplying explosive items. Using best practices in the private sector as a benchmark, the system was operating by December 1998 on a prototype basis.

With the system fully operational, between October 1999 and December 2000, 5,808 orders were placed from 47 bases in the United States and from 16 ships. The re-supply process not only greatly reduced the ordering time but reduced the Fleet burden of paperwork by about 46 work-years per year. The paperless and nearly effortless process saved the Navy about $3.2 million each year. Savings were realized by the reduction in time required by the customer to put in the order, the elimination of a Weapons Department role in requisition, and efficiencies in the new pick-pack-and ship operation at the stock facility.

In many less striking ways, computer practices at the division changed daily life. For example, the base phone book, often out-dated as soon as it was delivered, was no longer published in print form, but made available on-line. Basic reports, such as command histories and management plans, were available in read-only versions from the division's net. The technical library got less physical use as more and more publications were downloadable from the net. As noted by Smith as early as 1991, e-mails had replaced memoranda. By 2000, several technical manuals authored at Indian Head were mounted on the web and available in electronic form, readily updated at lower cost than print. An innovation being developed was a set of training practices, using "virtual classrooms" and "virtual apprenticeships." [43]

ENERGETICS AND THE ENVIRONMENT

A major driver of ordnance technology through the 1990s was concern with the environmental impact of weapons. By their nature, the production, storage, and testing of energetic substances could be disruptive to the environment. And of course, their operational deployment is designed and intended to disrupt, if not destroy, the target environment. Many projects, both large and small, at Indian Head through the 1990s and into the next century reflected the effort to mitigate or eliminate the undesirable environmental and health effects of producing, storing, and testing ordnance without degrading the warfighting requirements.

One study that serves as an example of the new sorts of programs generated by heightened environmental concern focused on the impact of underwater explosive shock testing of the new nuclear submarine, the *Seawolf*, on marine-mammal hearing. The purpose of the test was twofold. One objective was to develop a more efficient method for calculating potential acoustic damage to marine mammal hearing, and the other was to perform calculations for use in preparing an environmental impact statement for the *Seawolf*. Using human, in-air data regarding acoustic damage to hearing, a very conservative Temporary Threshold Shift, indicating discomfort or annoyance from a noise, was established. Extensive literature search, including studies of the hearing of whales and other marine mammals, revealed some baseline information. For example, whales regularly and repeatedly produced sounds in the level of 180 to 185 decibels in lower frequencies (between 1 and 10 Hz). Using these factors and others, an interim criterion line could be established, below which volume and frequency could be regarded as not giving discomfort.[44]

In a massive study conducted at the end of the decade, the Indian Head Division summarized pollution abatement ashore programs regarding ordnance, throughout both Navy and Marine Corps shoreside facilities. A survey with responses from over 200 people at 83 facilities covered a cross-section of professional disciplines involved in the management of energetics and ordnance items. The survey identified 69 concerns, many of them held by several or many respondents. The concerns focused on six major environmental areas: process manufacturing, open burning and open detonation, remediation, wastewater treatment, recycling and reuse, and inerting. A literature survey identified the currently available technologies applicable to the areas of concern, and specific technologies were matched to the concerns. One result of the study was to reveal that existing procedures and research, sometimes with modifications, already existed to deal with many of the repeated issues that came up in the survey. Indeed, some utilized existing, off-the-shelf technologies. The study showed the drastic need for communicating and educating administrators and other responsible individuals about the vast amount of technology and techniques being developed in the environmental area. The study, distributed back to the survey respondents, among others, included useful information such as the telephone numbers of experts who were knowledgeable about specific remedial techniques and new research, as well as bibliographies of reports and studies that addressed some of the concerns.[45]

One promising method for the disposal of surplus or over-age propellants, explosives, and pyrotechnics was the use of these materials to supplement boiler fuels. Under funding from the Strategic Environmental Research and Development Program, Indian Head cooperated with the U.S. Army Environmental Center in studying the processing of energetic materials for destruction. A process suggested by the Army involved dissolving the energetic materials in a solvent, such as toluene, producing a single-phase solution, that is, a true solution, rather than a slurry. This material would then be blended with fuel oil, and the resultant single-phase mixture would be fed to a boiler for combustion. Prior experiments had been limited to dissolving TNT in toluene, but both the Army and Navy wanted to follow up with experiments with Composition B, nitrocellulose, nitroguanadine, High Bulk Density Nitroguanadine, RDX, and Otto Fuel as well as TNT. The various energetic materials were processed with No. 2 fuel oil and burned, with the stack emissions and residues measured and characterized. Safety risks were minimized by reducing the size and concentration of the explosive particles and by keeping the solution in a dynamic or agitated state to prevent any settling. Using a conceptual design of a plant with an existing boiler, cost estimates showed that the method could be quite economical. The study reviewed the necessary permits that would be required under Maryland law to set up such a system.[46]

Responding to official orders and to the changed political environment, Indian Head worked on the topic of "Green Energetics," reducing the toxicity in the final products and in the manufacturing process for propellants and explosives. Sponsored by the Strategic Environmental Research and Development Program, the division conducted an environmental cost analysis for gun propellant processing, investigating the use of supercritical carbon dioxide as a substitute solvent for chemical synthesis of reactive materials, an environmentally benign alternative to organic solvents. The first phase of the experiment used computer modeling, demonstrating the use of process simulation and other models and databases to estimate the costs of the substitution, using a known gun design. The 5-inch, 54 caliber gun system requirements were fed into the programs and databases and two different formulations were evaluated and compared, showing profitability and costs. Following the computer analysis with laboratory tests, the new solvents were used in the preparation of 2,3-dimethyl-2,3-dinitrobutane, a required additive used for marking all plastic explosives. Another test series examined the possibility of using supercritical CO_2 in the production of TAIW, a precursor chemical in the manufacture of the explosive CL-20. The TAIW had low solubility in the supercritical carbon dioxide.[47]

In other studies, experimenters worked with alternatives to the use of lead salts in gun propelling charges. Traditionally, lead had been added because it facilitated the reduction of copper build-up on the interior walls of the gun barrel, when copper was sheared from the rotating bands of shells. A literature survey showed that the process was at first understood to be the formation of an alloy of lead and copper, but that more recent research had demonstrated that the lead melted and dissolved the copper in solution, which was blown from the barrel on the next round. Using that theory, a Canadian researcher had suggested in 1975 that bismuth would represent a good substitute for lead. Protecting workers in the production of NACO gun propellant from exposure to lead salts under modern health and safety rules would entail expensive procedures; if bismuth (often used in medicines) could be substituted, no such added worker protection would be required. Under funding from the Crane Division of NSWC, Indian Head and Dahlgren researchers cooperated in studying the effect of bismuth/tin foil in de-coppering a barrel. Researchers struggled to establish a controlled experiment that would compare the effect of the bismuth/tin foil in de-coppering a barrel, using a specially manufactured batch of NACO propellant without lead. Repeated firings with that powder were needed to get any copper build-up, before a new batch of firings could be held with the foil added to the lead-free powder to see if it would reduce the copper deposit. Despite difficulties in establishing a comparison baseline that seemed similar to issues faced by Dashiell a century before, the researchers felt confident in recommending bismuth/tin foil addition to gun rounds, or eventual mixing of bismuth into the NACO propellant and the elimination of the lead.[48]

CONTINUITIES AND CHANGES

Despite the plethora of dramatic forces for change impinging on the broader defense establishment and the ordnance community, and on Indian Head at the microcosm level, the decade-plus from 1989 through 2000 saw several historical patterns continue. As in prior years, leadership struggled to manage a complicated facility with limited resources while defending the institution and constantly redefining its role and agenda. Technical specialists, including by the end of the decade a wide variety of scientists, engineers, computer specialists, testing technicians and production personnel, found their work shaped by the Navy's changing requirements and by broader developments in the scientific and technical world. As in past decades, the lessons of the last war or two drove a few innovations. Following Vietnam and Desert Storm, the Navy and Marines sought the development of insensitive munitions, mine-clearing devices for shallow water, and methods to clear antipersonnel obstacles. The Navy's awareness of national priorities, as in the past, shaped other programs. Emphasis on protecting the environment and improving worker safety drove new work and produced numerous innovations.

Increasingly, the computer served as a research tool, just as the external invention of other tools in the past had changed the research structure at Indian Head. With the computer as a tool, not only could internal databases be maintained and outside data be more effectively "mined," but also complete experiments could be conducted through the development of simulation codes. Virtual testing of equipment like the shallow water anti-mine arrays allowed improvements in design without the expense of physical construction and testing of dozens of preliminary and intermediate models. Although new and cutting-edge, such application of the computer, as an externally-developed tool, to the work at Indian Head was in the long established pattern of two-way technology transfer between the civilian and military spheres that had enriched American technical life for two centuries.

The struggle to work across practical and theoretical disciplines to produce synergistic results through teamwork, rather than through individual innovation, was itself a century-old heritage at Indian head. Dashiell had at once represented the lone inventor and at the same time had sown the seeds of institutional research and the application of scientific method to ordnance problems. Those rich traditions provided continuities as Indian Head faced further changes at the beginning of its second century.

EPILOGUE

Retrospectives and Prospects

FROM THE DAYS when Ensign Robert Dashiell sat hunched over his table, working by candlelight to perfect a breech mechanism, to the modern, bustling engineering and production activity of Indian Head a century later, a range of technological and management questions have recurred. As the process of innovation in naval weaponry became more complex, out of the growth of both technological sophistication and bureaucratic size, each generation faced similar issues. Each answered them slightly differently.

In Dashiell's day, innovation would come from the hand of a single inventor, backed up by drafting specialists who would convert an idea to plans and a workshop which would turn out the product. Personnel management consisted of direct orders and close supervision of both employees and contractors from the Chief of the Bureau of Ordnance. The relationship with the private sector evolved as the Bureau sought to stimulate the invention of products in its own shops and laboratories and by its own employees, either in or out of uniform. The Navy sought to avoid paying a premium to outside manufacturers or to foreign companies. Secretaries of the Navy like William C. Whitney and Benjamin Tracy could stimulate and even force development of new alloys and new production facilities through contracts. It was through such pressure that Folger got Carnegie and Bethlehem to produce Harveyed steel. As a testing facility, the Proving Ground evaluated the conformity of privately manufactured powder, shot, armor, guns, and mounts to Navy-established specifications.

As an innovator, Dashiell was something of a genius. However, his personally aggressive style made him less than ideal when it came to seeing to the conditions of workers, administering a growing facility, or responding to the arguments of private suppliers. The apparent conflict of role between inventor-businessman and objective judge of products remained unresolved. His immediate successors brought a more moderate style, and continued some of the original testing procedures which he had brought from his earlier assignment at Annapolis and which he had perfected at Indian Head.

The next decade saw the Bureau of Ordnance turning to Indian Head to produce smokeless powder in sufficient quantity to supply the Navy in peacetime, to free it of complete dependence on private manufacturing. The commercial price would be carefully checked against the factory's direct costs to determine whether or not the Navy was being gouged.

The growth of a large Navy-owned, civilian-operated chemical plant went hand in hand with the appearance of new categories of workers—Navy-paid civilian researchers and managers. Partly because such civilians tended to stay on while the officers left, and partly because chemical plant administration held out no opportunity for an ordnance officer to improve his career, the civilians by default began to manage the day-to-day affairs of the factory side of the facility. Dr. George W. Patterson developed a mastery of chemical engineering and convinced commanding officers to make improvements in process and to experiment with varieties of equipment.

Innovation, as a process, began to change as well. In addition to the officers, some of whom experimented with gunnery and steel, the new category of civilian chemists became part of a slowly emerging naval research and development team, seeking more and more to pursue "experimental work." The Navy had recognized since the 1880s that experimentation was essential to a competitive posture in defense. By 1910 the Secretary of the Navy began to include in the budget small amounts of funding especially set aside for ordnance experiments. Much of that money flowed to Indian Head.

As production of smokeless powder picked up, the powder factory could be used to demonstrate that the prices charged by the "powder trust" were excessive. Evidence of the costs of production at Indian Head became part of the argument for dividing the holdings of du Pont and for creating Hercules Powder by court order. However, the issue of exactly what role the federal production capacity should play in a society based upon the private enterprise system remained unresolved and sometimes became the subject of a partisan political debate.

The first two decades at Indian Head saw the facility perform several quite different functions in the emerging procurement complex. The Spanish-American War had demonstrated that the private sector could not supply sufficient smokeless powder for the modern guns of the Navy, and the powder factory at Indian Head was built to fill that need. The Proving Ground still tested private products including armor, guns, shells, and powder for conformity to specifications. In a small way, some of the chemical engineering work done at Indian Head provided an early example of another role: that of process engineering which would be transferred to the private sector. Over the following decades, Indian Head would at one time or another repeat all of the roles which it had played from 1900 to 1912: supplier in emergency, yardstick to evaluate private pricing, testing ground for processes, and federal experimental and testing facility to ensure quality.

The first decades of the station saw several significant successes in innovation. Although an individual invention, the Dashiell breech was also the first developmental project to come from the new facility. It was followed by a rapid succession of improvements, some small and some major. Patterson standardized diphenylamine-stabilized smokeless powder and developed what a later generation would call the base-line statistics for comparing stability tests. The final Navy standard stability test was Patterson's own contribution. Numerous improvements in the multiple steps of making smokeless powder, sometimes initiated at du Pont, sometimes at Indian Head, advanced that industrial process.

World War I brought profound changes. The increase in gun range and rate of testing made the old proving ground completely inadequate, and wartime appropriations allowed for relocation down-river. Specialization at ordnance facilities began a process which eventually led to Indian Head spawning two offspring—the Experimental Ammunition Unit (EAU), which later merged with other groups in the Naval Ordnance Laboratory, and the Lower Station, which became the Dahlgren Proving Ground. The war led to branching out from the production of propellants to other "energetic materials" including warhead explosives and to the opening of the Explosive D Plant.

With laboratory specialization came a more faceless approach. The 1920s saw the appearance of corporate-modified products formally designated with "Mark" and "Mod" numbers, rather than innovations like the "Dashiell breech" or "Harveyed steel" remembered by the names of their inventors. A working agenda of problems in the explosives, propellant, and pyrotechnic areas began to drive the tasks at the EAU at Indian Head in the postwar years. Again, the quiet technical contributions of the station contributed to the Navy's strength, with studies led by H. J. Nichols eventually resulting in the improvement of a time-delay, armor-piercing fuze which became the standard in World War II. Indian Head participated in developmental work on the 5-inch gun which also became a reliable standby in that war.

Specialization at headquarters, with the creation of research desks and powder desks at the Bureau of Ordnance by 1920, foreshadowed the later complexity of dealing with different authorities in the naval purchasing structure. In the interwar years, Indian Head took more of its orders from the "F" desk, which controlled powder. The EAU and Dahlgren began to do more of the work for the research, or "Q" desk, now headed by F. F. Dick, a former Patterson understudy at Indian Head. After 1924, the EAU moved to the Washington Navy Yard, and the Indian Head laboratory concentrated more on chemical production problems than on innovative ordnance questions. Until 1932 Dahlgren remained a branch of the Indian Head station, and close ties between all three facilities during the interwar years often meant that Dahlgren, Indian Head, and the EAU worked on different aspects of the same ordnance agenda items.

National resentment at munitions makers after World War I and an isolationist foreign policy tended to reduce the funding to all of the emerging defense establishment. As a result of those cutbacks, Indian Head went into a period of quiet survival and more gradual growth. During these decades the station provided the fleet with powder, picking up production in the 1930s to outfit the new ships built during the Franklin Roosevelt administrations. In the years between the wars, living conditions for workers at the factory slowly improved, with stores, schools, churches, and housing springing up to make a small, isolated, yet more liveable community. When World War II came, improvements in roads, telephones, and other utilities suddenly made Indian Head a major component in the war effort. Maryland Highway 210, a straight link between Indian Head and Washington, began to change the community and the station.

A push to expand in the direction of research came with World War II. Not only were the chemical experts of Indian Head called upon to participate in explosives and propellant research but, specifically, the development of the extrusion process for rocket grains put Indian Head in on the beginning of the new age of rocket propulsion. The extrusion work required new testing facilities, involving temperature cycling, static firing, and x-ray machinery, for purposes of safety and quality control. The small research laboratory of the chemical department began to augment its role and interests, splitting off to form the Research and Development Department.

During the years of World War II, Indian Head once again gave birth to new institutions. The staff at the Jet Propulsion Laboratory at Indian Head moved on to China Lake, joining others from the California Institute of Technology in establishing the Naval Ordnance Test Station there. The Explosives Investigation Laboratory formed the basis of the Explosive Ordnance Disposal Technical Center (EODTC), remaining as a tenant activity at Indian Head but under a separate administration from the Powder Factory.

In the postwar decade, national differences of opinion between those devoted to pure and applied research were reflected in a small way at Indian Head. Some scientists there hoped to extend their role into research and development and to study the basic physics and chemistry of propellants, propellant combustion, and related unresolved scientific issues. The massive growth of government-owned, contractor-operated laboratories and the expansion of private firms in the weapons business created a hectic world of competition for money in research, development, engineering, production, testing, and evaluation. Modified propellants and Jet Assist Take-Off, or JATO, motors were developed at other laboratories, providing test work and production engineering tasks for Indian Head. As defined by leaders both locally and at headquarters, the main "missions" of Indian Head— to be involved in production and in engineering of the weapons—led the facility away from the research direction.

Research at Indian Head somewhat declined in prestige, skill, and resources in the early 1950s while the Navy expanded the production facilities by building the Biazzi continuous nitroglycerin production plant and other plants. Through the 1950s Indian Head found its mission not in competition with the private sector but in a series of specialized roles. It could provide an independent and impartial facility for the scaling up of laboratory products to full-scale production, and then provide that know-how to the various private sector firms which would competitively bid on the contracts to produce the final product. The concept of taking propellants and explosives "from glassware to steel containers" began to emerge more clearly as the role with the expansion of pilot plants in the late 1950s. The place of Indian Head in the burgeoning system received more definition, and that change was reflected in the name change from Powder Factory to Propellant Plant in 1958.

In the mid- and late 1950s, the drive to build the Polaris missile, through the Special Projects Office (SPO), provided energy and programs for Indian Head. New management techniques, a commitment to explicit goals and deadlines, and an influx of funding began. At Indian Head, Polaris led directly to much-needed modernization of facilities and a burst of recruitment. Joe Browning, one of a young, postwar-educated group of chemists and engineers, thrived on the new program, working for over two years at SPO itself before returning with a new sense of mission to Indian Head.

By the end of the 1950s Browning's attitude, which was one of practical adaptation to the expanding world of missiles, had begun to take hold. Browning's recognition of the need to reform management to better tap the talents of youth inspired the newly recruited generation at the station. The pattern, reforms, and collective memory he left of that concept continued as a strength of the institution for a generation.

As technical director, Browning brought a style of management which won him many loyal followers. Particularly advanced was his participative management principle, giving younger members of the staff opportunities to engage in running programs and solving problems, either individually or through the Assistant Management Board. In addition, he had a quietly charismatic personality which helped sell the services of the institution in Washington, winning program after program to the facility.

In the flush times of the 1960s, Indian Head elaborated its role in several directions, taking on new tasks in chemistry, in development, and in support of other institutions in the complex missile establishment. Technical contributions and achievements came thick and fast, with work on JATOs, with NACO ("Navy Cool") gun propellant, with production of the Mighty Mouse and the Zuni, and with an emerging role in propellant and cartridge actuated devices. But with the decline in the 1970s of money for research, the Station shrank once again, to a static level of about two thousand employees. The facility faced the post-Vietnam years with a rich history but with a need to define its role and to modernize its management.

By the mid-seventies, several changes in naval material procurement brought contraction. Research and Development was now officially declared by the Navy not to be part of the mission of Indian Head. Instead, the Station was to focus on production and engineering. But in many areas of production, the private sector was firmly entrenched. Finding the appropriate niche became a more difficult task as the procurement process became more complex and the number of specialized ordnance products expanded. Inevitably, the private sector sought and held the more profitable pieces of the work, and Indian Head's production opportunities frequently expanded into specialized, short-run, dangerous, or low-margin product lines. Marketing and "customer relations" with the program management offices at the systems commands which provided funding became ever more crucial to the success of the Station.

Indian Head lost administrative control of the Research and Development Department to the Naval Ordnance Laboratory. The ordnance computer center was also administratively severed from the Station during the reorganizations. However, like the EODTC, the Sea Automated Data Systems Activity (SEAADSA) and the Research and Development group remained geographically at Indian Head as "tenant activities."

Despite such changes, Technical Director Dave Lee continued the marketing effort, winning designated agency roles for Indian Head in the engineering work on naval weapons systems. Reorganization of the systems commands at headquarters brought new opportunities. Recognizing the need to organize the marketing of the Station's products and services to customers, he set up a system of Program Management to improve the responsiveness of the Station to the new systems command Program Managers. Engineering required less personnel than production, and Lee sought to maintain a healthy ratio between the two kinds of work.

Through the 1970s and early 1980s, the complex and interacting issues of service to clients, safety, personnel management, quality control, programmed innovation, and environmental impact remained. Technical successes continued to flow, with work on Long Range Bombardment Ammunition, the use of NOSOL (No Solvent) propellants, and the consolidation of the cartridge actuated device work at Indian Head.

In his tenure at Indian Head as technical director, Dominic J. Monetta displayed a driving, productive approach which reshaped the program management tools established in the prior decade and which, he hoped, would better address the continuing issues. Consciously using the excitement and successes of the early Browning years as a model, Monetta hoped to inspire a commitment to excellence. He worked to instill a management vocabulary reflecting a hard-headed model of human behavior in organizations, a structure consciously based on recognition of career motives and dedication to quality.

By 1989 the Station regularly served as a resource for the Navy. Indian Head provided innovative work on the systems of the next decade, many of them classified. Special teams worked on timely issues such as the investigation into the gun turret explosion and resultant tragic loss of forty-seven lives which occurred aboard USS *Iowa* on 25 April 1989. Projects for the National Aeronautics and Space Administration, for the Drug Enforcement Administration, and for other agencies promised to diversify the client base beyond the Department of Defense.

When Monetta departed to take a position at the Department of Energy in mid-June 1989, he left behind well-established strengths in management, personnel, and facilities at Indian Head. The staff was strong, with continuity of service at the mid-management level one of the Station's greatest assets. As under Joe Browning, younger staff could see an attractive career there; as in earlier exciting periods, working in technology at Indian Head held promise of being at the cutting edge. As a consequence, several senior managers had stayed on for decades, providing the benefit of their rich experience to the facility.

Capt. Fritz Wendt finished his tour of duty in the summer of 1989, and the Naval Ordnance Station saw a complete change of leadership with both a new commanding officer and a new technical director coming aboard within a few months of each other in June and October 1989. The new commanding officer, Capt. Ed Nicholson, immediately examined the successes and problems he inherited and set to work. Under the system of Naval Industrial Funding, increases in management expenses and in other nontechnical work had been reflected in the cost of products to clients. Nicholson immediately began to rein in expenses where they could be cut. He anticipated that one of the most serious problems for the Station over the coming decade would be insuring environmental protection.

The new technical director, Roger Smith, came aboard in October 1989. Smith's career path and his orientation was somewhat different from those of his three predecessors. Unlike Browning, Lee, and Monetta, Smith was not an engineer. His training was in physics, and his career had been entirely as a civilian in weapons systems. He had spent twenty-five years working in various program management positions at the Naval Sea Systems Command and the predecessor offices, with a primary focus on undersea warfare.

As Smith saw it, that different background could have two significant impacts on the direction of the Station. First, since he would not be the "star player" in the technical area, it would provide an opportunity for experienced local personnel to take further leadership roles. Secondly, his experience as a Program Manager gave him knowledge of how Indian Head's "customers" thought. Smith intended to use that knowledge to improve further Indian Head's reputation with those who bought its products and services.

Looking ahead at the tasks facing him, Smith anticipated that one important job would be to retain a strong production and industrial base. He noted that the Navy's own workload, like the American economy, had been shifting from production to service, with some of the same risks and ill effects. He believed that this shift could lead to an eroding of the Navy's capabilities in time of emergency. It was the solid industrial facility, the hardware, which provided the reason for Indian Head's continued existence. If that was neglected, the paper functions could be relocated to any facility.

AFTER SEPTEMBER 11, 2001

As the author completed the second edition of this book, the tragic events of September 11, 2001, unfolded. The response of the United States of America to the unprecedented attack on its soil was immediate, profound, and heralded a new era. In heartfelt memorial services, in the displays of American flags, and in bipartisan Congressional support for a military response to the attacks, the nation demonstrated a new resolve. Not only was the nation united in a dramatic way by the acts of terrorism at the World Trade Center in New York and at the Pentagon in Washington, but it soon became clear that the nation now faced a new kind of enemy, requiring new methods of warfare.

Since 1890, when Dashiell first opened the Proving Ground at Indian, political and popular backing for the military rose and fell in repeated cycles. Between wars, it seemed that public and political support for long-term investment in defense diminished. In the 1920s and in the 1980s, political support for defense waned. In World War I and World War II, and again briefly during the Gulf War the Navy's shore establishment found itself called upon to support the rapid arming of ships and aircraft. But after each war, the defense establishment received less attention, less funding, less political support.

The decade of the 1990s following the Gulf War had been marked by many of those same signs of decline. At Indian Head, administrative issues had centered around RIFs and diminished workload, consolidation of shore establishments, increasing proportion of overhead, and the struggles to maintain competency to meet defined missions and to find new customers in the defense world. Roger Smith and then Mary Lacey had found pathways through these difficult times and steered the Indian Head Division of the Naval Surface Warfare Center through the rough waters.

Following President George Bush's marshalling of national will to "bring to justice, or to bring justice to, the perpetrators" of the September 11 attacks, Indian Head sensed a change in NavSea headquarters, at the Department of Defense, and in the Congress. New work appeared rather suddenly on the agenda of the Indian Head engineers, scientists, and production workers. And as some employees noticed in their lives outside of the gates, one could once again be publicly proud to work for a branch of the Defense Department. One immediate administrative effect was that attempts to target the least efficient facilities for future closing were postponed. The next round of base closures and all the requirements for analysis and justification that such a round of closure entailed appeared to give way to more pressing priorities.

Rather suddenly, the Research, Development, Test, and Evaluation facilities of the United States Department of Defense found themselves thrust into a new world of demands and pressures. At an All-Hands meeting on November 8, 2001, Executive Director Mary Lacey pointed out that the Indian Head Division was hiring about 125 new scientists and engineers. There appeared to be at least $40 million in additional work immediately scheduled for the division. Areas of emerging requirements included work in propulsion, CAD/PADs, procurement support and loading of warheads, development of thermobaric (sustained temperature and pressure) weapons and high-temperature incendiaries, and logistics management for new chemical and biological warfare equipment. Future research would entail looking for ways to destroy chemical and biological weapons factories on the ground without causing the spread of dangerous chemicals and biologicals. Support for official DoD recognition of Indian Head as the National Center for Energetics appeared to be growing.

As the United States faced the unfolding events of the battle against terrorists and the governments that gave sanctuary to the terrorists, Indian Head remained ready to work with the energetic materials in aircraft safety devices, in the propellants of the weapons those planes carried, and in the warheads of weapons carried by ships, planes, and submarines.

The mission that had evolved over the eleven decades since the founding of the Naval Proving Ground in 1890 continued to adapt as the nation faced new challenges to its national security.

APPENDIX 1

Tours of Duty
Inspectors of Ordnance in Charge and Commanding Officers, Indian Head

[For biographic information, see Appendix 2]

Tour of Duty	Name	Rank as CO	Rank Attained
1890–1893	Dashiell, Robert B.	ENS	LT
1893–1896	Mason, Newton E.	LT	RADM
1896–1900	Couden, Albert R.	LCDR-CDR	RADM
1900–1902	Strauss, Joseph	LT	ADM
1902–1903	Patton, John B.	LT	CDR
1903–1906	Dieffenbach, Alfred C.	LT	CDR
1906–1908	Strauss, Joseph	LCDR	ADM
1908–1910	Jackson, Richard H.	LCDR	ADM
1910–1913	Holden, Jonas H.	LCDR	ADM
1913–1916	Hellweg, Julius	LCDR	COMO
1916–1917	Earle, Ralph	CDR	RADM
1917–1920	Lackey, Henry E.	CDR	RADM
1920–1923	Greenslade, John W.	CAPT	VADM
1923–1923	Bloch, Claude C.	CAPT	ADM
1923–1925	Pickens, Andrew C.	CDR-CAPT	RADM
1925–1928	Stark, Harold R.	CDR-CAPT	ADM
1928–1931	Leary, Herbert F.	CAPT	VADM
1931–1932	Schuyler, Garret L.	CAPT	RADM
1932–1935	Johnson, Lee P.	CDR	COMO
1935–1938	Wilson, William W.	CDR-CAPT	CAPT
1938–1940	Haines, Preston B.	CAPT	CAPT
1940–1943	Hersey, Mark L., Jr.	CAPT	COMO
1943–1946	Glennon, James B.	CAPT	CAPT
1946–1948	Hanlon, Byron H.	CAPT-RADM	ADM
1948–1948	Gallery, Philip D.	CAPT	RADM
1948–1952	Voegeli, Clarence E.	CAPT	CAPT
1952–1952	Scanland, Francis W., Jr.	CDR	CAPT
1952–1955	Benson, William H.	CAPT	CAPT
8/01/55–6/30/58	King, George E.	CAPT	CAPT
7/01/58–4/30/60[1]	Atkins, Griswold T.	CAPT	CAPT
5/09/60–9/02/60	Galvani, Amedeo H.	CDR	CAPT
9/02/60–7/31/63	Wesche, Otis A.	CAPT	CAPT
7/31/63–8/30/65	Dreyer, Oscar F.	CAPT	CAPT
8/30/65–7/31/69	Olsen, Leslie R.	CAPT	CAPT
7/31/69–6/30/72	Frese, Bernard W., Jr.	CAPT	CAPT
6/30/72–5/28/76	Gary, Stanley P.	CAPT	CAPT
5/28/76–6/23/80	Warren, Thomas C.	CAPT	CAPT
6/23/80–6/24/83	Underwood, Fred S.	CAPT	CAPT

Tour of Duty	Name	Rank as CO	Rank Attained
6/24/83–7/31/87	Tadlock, James D.	CAPT	CAPT
7/31/87–7/21/89	Wendt, George F.	CAPT	CAPT
7/21/89–6/2/92	Nicholson, Edwin P.	CAPT	CAPT
6/2/92–7/28/94	Maxwell, David G.	CAPT	CAPT
7/28/94–7/18/97	Newton, Wayne J.	CAPT	CAPT
7/18/97–7/14/00	Walsh, John J.	CAPT	CAPT
7/14/00–	Seidband, Marc	CAPT	

[1]Commander Brainard Belmore served as Commanding Officer for nine days between 4/30/60 and 5/09/60.

Biographical Information
Officers in Charge and Commanding Officers, Indian Head

Robert B. Dashiell   *Indian Head 1890–1893*

Appointed at-large to U.S. Naval Academy (USNA) in Class of 1877; various duties including service aboard USS *Essex* and USS *Pensacola* (1877–88); assigned to Naval Ordnance (1888); in charge of establishing Naval Proving Ground (NPG) (opened 1890) including supervision of building and acquisition of staff and materials; specialist in gun mechanical design; invented Dashiell rapid-fire breech (1890); in charge of NPG (1890–93); assigned to Philadelphia Navy Yard (1893); served aboard USS *New York*; promoted to Assistant Naval Constructor (1895); died in 1899, probably of malaria.

Newton E. Mason   *Indian Head 1893–1896*

Born 14 October 1850 in Moaretown, Pennsylvania; graduated from USNA in 1869; served on USS *Sabine*, USS *Wabash*, USS *Kansas*, USS *Ossipee*, Irish famine relief ship *Constellation*, USS *Monocacy*, and USS *Pensacola* between 1869 and 1883; stationed at Washington Navy Yard for ordnance duty (1883–84); served in Bureau of Ordnance (1885–89); was in charge of NPG at Indian Head (1893–96); appointed Lt. Commander of USS *Brooklyn* during Spanish-American War; promoted to Commander of USS *Cincinnati* (1902–04); served as Chief of the Bureau of Ordnance at Navy Department with rank of Rear Admiral; member of General Board until his retirement in 1912; recalled to active duty during World War I to serve on War Industries Board.

Albert R. Couden   *Indian Head 1896–1900*

Born 30 October 1846 in Michigan City, Indiana; appointed to USNA in 1863 and graduated 1867; various duties, including active service on USS *Marion* until 1896; in charge at Indian Head (1896–1900); commanded USS *Wheeling* (1900–01) and USS *Marion* (1901–02); commanded Naval Station at Cavite, Philippine Islands (1902–04); served as General Inspector of Ordnance for the Navy (1904–06); commanded USS *Louisiana* (1906–07); later was President of Board of Naval Ordnance until his retirement in 1908; recalled to active duty with Bureau of Naval Ordnance during World War I.

Joseph Strauss   *Indian Head 1900–1902, 1906–1908*

Born 16 November 1861 in Mt. Morris, New York; graduated from USNA in 1885; participated in Spanish-American War; served as Inspector of Ordnance at Indian Head; with rank of Admiral served as Chief of Bureau of Ordnance (1913–16); commanded battleship *Nevada* (1916–18); commanded Atlantic Fleet mine force

(1918–19); member of the General Board (1920); Commander-in-Chief of the Asiatic Fleet (1921–22); after retirement in 1925 served on Safety and Salvage of Submarines Board and the Battleship Design Board.

John B. Patton   *Indian Head 1902–1903*

Appointed Naval Cadet in 1885; served aboard the *Pensacola* and the *Baltimore* (1889–91); promoted to Assistant Engineer in 1891; detailed to the Navy Yard in Portsmouth, New Hampshire, in 1892; served on the *Concord* (1893–96); appointed Inspector of Steel in Pittsburgh, Pennsylvania, in 1896; sailed with the *Culgoa* (1898–1901); Commanding Officer of Indian Head (1902–03); served on the *Florida* (1903–05); sailed around the world with the Great White Fleet (1907–09); retired in 1914.

Alfred C. Dieffenbach   *Indian Head 1903–1906*

Appointed Cadet Engineer in 1881; promoted to Ensign in 1887; served on the *Omaha* (1888–90); detailed to the Naval Ordnance Proving Ground at Indian Head in 1891; appointed Inspector of Ordnance at Hartford, Connecticut, in 1892; commissioned as Lieutenant in 1896; sailed on the *Machias* (1897–98); appointed to the Bureau of Ordnance (1898–1900); served on the *Buffalo* and the *Concord* (1900–02); Commanding Officer of Indian Head (1903–06).

Richard H. Jackson   *Indian Head 1908–1910*

Born 10 May 1866 in Tuscumbia, Alabama; graduated USNA in 1887; served in Spanish-American War; navigator of USS *Colorado* (1905–07) and executive officer (1907–08); in charge of NPG at Indian Head (1908–10); Naval Station, Cavite (1910–11); duty with General Board Navy Department (1913–15); special representative from Navy Department to Ministry of Marine, Paris, and naval attache, Paris (June 1917–November 1918); Rear Admiral commanding Battleship Div. 3 (1922); Assistant Chief of Naval Operations (1923); Vice Admiral commanding Battleship Divisions (1925); Admiral, Commander-in-Chief of Battle Fleet (1926); member General Board (1927–30).

Jonas H. Holden   *Indian Head 1910–1913*

Graduated from USNA 1896; served on the *Columbia*, the *Maine*, and the *Scorpion* (1896–98); commissioned 11 May 1898; returned to sea and served aboard the *Solace*, the *Olympia*, the *Oregon*, and the *Brooklyn* (1898–1900); ten months duty as Aide to the Commander-in-Chief (April 1900–March 1901); received instruction in torpedoes and electronics at the Newport Torpedo Station; detached to the Naval Academy (1901–02); additional service at sea on the *Maine* and the *Lancaster* (1902–04); Commanding Officer at Indian Head (1910–13).

Julius F. Hellweg   *Indian Head 1913–1916*

Rose through the ranks to Commodore; served as supervisor of naval overseas transportation service in Norfolk (1919); Captain of Port Bordeaux, France, and senior naval officer in France (1919–20); in charge of bringing five former German warships to the United States (1920); Commander 14th Destroyer Squadron (1924–26); superintendent of the Naval Observatory in Washington, D.C., until retirement in 1930.

Ralph Earle   *Indian Head 1916–1917*

Born 3 May 1874 in Worcester, Massachusetts; appointed to USNA 1892; commanded Chemical Laboratory, Naval Station, Puerto Rico; Inspector of Ordnance in Charge of NPG, Indian Head (1916–17); Chief of Bureau of Ordnance with rank of Rear Admiral and a principal proponent of the North Sea mine barrage (1917–21); Inspector of Ordnance in Charge, Naval Torpedo Station (1923–27); retired from the Navy and was President of Worcester Polytechnic Institute in Worcester, Massachusetts, until his death in 1939.

Henry Ellis Lackey   *Indian Head 1917–1920*

Born 23 June 1876 in Norfolk, Virginia; appointed to USNA by President of the U.S. in 1895; in Battle of Santiago aboard USS *New York*; duty at NPG (1908–10); awarded the Navy Cross for command of NPG, Indian Head (1917–20); Director of Ships' Movements Division, Office of Naval Operations (1921–24); Command of Naval Training Station, Norfolk (1927–30); Commander Cruiser Division Four, Scouting Force (1933–35); Director, Shore Establishments (1935–37); Commander Squadron 40-T (1937–39); President, General Court Martial (1939–40).

John Wills Greenslade   *Indian Head 1920–1923*

Born 11 January 1880 in Bellevue, Ohio; USNA Class of 1899; Instructor, Physics and Chemistry at USNA (1905–06, 1909–11) and Electrical Engineering and Physics (1915–17); commanded USS *Housatonic*, laid 9,200 mines in North Sea (1917–19); Command of NPG, Indian Head (1920–23); commanded Mine Squadron One, Control Force (1923–25); Naval War College as student then Head of Operations (1925–28); member of General Board (1930–32, 1936–37, 1939–41); Commander, Battleships, Battle Force (1938–39); Commandant, Twelfth Naval District, San Francisco, with additional duty as Commander of Pacific Southern Naval Coastal Frontier (1941–44); Pacific Coast Coordinator of Naval Logistics, San Francisco (1944–45).

Claude Charles Bloch   *Indian Head 1923–1923*

Born 12 July 1878 in Woodbury, New Jersey; graduated USNA in 1899; served on USS *Iowa* during Spanish-American War; served on USS *Newark* and USS *Wheeling* during the Philippine and Boxer campaigns; with Bureau of Ordnance in Washington (1904–06, 1913–16); stationed at NPG, Indian Head (1909–11); Assistant Chief of Bureau of Ordnance (1918–21) and Chief (1923–27); served as Commanding Officer at Indian Head for three to four months in 1923; Judge Advocate General of the Navy (1934–36) and Commander of the Battle Force with rank of Admiral (1937); Commander-in-Chief of U.S. Fleet (1938–40); subsequently Commandant of 14th Naval District, Hawaii.

Andrew Calhoun Pickens   *Indian Head 1923–1925*

Born 8 December 1881 in Mobile, Alabama; detached from USNA 1904; Bureau of Ordnance, Navy Department (1918–22); Commander of NPG, Dahlgren, Virginia, and NPF, Indian Head (1923–25); Command USS *Nitro* (1925–27); Assistant to Chief of Bureau of Ordnance (1927–30); Commander Destroyer Squadron Three (1934–35); Chief of Staff and Aide to Commander-in-Chief, U.S. Fleet (1936–37); commanded Cruiser Division Seven (1939–41); President, Naval Examining Board (1941–42); retired from active service in 1942.

Harold R. Stark   *Indian Head 1925–1928*

Born 12 November 1880 in Wilkes Barre, Pennsylvania; appointed to USNA Class of 1903; staff of Commander, U.S. Naval Forces in European Waters (1917–19); Naval Inspector of Ordnance in Charge of the NPG, Dahlgren, and NPF, Indian Head (1925–28); Aide to the Secretary of the Navy (1930–33); Chief of the Bureau of Ordnance (1934–37); Command of Cruisers Battle Force (1937–38); in August 1939 succeeded Admiral Leahy as Chief of Naval Operations with accompanying rank of Admiral; faced with the possibility of U.S. involvement in WWII was able to convince Congress to achieve a high state of readiness; awarded the Distinguished Service Medal for this service; Commander of U.S. Naval Forces in Europe and Commander Twelfth Fleet (1942–45); Office of Chief of Naval Operations until retirement in April 1946.

Herbert Fairfax Leary   *Indian Head 1928–1931*

Born 31 May 1885 in Washington D.C.; appointed to USNA from Maryland and graduated 1905; August 1905 detailed at Portsmouth, New Hampshire, during Russo-Japanese Peace Conference; postgraduate ordnance instruction at Indian Head and elsewhere; Navy Cross for work on staff of Commander Battleship Force One during WWI; Aide on staff of Commander Destroyer Squadrons, Battle Fleet (1926–28); Naval Inspector of Ordnance in Charge at NPG, Dahlgren, and NPF, Indian Head (1928–31); fitted and commanded USS *Portland* (1932–34); staff of Commander-in-Chief, U.S. Fleet (1938–39); after Pearl Harbor commanded Naval Forces in Australia-New Zealand Area, then Commanded U.S. Naval Forces, Southwest Pacific; then Commander Task Force ONE, Pacific Fleet; May 1943 reported as Commandant, Fifth Naval District, Norfolk; Commander Eastern Sea Frontier (1943–46).

Garret Lansing Schuyler   *Indian Head 1931–1932*

Born 18 March 1885 in New York, New York; entered USNA 1903; Proof Officer at Indian Head (1912–16); in London as Assistant Naval Attache during WWI; Navy Cross for duty at U.S. Railway Battery at St. Nazarre; Gunnery Officer, USS *Mississippi* (1920–23); Bureau of Ordnance, Navy Department (1923–31); Commanding Officer at NPF, Indian Head (1931–32); Inspector of Ordnance in Charge at NPG, Dahlgren (1932–34); returned to Bureau of Ordnance; awarded the Legion of Merit for achievements in science and development of ordnance including delay coils and bomb displacers.

Lee Payne Johnson   *Indian Head 1932–1935*

Born 26 October 1886 in Concord, North Carolina; appointed to USNA 1905; USS *Connecticut* during occupation of Vera Cruz, Mexico (1914); Commanding Officer, USS *Balch* (1918–19) for which he was awarded Navy Cross; duty at NPG, Dahlgren (1921–24); Gunnery Officer of USS *Utah* (1924–27); Head of Explosives Section, Bureau of Ordnance (1927–30); commanded mine layers (1930–32); in charge of NPF, Indian Head (1932–35); Executive Officer, *California* (1935–37); in charge of Naval Ammunition Depot, Balboa, Canal Zone (1937–39); Senior Course at Naval War College, Newport, Rhode Island (1939–40); commanded cruiser *Tuscaloosa* (1940–42); Chief of Staff and Aide to Commander Amphibious Force, Atlantic Fleet, where he participated in amphibious operations in Southern Europe including the Sicilian invasion; awarded the Legion of Merit.

William W. Wilson   *Indian Head 1935–1938*

Born 10 March 1886 in New Castle, Delaware; entered USNA in 1904; on USS *North Dakota* participated in Mexican Campaign (1914); assisted fitting out USS *Pennsylvania* and served aboard until 1918; assigned to duty in Office of the Chief of Naval Operations, Washington D.C.; command of the USS *Maury* (1921–24); served in Aeronautic Organization in the Bureau of Ordnance (1924–27); Aide on Staff of the Commander-in-Chief, U.S. Fleet (1929–30); Executive Officer of the USS *New York* (1933–35); in charge of NPF at Indian Head (1935–38); assisted in fitting USS *Nashville* and commanded upon her commission (1938–40).

Preston Bennett Haines   *Indian Head 1938–1940*

Born 22 August 1886 in Peekskill, New York; appointed to USNA from New York 1905; Philadelphia Navy Yard (1916–19); served in Bureau of Ordnance (1922–24, 1927–30, 1932–35); Commanding Officer, USS *McFarland* (1920–22), USS *Preston* (1924–27); Aide and Force Gunnery Officer on the Staff of Commander Battle Force (1935–37); commanded NPF at Indian Head (1938–40).

Mark Leslie Hersey, Jr.   *Indian Head 1940–1943*

Born 8 July 1888 in Whipple Barracks, Arizona; appointed to USNA from Maine (1905); on USS *Florida* and participated in the occupation of Vera Cruz (1914); had consecutive duty at the NPG, Indian Head, and NPG, Dahlgren (1922–23); Commandant, Naval Station, Guantanamo Bay, Cuba (1936–38); commanded USS *Raleigh* (1938–40); Commanding Officer, NPF (1940–43); Commander, Naval Bases, South Solomons, Sub-Area (1944–45), for which he was awarded the Legion of Merit; Commandant at Naval Base, Manila-Subic Bay, Philippine Islands (1945–46).

James Blair Glennon   *Indian Head 1943–1946*

Born 4 November 1888 in a Navy family; entered USNA 1905; studied ordnance at the Postgraduate School, other naval facilities including Dahlgren, and private plants; served on USS *Alabama* and USS *Arizona* during WWI; served as Executive

Officer at Dahlgren (1925–27); Officer in Charge of Naval Mine Depot, Yorktown, Virginia (1935–37); in charge of Naval Ordnance Laboratory, Navy Yard, Washington (1942–43); in charge of NPF (1943–46)—personnel rose to 5,000 and monthly powder output reached its peak; credited with great advances in the Naval Mine Program.

Byron H. Hanlon   *Indian Head 1946–1948*

Born in Rocklin, California; entered the USNA in 1917; ordnance engineering at Naval Postgraduate School and Lehigh University; at Pearl Harbor on December 7, 1941; Officer in Charge of Planning Division at Naval Gun Factory (1942–44); Commander Underwater Demolition Teams, Amphibious Force Pacific (1944–45); commanded USS *North Carolina* (1945); Commanding Officer of NPF (1946–48); Superintendent of the Naval Gun Factory (1950–52); during Korean hostilities, Commander-in-Chief, U.N. Command Representative on the Combined Economic Board, Korea; 1955 named Commander Joint Task Force SEVEN and served as Representative of the Atomic Energy Commission at the Pacific Proving Ground during Operation Redwing; Commander Amphibious Force, U.S. Atlantic Fleet, from November 1957 till retirement when advanced to Admiral on basis of combat awards.

Philip D. Gallery   *Indian Head 1948–1948*

Born 17 October 1907 in Chicago, Illinois; entered USNA in 1924; instruction at the Postgraduate School in Annapolis; Executive Officer of the USNA Preparatory School, Naval Training Station, Norfolk; commanded the Antiaircraft Training and Test Center, Dam Neck, Virginia, where he received the Legion of Merit; commanded USS *Jenkins* during Philippines and other campaigns and earned the Gold Star; commanded Destroyer Division Seventy-Two during atomic tests at Bikini Atoll in 1946; detached to serve as Executive Officer of NPF and later Commanding Officer till 1948; Commander Surface Antisubmarine Development Detachment, U.S. Atlantic Fleet; on staff of Commander Operational Development Force until retirement in 1958.

Clarence E. Voegeli   *Indian Head 1948–1952*

Born 24 April 1900 in Fountain City, Wisconsin; enlisted in U.S. Navy the day the United States declared war on Germany in 1917; June 1918 entered USNA; studied Ordnance Engineering at Annapolis Postgraduate School and completed the course at Washington Navy Yard; commanded USS *Highlands*, USS *Dane*, and USS *Los Angeles*; served as President of the Board on Naval Ordnance; commanded NPF until his retirement (1948–52).

Francis W. Scanland, Jr.   *Indian Head 1952–1952*

Born 12 September 1912 in Baltimore, Maryland; appointed to USNA in 1930 from Louisiana; duty on staff of Commander Submarines, Southwest Pacific; took command of USS *Hawkbill* (submarine), earned Legion of Merit and Gold Star among others for his service; duty at Submarine Base in Key West; commanded Submarine Division Fifty-One; served as Executive Officer and Commanding Officer of NPF at Indian Head (1951–53); served as Chief of Staff and Aide of Commander Carrier Division TWENTY; Commander Destroyer Squadron NINETEEN; commanded Naval Ammunition Depot, Charleston, South Carolina, and Naval Weapons Station, Yorktown, Virginia.

William H. Benson   *Indian Head 1952–1955*

Graduated from USNA in 1925; studied general ordnance at the USNA Postgraduate School in 1931; served as gunnery officer aboard the cruiser *Augusta*; in 1942 was appointed Armaments Officer at the Naval Proving Ground in Dahlgren, Virginia; commanded the attack transport *George Clymer* (1945); returned to Dahlgren in 1946 as Deputy Commanding Officer; attended the Naval War College in 1949; served in Korea under Vice Admiral C. Turner as Assistant Chief of Staff for Plans, Operations, and Intelligence (1951); commanded *Destroyer 20* (1951–52); awarded the Legion of Merit; Commanding Officer at Indian Head (1952–55).

George E. King   *Indian Head 1955–1958*

Born 15 April 1904 in Normal, Illinois; appointed at-large to the USNA in 1924; received instruction at the Naval Torpedo Station and the Postgraduate School at Annapolis; served at the Naval Gun Factory and assigned to the Bureau of Ordnance (1947–49); attended the advanced management program at Harvard Business School (1949); Program Director, Planning and Progress Division at Bureau of Ordnance; Commander at Indian Head (1955–58).

Griswold T. Atkins   *Indian Head 1958–1960*

Born 28 December 1909 in Annapolis, Maryland; USNA Class of 1930; Postgraduate School training in ordnance (1936–39); assigned to battleship *Tennessee* (1939–42) (*Tennessee* at Battleship Row for Pearl Harbor); Plate Battery Officer at NPG, Dahlgren, as well as Head of the Armor, Projectiles, Bombs, Rockets, Warheads and Ballistics Section in Bureau of Ordnance; ordered to duty at NPF February 1958; designated Commanding Officer of Naval Propellant Plant August 1958.

Amedeo Henry Galvani   *Indian Head 1960–1960*

Born 28 August 1920 in Plymouth, Massachusetts; appointed to USNA 1939; earned M.S. in Chemistry from Lehigh University; assistant at Naval Ammunition Depot, Hawthorne, Nevada; Commander of USS *John Paul Jones* (DD-932); at Indian Head served as Executive Officer and temporary Commanding Officer; later commanded the Naval Ammunition Depot, McAlester, Oklahoma.

Otis Albert Wesche   *Indian Head 1960–1963*

Born 28 April 1914 in Ridgeville Corners, Ohio; was appointed to the USNA from Ohio in 1934; during WWII served on the USS *Idaho* and *Wasp*; received instruction in Ordnance Engineering at Postgraduate School in Annapolis (1944–46); Gunnery Officer on the USS *Missouri*; assumed command of USS *Southerland* May 1952;

was awarded the Bronze Star Medal for service in Korea; took command at Indian Head May 1960 and on 23 April 1963, transferred to the Bureau of Naval Weapons.

Oscar F. Dreyer   *Indian Head 1963–1965*

Born 22 June 1917 in St. Louis, Missouri; enlisted in the U.S. Navy in 1935 and entered the USNA in 1937; aboard the USS *McCall* and was awarded the Bronze Star Medal for action against Japanese forces; attended Postgraduate School; earned an M.S. in Chemical Engineering from Cornell University (1946); commanded destroyers and later a Destroyer Division; Ordnance Officer for three years at Naval Ammunition Depot, McAlester, Oklahoma; became Assistant Manager for the Bureau of Naval Weapon's government-owned and -operated plants in 1959; July 1963 became Commander of NPP, Indian Head, for two years; later Commanding Officer at Port Hueneme, California.

Leslie R. Olsen   *Indian Head 1965–1969*

Born 6 November 1919 in Brigham City, Utah; appointed to USNA class of 1942 (which graduated in 1941) from Nevada; on USS *Porter* (DD-356), sunk in Battle of Santa Cruz in October 1942; studied metallurgy at the Naval Postgraduate School; detached to California Institute of Technology, graduated 1946; served with Bureaus of Aeronautics and Ordnance; in action in Korean waters under U.N. Command; February 1956 Head of Guided Missile Systems Section in the Office of the Chief of Naval Operations; was Commanding Officer on USS *Sierra* before duty as Commanding Officer at Indian Head.

Bernard W. Frese, Jr.   *Indian Head 1969–1972*

Born 3 January 1920 in Cincinnati, Ohio; USNA Class of 1943, B.S. in Electrical Engineering (EE) (graduated June 1942 in accelerated class for National Emergency); attended Postgraduate School in Ordnance Engineering (1946–47); earned M.S. (EE) at Massachusetts Institute of Technology (MIT) (1949); attended Naval Guided Missile School (1951); qualified to command destroyers (1956); studied at the Industrial College of the Armed Forces (1960–61); promoted to Captain (September 1962); served as Commanding Officer, Indian Head (1969–72).

Stanley P. Gary   *Indian Head 1972–1976*

Graduated from USNA in 1946; served as Fire Control Officer on the USS *John R. Craig*; attended Naval Mine Warfare School; commanded USS *Mocking Bird*; earned M.S. (EE) from MIT (1954); served aboard the USS *Iowa* (1954–56); assigned to the guided missile system project in 1956; became Technical Officer, Test Officer, and finally Commanding Officer of the U.S. Naval Ordnance Missile Testing Facility, White Sands Missile Range; awarded Legion of Merit for his work on the Polaris/Poseidon research project; before coming to Indian Head was Manager of the Missile Branch, Strategic Systems Project Office, Chief of Naval Material.

Thomas C. Warren   *Indian Head 1976–1980*

Born 15 July 1931; graduated from Trinity College (1949) and USNA (1953); received MS (EE) from the USNA Postgraduate School (1960); served as Gunnery Officer aboard the USS *Estes* and USS *Taussig* and Missile Officer aboard the USS *Oklahoma City*; appointed Terrier Fire Control Manager with the Surface Missile Systems Project Office at BuWeps/NavOrd; became Combat Systems Officer at the Norfolk Naval Shipyard in 1970; assigned duty as Missile Branch Head, Strategic Systems Project Office, in 1972, where he worked on the Trident I, Poseidon, and Polaris missile systems; appointed Commanding Officer at Indian Head in 1976.

Fred S. Underwood   *Indian Head 1980–1983*

Born 8 October 1933; graduated from USNA (1955); received M.S. (EE) from the USNA Postgraduate School (1963); served as Gunnery Officer aboard the USS *Heerman* and the USS *Mullinnix*, Executive Officer on the USS *Pinnacle*, and Operations Officer on the USS *Luce*; also served on the staff of COMSERVPAC in Pearl Harbor; designated Officer-in-Charge of the Naval Shore Electronics Activity, Guam; assigned to the Combat Systems Coordination Office at the Naval Ships Engineering Center in Hyattsville, Maryland; later assigned to the Naval Sea Systems Command Headquarters in Washington, D.C.; appointed Commanding Officer of Indian Head in June 1980.

James D. Tadlock   *Indian Head 1983–1987*

Graduated from Auburn University with a B.S. in Aeronautical Engineering; commissioned as a Navy line officer in August 1960 and began service as Assistant Electronics Material Officer aboard the USS *Soley*; attended the Naval Destroyer School; served in Vietnam with the Naval Advisory Group; received an M.B.A. and an M.S. in Ordnance Engineering from the USNA Postgraduate School in 1973; assigned as Assistant Ordnance Officer at Naval Weapons Station in Yorktown, Virginia; became Program Director for the Naval Material Command Headquarters' Missile Propulsion Technology Programs; appointed Director of the Gun Fire Control Division in Naval Sea Systems Command; took command of Indian Head in 1983.

George F. Wendt   *Indian Head 1987–1989*

Born 28 August 1936; completed a B.S. in Civil Engineering at Ohio University (1959); appointed Electronics Officer aboard the USS *Meredith* after attending the Officer Candidate School in Newport, Rhode Island; studied at the Destroyer Force Engineering School and served as Engineering Officer on the USS *Barton*; received an M.S. in Physics from the USNA Postgraduate School; assumed command of River Division 573 in Vietnam; attended the Guided Missile School in Dam Neck, Virginia (1968); appointed Ordnance Engineering Officer at the Mare Island Naval Shipyard (1973); designated Director of Engineering, Naval Ship Weapons Systems Engineering Station, Port Hueneme, California (1983–85); in 1985 became Commanding Officer of the Naval Sea Support Center, San Diego,

California; Commanding Officer of Naval Ordnance Station, Indian Head (1987–89).

Edwin P. Nicholson *Indian Head 1989–1992*

Graduated from the NROTC program at the University of North Carolina in 1964; graduated from the Destroyer School in 1967 and served two tours of duty in Vietnam on the USS *Picking*; earned a B.S. in Chemistry from the Naval Postgraduate School in 1973; served as Chief Engineer on the USS *St. Louis* (1973–1976); served as Executive Officer on the USS *Dyess* (1976–1978); appointed Deputy Head of Research and Technology at the Naval Surface Weapons Center in White Oak, Maryland (1978–81); commanded the USS *Vogelgesang* and the USS *Connole* (1981–84); graduated from the Naval War College in 1986; served as Commanding Officer of Naval Weapons Station Earle (1986–87); awarded the Bronze Star; became Commanding Officer of Naval Ordnance Station, Indian Head, in July 1989.

David G. Maxwell *Indian Head 1992–1994*

Graduated from USNA in 1968; master's degree in operations research from the Naval Postgraduate School; served as Communications and Operations Officer aboard USS *Bradley* and as Operations Officer aboard USS *Robinson*, off the coast of Vietnam. In 1979 served as Scheduling and Naval Tactical Data Systems Officer on Cruiser Destroyer Group Eight staff. In 1981, served as Executive Officer aboard USS *Dahlgren*. Project Manager (1983-85) for Naval Tactical Data System program development aboard CGN-993, DDG-993, CG-26, and CGN-9 class ships. Served as Commanding Officer (1986-1988) of the USS *William V. Pratt* (DDG-44). In 1988-1992, Commanding Officer, Naval Ordnance Missile Test Station at White Sands, New Mexico. Appointed Commanding Officer, Indian Head, June 2, 1992.

Wayne J. Newton *Indian Head 1994–1997*

Graduate of University of California, San Diego, commissioned an ensign 1974. M. S. degree in telecommunication management, Naval Postgraduate School. Served aboard ballistic missile submarines USS *John Marshall* and USS *Nathan Hale*. Strategic Forces Analyst and Action Officer with Joint Chiefs of Staff. Executive Officer, Naval Electronics System Engineering Center Portsmouth, 1986-1989.Combat Systems Officer, Charleston Naval Shipyard, later Production Resources Officer, 1990. Completed the program manager curriculum at the Defense Systems Management College in 1992. Head, Submarine Communication Systems Engineering Division, Submarine Communications Program Office, SPAWAR, 1993. Appointed Commanding Officer, Indian Head, 28 July 1994.

John J. Walsh *Indian Head 1997–2000*

Born in Waterbury, Connecticut 17 October 1948. Enlisted in the Navy 1967. Upon discharge in 1970, attended University of Massachusetts, and earned a B.S. in Management, 1974. Commissioned an ensign 1974. Enlisted service aboard USS *Thor*; served as combat information and Electronic Warfare Officer aboard USS *Austin*,

Operations Officer and Diving Officer USS *Petrel*, Chief Engineer USS *Ortolan*, and Officer in Charge, Explosive Ordnance Disposal Mobile Unit TWO, USS *Santa Barbara*. Shore duty assignments included Commanding Officer, EOD, Indian Head, and Chief Staff Officer, Keyport Division, Naval Undersea Warfare Center. Qualified as a Deep Sea Diving Officer, Saturation Diving Officer, and Master Explosive Ordnance Disposal Officer. Appointed Commanding Officer, Indian Head, 18 July 1997.

Marc A. Siedband   *Indian Head 2000–*

Graduated with a B.S. in Mechanical Engineering and commissioned as an ensign from the USNA in 1975. After nuclear power and surface warfare training, served as Mechanical Division Officer aboard USS *Enterprise*. After serving as Leading Officer of the Watch at the Naval Reactor School in Idaho, he served aboard USS *Jesse L. Brown* and USS *Carl Vinson*. In 1987, earned a master's degree in Mechanical Engineering at the Naval Postgraduate School, and then served as senior ship superintendent for the overhaul of the USS *Arkansas*. After service with COMNAVAIRPAC staff, he served as Chief Engineer aboard USS *Constellation*. In 1995, served as Branch Head for Ship Maintenance and Modernization in the Office of the CNO. Appointed Commanding Officer, Indian Head, 14 July 2000.

APPENDIX 3

Oral History Interviews Conducted for this Work

Interviewee	Interviewer*	Date
John W. (Bill) Jenkins	RC	7-26-88
Vincent C. Hungerford	RC	1-17-89
John Murrin	RC	1-17-89
Mr. and Mrs. B. Bledsoe	JG	1-24-89
Dr. Frank Susan	JG	1-24-89
Jay Thornburg	RC	2-9-89
John P. McDevitt	RC	2-9-89
Jay Carsey	RC, EC	3-3-89
Joe L. Browning	RC, VH	2-17-89
Opal Willis	RC	3-30-89
Warren Bowie	RC	3-30-89
Al Camp	RC	4-13-89
Dr. Dominic J. Monetta	RC	4-15-89

Interviews conducted by the author for Second Edition:

John Trick	RC	11-16-00
James Colvaard	RC	12-13-00
Capt. Marc Seidband	RC	1-25-01
Stephen Mitchell	RC	2-9-01
Al Meaders	RC	4-19-01
Mary Lacey	RC	5-8-01, 5-24-01

Tapes and abstracts on deposit at Indian Head Technical Library, Naval Ordnance Station, Indian Head, Maryland.

*RC = Rodney Carlisle; JG = James Gilchrist; EC = Edith A. Chalk; VH = Vincent C. Hungerford.

For the second edition the author conducted other informal interviews and discussions with Joe Shannon, Chris Adams, Dennis Chappell, Patsy Morgan, Jane Woods, Teresa Ferrero, Tara Landis, and Capt. Ed Nichols (Ret.).

APPENDIX 4

Milestones in Technology and Management

1890	Naval Proving Ground opened at Indian Head
1890	Ensign Robert Dashiell, officer in charge, patented gun breech [see Appendix 2]
1891	First shots fired at Indian Head
1891	Nickel-steel armor tested at Indian Head
1892	Harveyed steel armor tested
1890–93	Gun and mount design changes instituted from Indian Head testing
1893	Lt. Newton Mason began housing, medical program
1890–99	Experimental work conducted at Newport with smokeless powder; George Patterson worked under Charles Munroe
1898	Dashiell guns fired in Spanish-American War
1900–02	Lt. Comdr. Joseph Strauss was officer in charge, later Chief, Bureau of Ordnance 1913–16, and Commander, North Sea Mine Barrage, 1917–18 [see Appendix 2]
1900	Smokeless powder production began at Indian Head; George Patterson was chief chemist
1906	Sulfuric acid plant constructed
1905–08	Patterson utilized rosaniline as stabilizer in smokeless powder
1908	Patterson shifted to diphenylamine stabilizer, standard since then
1909–11	$100,000 experimental budget produced over 100 separate studies
1911–14	Program of experimental ordnance work conducted
1912	Du Pont divided by court order, Hercules set up as consequence of Indian Head cost analysis
1913–16	Indian Head powder production capacity quadrupled to replace private sector
1913	Gunnery officer fired past President Wilson's yacht, criticized for zeal
1914	World War I began
1917	United States entered World War I
1917	Explosive D plant opened
1918	World War I ended
1919	Government-built housing opened
1919	Railroad link opened
1919–24	Experimental Ammunition Unit operated at Indian Head, later incorporated into Naval Ordnance Laboratory
1921	*Ostfreisland* bombing tests conducted
1921	All gun testing shifted to Dahlgren, operated until 1932 as Indian Head's Lower Station
1922	Army Navy Munitions Board consolidated propellant and explosive purchasing
1923	Name changed to Naval Powder Factory
1923	Specifications set for density of Explosive D

1924–27	Variable Delay Armor Penetration Fuze work conducted
1925–28	Harold R. Stark was officer in charge, later Chief of Naval Operations, 1939–42 [see Appendix 2]
1935–40	Powder production increased as fleet expanded under Franklin Roosevelt
1936	German and British progress in powder studied
1939	Germany invaded Poland, World War II began
1940	George Patterson retired; 50 years of service
1941	Production of smokeless powder and Explosive D increased; plant expanded
1941	Pearl Harbor attacked: United States in World War II
1942	Extrusion presses set up for rocket grains
1942–44	Section of Jet Propulsion Laboratory operated at Indian Head, later shifted to JPL, Pasadena
1942	Explosives Investigation Laboratory consolidated at Indian Head, origin of Explosive Ordnance Technology Center
1944	First testing of Jet Assist Takeoff Units (JATOs); large static test stands built
1945	World War II ended
1946	Research and Development Department established
1947	Bureau of Ordnance approved pilot plants for Indian Head
1947	Explosives Investigation Laboratory converted to tenant activity
1950	Korean War began
1952	2.75″ rocket grain production increased for Korean War
1954	Biazzi nitroglycerin plant began production
1954	Casting plant opened
1954	Terrier booster grains produced
1954	Korean War ended
1958	First Polaris work at Indian Head
1958	Commander Jim Dodgen appointed as technical director
1958	Name changed to Naval Propellant Plant
1958	Plastisol nitrocellulose produced
1959	Bureau of Ordnance, Bureau of Aeronautics merged in Bureau of Weapons
1959	Construction of 23 new buildings for Polaris started
1959	Bullpup sustainers produced and shipped
1959	Produced high-bulk density nitroguanidine
1959	Assistant Management Board established in R and D Department
1960	Patent granted on Otto Fuel II
1960	Polaris base grain facilities completed
1960	Sidewinder propulsion and gas units produced
c1960	Terrier, Talos, Tartar produced as cast motors
1961	RCA 501 computer on line for ballistic evaluation
1961	Production start-up of nitroplasticizers for Polaris
1962	Joe L. Browning appointed first civilian technical director
1962	Produced high-energy casting powders for advanced rockets
1962	Retooled Biazzi plant for production of Otto Fuel

1962–63	Propellant Plant Assistant Management Board initiated
1963	Bureau of Weapons split: NAVORD, NAVAIR
1963	Inert Diluent Plant construction started
1962–66	X-259 (Athena), X-248 (Scout) experimental space rocket work conducted
1964	Gulf of Tonkin Incident; Vietnam war intensified
1965	Plastic explosive C-3 produced in emergency
1965	2.75″ rocket production revived for Vietnam
1966	Name changed from Naval Propellant Plant to Naval Ordnance Station
1966	Improved Zuni rocket produced
1967	Inert Diluent Plant opened
1967	Project Gunfighter initiated, long-range ordnance studied
1967	Poseidon casting powder C3 produced
1967	Sidewinder 1C propellent processed by pneumatic mix
1967	Production of NACO gun propellant started
1968	Gunfighter projectile 8″/55 hits 500-yard target at 43 miles
1968	Rocket Assisted Projectiles supplied to fleet
1969	"Peacetime retrenchment" 3,500 to 2,000 employees
1970	SCAMP antimaterial projectile developed
1970	Research and Development transferred to Naval Ordnance Laboratory, remained as tenant
1971	Adoption of NOSOL, or no-solvent, propellant
1972	Long Range Bombardment Ammunition shipped to fleet
1973	Liquid gun propellant research begun
1973	Cartridge Actuated Device and Propellant Actuated Device work consolidated for three armed services at Indian Head
1974	NAVSHIP and NAVORD combined in NAVSEA
1974	Training and simulator missiles constructed
1974	Unsymmetrical dimethylhydrazine produced, replacing private source
1975	David Lee appointed as second civilian technical director
1975	Ordnance Environmental Support Office established at Indian Head
1976	Program management system installed
1977	SEAADSO established at Indian Head as tenant, combining ship and ordnance computer systems; later SEAADSA
1981	Personnel Development Council re-established, successor to Assistant Management Board
1985	David Lee resigned as technical director
1986	Dr. Dominic J. Monetta appointed as third civilian technical director
1987	Program management shifted to functional departments; support departments shifted to civilian management
1987–89	Six centers of excellence designated by Navy at Indian Head: Cartridge and Propellant Actuated Devices (CAD/PAD) Propulsion (for guns, missiles, and rockets) Warhead Explosives Energetic Chemicals Missile Simulators Explosive and Environmental Safety

1989	Dr. Monetta resigned
1989	Roger Smith appointed as fourth civilian technical director
1990	Indian Head celebrates 100th anniversary
1991	Roger Smith receives Navy Superior Civilian Service Award
1992	Name is changed to IH Division, Naval Surface Warfare Center (IHDIV)
1992	Indian Head is designated as Tri-Service Agent for the 2.75 inch rocket
1993	White Oak personnel (250 +) join IHDIV, result of BRAC 93
1994	CRADA with University Maryland Technical Extension Service is announced
1994	IH becomes center of excellence for energetics
1995	Cast Assembly facility opens
1996	CAD/PAD manufacturing rework facility ribbon cutting
1996	Nine MILCON projects at IH announced, totaling over $54 million
1997	Defense Environmental Quality Award and Natural Resources Conservation Award for Small Institutions
1997	IHDIV receives Hammer Award from VP Gore
1998	The CAD/PAD JPO stands up
1998	Yorktown, Earle, Concord, and Seal Beach detachments are added to Indian Head, bringing 240 personnel into workforce
1998	Functional Ground Test Facility for the Tomahawk missile demonstrated
1999	Roger Smith dies; Mary Lacey appointed as fifth technical director
1999	EOD departs to Eglin Air Force Base, Florida
1999	Chemical Biological Incident Response Force relocates to IH as tenant
2000	Anh Duong of IH, receives ONR Arthur Bisson Prize for Naval Technology Achievement for transition of insensitive PBX into 16 weapons
2000	Ground breaking for Continuous Processing Scale-UP Facility, to include $6.6 million in equipment
2000	Green Energetic Material gun propellant developed
2001	David Packard Excellence in Acquisition Award to CAD/PAD Reengineering Team

APPENDIX 5

Chiefs of the Navy Bureau of Ordnance and its Successors: 1881–2001

Bureau of Ordnance: 1881–1958

1881–1890	COMO Montgomery Sicard
1890–1893	CAPT William M. Folger
1893–1897	COMO William T. Sampson
1897–1904	RADM Charles O'Neil
1904–1904	RADM George A. Converse
1904–1911	*RADM Newton E. Mason
1911–1913	RADM Nathan C. Twining
1913–1916	*RADM Joseph Strauss
1916–1920	*RADM Ralph Earle
1920–1923	RADM Charles B. McVay, Jr.
1923–1927	*RADM Claude C. Bloch
1927–1931	RADM William D. Leahy
1931–1934	RADM Edgar B. Larimer
1934–1937	*RADM Harold R. Stark
1937–1941	RADM William R. Furlong
1941–1943	RADM William H. P. Blandy
1943–1947	VADM George F. Hussey, Jr.
1947–1950	RADM Albert G. Noble
1950–1954	RADM Malcolm F. Schoeffel
1954–1958	RADM Fredric S. Withington

Bureau of Naval Weapons: 1958–1966

1958–1962	RADM Paul D. Stroop
1962–1964	RADM Kleber S. Masterson
1964–1964	RADM Wellington T. Hines (acting)
1964–1966	RADM Allen M. Shinn

Naval Ordnance Systems Command: 1966–1973

1966–1969	RADM Arthur R. Gralla
1969–1972	RADM Mark W. Woods
1973–1973	RADM Roger E. Spreen

Naval Sea Systems Command: 1973–

1973–1976	RADM Robert C. Gooding
1976–1980	VADM Clarence R. Bryan
1980–1985	VADM Earl B. Fowler, Jr.
1985–1988	VADM William R. Rowden

1988–1991	VADM Peter M. Hekman, Jr.
1991–1994	VADM Kenneth C. Malley
1994–1998	RADM George R. Sterner
1998–	VADM George P. Nanos

*Previously served as commanding officer at Indian Head (see Appendix 1).

Notes

CHAPTER 1. *Origins of the Naval Ordnance Station*

1. Dashiell gives an account of this work in a deposition taken in his patent case (hereafter "Deposition"), Record Group (RG) 74, Entry 25, 2793/1892, National Archives and Records Administration, Washington, D.C. (hereafter NARA), p. 48; photograph of Dashiell: *1890 Naval Proving Ground, Naval Powder Factory, Naval Propellant Plant* (Indian Head, 1961), p. 4; shanty life: Dashiell to Captain, 10 March 1891, RG 74, Entry 25, 3338/1890, NARA.

2. John D. Alden, *The American Steel Navy* (Annapolis: Naval Institute Press, 1972); Dashiell's brilliant record: *A Narrative History of the Naval Ordnance Station, Indian Head and of the Gun Systems Division* (manuscript in Naval Ordnance Station collection, n.d., c. 1975), unpaged.

3. The standard gun could get off ten shots in five minutes, the Dashiell could fire ten in three minutes or less: Dashiell to Sampson, 16 February 1893, RG 74, Entry 25, 1007/1893, NARA; breech action: Deposition, RG 74, Entry 25, 2793/1892, NARA, p. 36; for relationship between smokeless powder and gunnery see Philip R. Alger, "Ordnance and Armor," *United States Naval Institute Proceedings* 27 (1901): 532; U.S. Department of the Navy, *Report of the Secretary of the Navy* (Washington, D.C.: U.S. Government Printing Office, 1890), p. 242 (hereafter *Report of the Secretary of the Navy*, for the year given); *Report of the Secretary of the Navy*, 1891, pp. 210–11; *Report of the Secretary of the Navy*, 1892, pp. 203–8, especially p. 208; *Report of the Secretary of the Navy*, 1894, p. 236; *Report of the Secretary of the Navy*, 1896, p. 276; *Report of the Secretary of the Navy*, 1897, p. 309.

4. Professional Notes, *United States Naval Institute Proceedings* 28 (1902): 415. William Hovgaard, in *Modern History of Warships* (U.S. Naval Institute, Annapolis, 1971), p. 428, shows the relationship between smokeless powder and the encouragement of the rapid-fire mechanism, stressing the fact that the new powder would not require the sponging-out step. The concept that the anticipation of one development, smokeless powder, could stimulate another development which would take place earlier, the rapid fire breech, may seem to fly in the face of logic. One assumes that effect should follow cause in a chronological sequence. However, Dashiell explicitly recognized that smokeless powder would need more rapid fire, and Folger's interest in improving the speed of handling apparently derived from anticipating the powder, which did not come into use for more than a decade. One specialist in the history of innovation has called such a phenomenon a "presumed anomaly." Ed Constant, in *The Origins of the Turbojet Revolution*, shows how the anticipated reduction of air-resistance of airplanes to allow for speeds in excess of those achievable with piston engines stimulated the search in Austria, Germany, and Britain for a power plant capable of higher speeds in the 1930s, leading to the near-simultaneous invention of the jet engine in several locations. Edward W. Constant II, *The Origins of the Turbojet Revolution* (Baltimore and London: The Johns Hopkins University Press, 1980).

5. Benjamin F. Cooling, *Gray Steel and Blue Water Navy* (Hamden, Conn.: Archon, 1979), pp. 85–109.
6. Rigged bids and the military-industrial complex: Ibid., pp. 120–37.
7. For Fiske see Clark G. Reynolds, *Famous American Admirals* (New York: Van Nostrand Reinhold, 1978), pp. 117–19; Paolo Coletta, *Admiral Bradley A. Fiske and the American Navy* (University of Kansas, 1979); Fiske's autobiography, *From Midshipman to Rear Admiral* (New York: Century Company, 1919); for Melville: William Ledyard Cathcart, "George Wallace Melville," *JASNE* 24 (1912): 477–511; for Taylor: David K. Allison, Ben G. Keppel, and C. Elizabeth Nowicke, *D. W. Taylor* (manuscript done at David Taylor Naval Ship Research and Development Center, 1987).
8. Dashiell's service record is in the Deposition, RG 74, Entry 25, 2793/1892, NARA, p. 58.
9. For Annapolis conditions: Austin M. Knight to Chief of the Bureau of Ordnance, 14 May 1886, RG 74, Entry 25, 848/1886, NARA.
10. For importance of the barge delays see *Report of the Secretary of the Navy*, 1890, p. 253; Greensbury's Point: Dayton to Chief of the Bureau of Ordnance, 1 August 1891, RG 74, Entry 25, 3919/1891, NARA; for general conditions at Annapolis: Knight to Chief of the Bureau of Ordnance, 14 May 1886, RG 74, Entry 25, 848/1886, NARA; Knight to Chief of the Bureau of Ordnance, 17 March 1886, RG 74, Entry 25, 660/1886, NARA.
11. Deposition, RG 74, Entry 25, 2793/1892, NARA, pp. 31, 46, 59. The Deposition contains the references to: the trolley car, pp. 31, 59; "get up a gun like that," p. 46; "shove along with it," p. 59.
12. Ibid., p. 30.
13. Ibid., p. 35.
14. "Utterly worthless contrivance": Dashiell to Ensor, 23 August 1892, RG 74, Entry 25, 2793/1892, NARA.
15. For Russian gunboat visit, discussion of breech mechanism, and Dashiell's goals in constructing it, see Deposition, RG 74, Entry 25, 2793/1892, NARA, pp. 52, 54, 37, 13.
16. Dashiell to Chief of the Bureau of Ordnance, 19 November 1891, RG 74, Entry 25, 6259/1891, NARA; documents pertinent to lawsuit are in a packet in RG 74, Entry 25, 2793/1892, NARA.
17. Dashiell using his own resources: Dashiell to District Attorney Ensor, 23 August 1892, RG 74, Entry 25, 2793/1892, NARA; Sandy Hook story: Folger to Dashiell, 13 October 1892, RG 74, Entry 25, 6219/1892, NARA; case as pending: Phillips to Sampson, 12 July 1894, RG 74, Entry 25, 2793/1892, NARA; Folger's description of suit: Folger to Attorney General, 31 December 1892, RG 74, Entry 25, 2793/1892, NARA; reversal at Court of Appeals: Phillips to Sampson, 14 February 1895, RG 74, Entry 25, 2793/1892, NARA; Supreme Court Decision, 162 U.S. 425, also detailed in RG 74, Entry 25, 2793/1892, NARA; Senate investigation: Senate, *Report of the Secretary of the Navy Relative to the Cost of Armor Plate for Vessels of the U.S. Navy*, 53d Cong., 3d sess., S. Doc. 1453, as cited in Cooling, *Gray Steel*, pp. 132–33.
18. Later use of gun: R. R. Ingersoll, *Text-Book of Ordnance and Gunnery—Compiled and Arranged for the Use of Naval Cadets, U.S. Naval Academy*, 4th ed. (Annapolis: U.S. Naval Institute, 1899), pp. 92–94; Dashiell testing his own gun: *Report of the Secretary of the Navy*, 1892, p. 224; report on gun: Dashiell to Sampson, 16 February 1893, RG 74, Entry 25, 1007/1893, NARA. While the royalty for the Dashiell was half that of the proposed Seabury deal, the Seabury guns would be delivered,

while the Dashiell guns were licensed for manufacture by the Navy at the gun factory. Although Dashiell had assumed that the existing tooling at the gun factory could turn out the breeches, in 1892 the Navy added twenty-four lathes and a variety of other metal-working tools including milling machines and boring mills for a total expense of over $73,000 in order to manufacture the breeches. Conditions at gun factory: Sampson to Commandant of the Navy Yard, 27 September 1892, RG 74, Entry 25, 4745/1892, NARA.

19. Dashiell's opinion on four- and five-inch guns: Dashiell to Chief of the Bureau of Ordnance, 5 June 1893, RG 74, Entry 25, 7390/1892, NARA; five-inch Dashiell guns aboard the *Olympia* as well as 159 aboard other ships: *Report of the Secretary of the Navy*, 1894, p. 249; the role of rapid-fire, smaller ordnance against unarmored portions: Hovgaard, *Modern History of Warships*, p. 391; for the role of the guns aboard the *Olympia* in particular, see Samuel S. Robison, *A History of Naval Tactics from 1530 to 1930: The Evolution of Tactical Maxims* (Annapolis: U.S. Naval Institute, 1942), p. 764. However, it should be noted that not all officers were happy with the results of the new five-inch guns: the after-action report of the *Baltimore* stated that the guns experienced frequent jamming, bent firing pins, and broken retractors. Ron Spector, *Admiral of the New Empire: The Life and Career of George Dewey* (Columbia: University of South Carolina Press, 1974), p. 60. It should be noted that the *Baltimore*'s guns were not fitted with the Dashiell breech.

20. Armor tests 322 feet from gun to target: Mason to Chief of the Bureau of Ordnance, RG 74, Entry 25, 8198/1896, NARA; for an early description of Indian Head: *Report of the Secretary of the Navy*, 1891, pp. 226–27; for new ships: *Report of the Secretary of the Navy*, 1890, pp. 237–38; for 8,000-yard range: Couden to Chief of the Bureau of Ordnance, 4 March 1898, RG 74, Entry 25, 211/1898, NARA.

21. Dock finished 12 December is recorded in Dashiell to Chief of the Bureau of Ordnance, RG 74, Entry 25, 2808/1890, NARA (various dates in this packet of letters, summary letter dated 5 March 1891); first shot at Indian Head and shutdown of Annapolis Proving Ground: *Report of the Secretary of the Navy*, 1891, p. 227.

22. Dashiell using military terms: Dashiell to Captain, 10 March 1891, RG 74, Entry 25, 3338/1890, NARA; Folger's request for goods from Dayton: Dashiell to Captain, 2 March 1891, RG 74, Entry 25, 985/1891, NARA; Dashiell requests and receives a horse: Dashiell to Chief of the Bureau of Ordnance, 9 September 1890, RG 74, Entry 25, 3535/1890, NARA; for scrounging, see Dashiell to Chief of the Bureau of Ordnance, 5 January 1891, RG 74, Entry 25, 72/1891, NARA; for more scrounging, see Dashiell to Chief of the Bureau of Ordnance, 27 January 1891, RG 74, Entry 25, 460/1891, NARA; for even more, see Dashiell to Chief of the Bureau of Ordnance, 16 February 1891, RG 74, Entry 25, 776/1890, NARA; for Dayton's perspective: Dayton to Chief of the Bureau of Ordnance, 18 September 1891, 4862/1891, NARA.

23. Firing of worker seeking paid holidays: *1890, Naval Proving Ground, Naval Powder Factory, Naval Propellant Plant* (Indian Head, 1961), p. 6; first aid: Dashiell to Folger, 4 April 1891, RG 74, Entry 25, 1636/1891, NARA; cowboy style: Dashiell to Captain, 1 March 1891, RG 74, Entry 25, 2251/1893, NARA.

24. *Report of the Secretary of the Navy*, 1892, pp. 223–24; telegraph line: Indian Head Report, September 1892, RG 74, Entry 25, 4890/1892, NARA; chronograph house: Dashiell to Chief of the Bureau of Ordnance, 31 March 1891, RG 74, Entry 25, 1549/1891, NARA; cottage: Dashiell to Folger, 29 October 1891, RG 74, Entry 25, 5777/1891, NARA.

25. Dashiell's form letter reports were emulated by Mason, Couden, and Strauss. Not until 1908 and 1910 do standardized, preprinted forms appear in the RG 74, Entry 25 collection.

26. "River pirates" and contractor delays: letters in Dashiell to Chief of the Bureau of Ordnance, RG 74, Entry 25, 2808/1890, NARA. Summary letter is dated 5 March 1891.

27. Dashiell was so prolific and creative that the files literally bulge with his reports, suggestions, and private correspondence. Most of his gun improvements appear to have been implemented, as revealed by handwritten notations by Folger on many of the memoranda, some of which are cited below.

28. Dashiell to Chief of the Bureau of Ordnance, 29 April 1891, RG 74, Entry 25, 2010/1891, NARA; Dashiell to Chief of the Bureau of Ordnance, 7 May 1891, RG 74, Entry 25, 2010/1891, NARA; Dashiell to Chief of the Bureau of Ordnance, 24 August 1891, RG 74, Entry 25, 4316/1891, NARA; Dashiell to Chief of the Bureau of Ordnance, 6 October 1891, RG 74, Entry 25, 5318/1891, NARA; Dashiell to Chief of the Bureau of Ordnance, 10 December 1891, RG 74, Entry 25, 5394/1891, NARA; Dashiell to Chief of the Bureau of Ordnance, 3 February 1892, RG 74, Entry 25, 1109/1892, NARA; Dashiell to Chief of the Bureau of Ordnance, 3 June 1892, RG 74, Entry 25, 4077/1892, NARA; Dashiell to Chief of the Bureau of Ordnance, 13 June 1893, RG 74, Entry 25, 1138/1893, NARA; Dashiell to Chief of the Bureau of Ordnance, 24 February 1893, RG 74, Entry 25, 1138/1893, NARA; Dashiell to Chief of the Bureau of Ordnance, 7 June 1892, RG 74, Entry 25, 4213/1892, NARA; Dashiell to Chief of the Bureau of Ordnance, 14 June 1892, RG 74, Entry 25, 4357/1892, NARA; Dashiell to Chief of the Bureau of Ordnance, 27 April 1893, RG 74, Entry 25, 2637/1893, NARA.

29. Dashiell to Chief of the Bureau of Ordnance, 25 February 1892, RG 74, Entry 25, 1732/1892, NARA.

30. For development of semi-automatics: Couden to Chief of the Bureau of Ordnance, 19 November 1897, RG 74, Entry 25, 8751/1897, NARA; Sampson to the Secretary of the Navy, 10 December 1897, RG 74, Entry 25, 8751/1897, NARA; ten rounds over again: Howell to Tracy, 18 April 1892, RG 74, Entry 25, 1732/1892, NARA. In 1897 the Navy decided on a Maxim-Nordenfeldt design as the standard semi-automatic design: Secretary of the Navy to Chief of the Bureau of Ordnance, 21 December 1897, RG 74, Entry 25, 8751/1897, NARA.

31. Dashiell to Chief of the Bureau of Ordnance, 23 April 1892, RG 74, Entry 25, 2966/1892, NARA. Two years later, the Driggs gun was retested under the supervision of Dashiell's successor; with some of the modifications which Dashiell had recommended, the gun performed somewhat better. By 1900 the Proving Ground could report that the still "experimental" Driggs semi-automatic had been fired three hundred times, with up to five shots in one four-second period. New Driggs gun report: Dashiell to Chief of the Bureau of Ordnance, 27 September 1894, RG 74, Entry 25, 2966/1892, NARA; Strauss to Chief of the Bureau of Ordnance, 8 March 1900, RG 74, Entry 25, 1645/1898, NARA.

32. Weight of armor: *Report of the Secretary of the Navy*, 1890, p. 253.

33. Nickel from Carnegie and Bethlehem: Cooling, *Gray Steel*; Dashiell to Chief of the Bureau of Ordnance, 12 September 1891, RG 74, Entry 25, 4817/1891, NARA.

34. Report of the Armor Board, 29 September 1891, RG 74, Entry 25, 2060/1892, NARA.

35. *Report of the Secretary of the Navy*, 1893, pp. 300–301; report from Dashiell, 15 June 1893, RG 74, Entry 25, 5720/1893, NARA, pp. 12–13.

36. Cooling, *Gray Steel*, pp. 84–109, passim.

37. *Report of the Secretary of the Navy*, 1893, p. 301.

38. This exchange is contained in Dashiell to Chief of the Bureau of Ordnance, 18 February 1892, Driggs to Dashiell, 27 February 1892, Dashiell to Chief of the Bureau of Ordnance, 2 March 1892, and Dashiell to Chief of the Bureau of Ordnance, 3 March 1892, all found in RG 74, Entry 25, 1544/1892, NARA.

39. *Report of the Secretary of the Navy*, 1892, pp. 209–11; Mason to Chief of the Bureau of Ordnance, 1 July 1896, RG 74, Entry 25, 5032/1896, NARA; Mason to Chief of the Bureau of Ordnance, 9 November 1896, RG 74, Entry 25, 8198/1896, NARA.

40. Dashiell to Chief of the Bureau of Ordnance, 5 June 1893, RG 74, Entry 25, 7390/1892, NARA.

41. In the last year of his life, Dashiell submitted blueprints for yet another rapid-fire gun design. Dashiell to Chief of the Bureau of Ordnance, 13 April 1898, RG 74, Entry 25, 4485/1898, NARA. Before he could see this invention through to completion, Dashiell died in 1899. Obituary reprint in *A Narrative History of the Naval Ordnance Station*, unpaged. *Report of the Secretary of the Navy*, 1893, pp. 306–7.

42. *Report of the Secretary of the Navy*, 1893, p. 307; *Report of the Secretary of the Navy*, 1896, p. 346.

43. *Report of the Secretary of the Navy*, 1893, pp. 306–7.

44. Professional Notes, *United States Naval Institute Proceedings* 26 (1900): 226–30.

45. P. R. Alger, "The High Explosives in Naval Warfare," *United States Naval Institute Proceedings* 26 (June 1900): 245–78. Despite Alger's preference, the Bureau continued to experiment with various types of explosives for loading armor-piercing shells. In 1900 the Proving Ground would report favorably on the use of "Thorite" for that purpose, an explosive containing mostly ammonium nitrate, fused with a small black powder charge. Strauss to Chief of the Bureau of Ordnance, 21 February 1900, RG 74, Entry 25, 3433/1898, NARA.

CHAPTER 2. *Early Powder Manufacturing and Ordnance Testing*

1. Presence of Dashiell guns aboard the *Olympia* and *Raleigh*: U.S. Department of the Navy, *Report of the Secretary of the Navy*, 1894 (Washington, D.C.: U.S. Government Printing Office, 1894), p. 249 (hereafter *Report of the Secretary of the Navy*, for the year given.)

2. L. G. Calkins, "History and Professional Notes," *United States Naval Institute Proceedings* 25 (1899): 270–72, as cited in Samuel S. Robison, *A History of Naval Tactics from 1530 to 1930: The Evolution of Tactical Maxims* (Annapolis: U.S. Naval Institute, 1942), p. 764.

3. Factory begins production: *Report of the Secretary of the Navy*, 1902, p. 517; Ralph Earle, "The Development of Our Navy's Smokeless Powder," *United States Naval Institute Proceedings* 40 (1914): 1051.

4. Lt. Joseph Strauss, "Smokeless Powder," *United States Naval Institute Proceedings* 27 (1901): 733–38; for Munroe's experiments: Report of the Naval Torpedo Station, Newport, RI, 15 September 1891, Record Group (RG) 74, Entry 25, 4830/1891, National Archives and Records Administration, Washington, D.C. (hereafter NARA), pp. 12–26; Munroe to Chief of the Bureau of Ordnance, 5 April 1886, RG 74, Entry 25, 820/1886, NARA; for literature and safety: Munroe to Jewell, 5 August 1891, RG 74, Entry 25, 4343/1891, NARA; Nobel's experiments: Munroe

to Jewell, 5 August 1891, RG 74, Entry 25, 4343/1891, NARA. On the general characteristics of smokeless powder in guns, see Hovgaard, *Modern History of Warships*, pp. 427–29.

5. The *United States Naval Institute Proceedings* digested ordnance and powder information from a wide variety of other journals: cf "Professional Notes" in vols. 25–27; availability of smokeless powder: Bureau of Ordnance to Secretary of the Navy, 6 September 1902, RG 80, Entry 19, 14498/1-3, NARA.

6. For basic testing: du Pont to Folger, 2 June 1891, and Dashiell to Folger, 30 May 1891, both in RG 74, Entry 25, 2622/1891, NARA; Dashiell to Chief of the Bureau of Ordnance, 16 December 1891, RG 74, Entry 25, 3710/1891, NARA; Dashiell to Chief of the Bureau of Ordnance, 29 April 1892, RG 74, Entry 25, 3261/1892, NARA; Dashiell to Chief of the Bureau of Ordnance, 10 December 1892, RG 74, Entry 25, 8295/1892, NARA; and du Pont to the Bureau of Ordnance, 11 August 1897; du Pont to Charles O'Neil, 11 June 1897; Couden to Chief of the Bureau of Ordnance, 9 June 1897; du Pont to Sampson, 25 May 1897, all in RG 74, Entry 25, 8352/1896, NARA; see also: Dashiell to Chief of the Bureau of Ordnance, 26 January 1892, RG 74, Entry 25, 855/1892, NARA; Dashiell to Chief of the Bureau of Ordnance, 22 March 1892, RG 74, Entry 25, 2328/1892, NARA; Dashiell to Chief of the Bureau of Ordnance, 24 March 1892, RG 74, Entry 25, 2392/1892, NARA; Dashiell to Chief of the Bureau of Ordnance, 25 March 1892, RG 74, Entry 25, 2431/1892, NARA.

7. "Professional Notes," *United States Naval Institute Proceedings* 25 (1899): 236–37.

8. For the Folger-Maxim story: Maxim to Secretary of Navy, 22 July 1891, RG 74, Entry 25, 2885/1891, NARA; Hiram S. Maxim, *My Life* (New York: McBride, Nast & Company, 1915), p. 242; for less biased version: du Pont to Chief of the Bureau of Ordnance, 1 August 1891, RG 74, Entry 25, 2885/1891, NARA. Du Pont tested Maxim's powder themselves and reported that Munroe's torpedo factory powder was better "because it has no mixture of nitroglycerine."

9. U.S. preference for guncotton, torpedo factory capacity, and du Pont: *Report of the Secretary of the Navy*, 1892, p. 208; British concern with perforations: "Professional Notes," *United States Naval Institute Proceedings* 28 (1902): 418–20.

10. Lt. Robert W. Henderson, "The Naval Torpedo Station," *United States Naval Institute Proceedings* 29 (1903): 194; need for safer location: Jewell to Chief of the Bureau of Ordnance, 15 April 1892, RG 74, Entry 25, 2948/1892, NARA.

11. Patterson's WPI background: Alumni Office, Worcester Polytechnic Institute, communication with author.

12. Lt. R. W. Henderson, "The Evolution of Smokeless Powder," *United States Naval Institute Proceedings* 30 (1904): 353–72.

13. For blueprints: McLean to Chief of the Bureau of Ordnance, 6 January 1898, RG 74, Entry 25, 184/1898, NARA; for Indian Head factory plans: Couden to Chief of the Bureau of Ordnance, 5 March 1898, RG 74, Entry 25, 184/1898, NARA.

14. Early experiments in Russian powder: report from H. F. Brown, 18 May 1896, RG 74, Entry 25, 4111/1896, NARA; for Bernadou's translation: Bernadou to Inspector in Charge, Torpedo Station, 24 October 1896, RG 74, Entry 25, 8352/1896, NARA; for published version: Lt. John B. Bernadou, trans., "Pyro-Collodion Smokeless Powder," *United States Naval Institute Proceedings* 28 (1897): 644–54, and continued in 29 (1898): 605–16.

15. Strauss, "Smokeless Powder," 734–35; Henderson, "The Evolution of Smokeless Powder," 353–72.

16. For changes in the chemical engineering industry see Eduard Farber, "Man Makes His Material," *Technology in Western Civilization*, vol. 2, ed. Melvin Kranzberg and Carroll W. Pursell, Jr. (London: Oxford University Press, 1967).
17. *Report of the Secretary of the Navy*, 1900, pp. 631–32.
18. *Report of the Secretary of the Navy*, 1903, p. 749.
19. *Report of the Secretary of the Navy*, 1901, p. 705.
20. *Report of the Secretary of the Navy*, 1901, p. 706.
21. Strauss to Chief of the Bureau of Ordnance, 7 August 1907, RG 74, Entry 25, 18953/10, NARA.
22. Strauss, memorandum, 22 April 1908, RG 74, Entry 25, 21816/1, NARA.
23. *Report of the Secretary of the Navy*, 1902, p. 517.
24. *Report of the Secretary of the Navy*, 1903, p. 676.
25. Officer training program: *Report of the Secretary of the Navy*, p. 406.
26. Strauss to Chandler, 16 December 1913, RG 74, Entry 26, Semi-Official Correspondence of the Bureau, Box 3, vol. 1, NARA.
27. *Report of the Secretary of the Navy*, 1902, p. 517.
28. Strauss to Chief of the Bureau of Ordnance, 9 June 1908, RG 181, Acc. 9959, 2103-1, Philadelphia National Archives Branch, Philadelphia, Pa. (hereafter PNAB), pp. 8–9.
29. *Report of the Secretary of the Navy*, 1902, p. 517.
30. Ibid., p. 516.
31. *Report of the Secretary of the Navy*, 1903, p. 747.
32. Cruickshanks departed in 1915 and Rainsford was recommended for his post: Sen. Blair Lee of Maryland to Strauss, 23 July 1915, RG 74, Entry 26, Semi-Official Correspondence of the Bureau, Box 3, vol. 2, NARA. After Olmstead resigned, Strauss argued that the Bureau should not rehire him because he had taken his experience to a private firm, Aetna Explosives Co., to get a better income: Olmstead to Strauss, 19 October 1915, RG 74, Entry 26, Semi-Official Correspondence of the Bureau, Box 3, vol. 2, NARA; Strauss to Hellweg, 21 October 1915, RG 74, Entry 26, Semi-Official Correspondence of the Bureau, Box 3, vol. 2, NARA.
33. Gering's longevity at Indian Head: typescript of Dashiell's "Report of Progress at Station," 15 June 1893, RG 74, Entry 25, 5720/1893, NARA; Gering's promotion from the ranks: typescript of "Annual Report for Fiscal Year 1922," 10 August 1922, RG 181, Acc. 9959, 2103-1-3, PNAB.
34. Dieffenbach to Chief of the Bureau of Ordnance, January 1907, RG 74, Entry 25, 18953/1, NARA.
35. Lab report: 21 January 1907, RG 74, Entry 25, 19265/23, NARA; Laflin and Rand of Wilmington adopt procedure: 14 February 1907, RG 74, Entry 25, 19265/23, NARA.
36. Jackson to Chief of the Bureau of Ordnance, 29 May 1909, RG 74, Entry 25, 22858/1, NARA.
37. Appropriation and sulfuric acid house: Jackson to Chief, 12 May 1910, RG 74, Entry 25, 22858/41, NARA; power house: Jackson to Chase, 4 June 1909, RG 74, Entry 25, 22858/1, NARA; Holden's Report: Holden to Chief of Bureau, 27 June 1911, RG 74, Entry 25, 21816/2, NARA.
38. *Report of the Secretary of the Navy*, 1901, p. 706.
39. Farnum's biographical information: undated public relations office materials, Indian Head (c. 1986), Naval Ordnance Station; Farnum re the Philippines: Farnum to Twining, 5 March 1912, RG 74, Entry 26, Semi-Official Correspondence of the Bureau, Box 2, NARA. Farnum wrote: "As you can readily understand, I have

greatly objected to this transfer as I am forced to go to an undesirable location at no increase of pay and it is well known that the pay at Indian Head like the Philippines has to be higher in order to get men to go there."

40. Petition from the merchants and citizens to the Secretary of the Navy, October 1914, RG 80, Entry 19, 6692/177, NARA.

41. *Report of the Secretary of the Navy*, 1903, p. 749.

42. *Report of the Secretary of the Navy*, 1903, p. 747.

43. Patterson to Inspector, 20 July 1907, RG 181, Acc. 9959, 2103-1, PNAB.

44. *Report of the Secretary of the Navy*, 1902, p. 516.

45. Patterson, Report of Action Taken . . . , 27 July 1911, RG 181, Acc. 9959, 2103-1, PNAB.

46. *Report of the Secretary of the Navy*, 1901, pp. 706–7.

47. "Annoyance": *Report of the Secretary of the Navy*, 1900, p. 631; acquisition: *Report of the Secretary of the Navy*, 1902, p. 515; marines: Holden report typescript, 30 June 1911, RG 181, Acc. 9959, 2103-1, PNAB, p. 11; purchase from Gaffields: *Federal Owned Real Estate Under the Control of the Navy Department* (Washington: GPO, 1937), p. 193.

48. *Report of the Secretary of the Navy*, 1900, pp. 630–31.

49. *Report of the Secretary of the Navy*, 1900, pp. 701–3.

50. *Report of the Secretary of the Navy*, 1901, pp. 703–4.

51. *Report of the Secretary of the Navy*, 1902, p. 515.

52. *Report of the Secretary of the Navy*, 1903, p. 745.

53. Alcohol and water drying problem: typescript annual report, 5 August 1907, RG 181, Acc. 9959, 2103-1, PNAB; 1908 stability tests: typescript report, 9 June 1908, RG 181, Acc. 9959, 2103-1, PNAB, p. 4; 1909 work: typescript report, 30 July 1901, RG 181, Acc. 9959, 2103-1, PNAB, pp. 7–8.

54. *Report of the Secretary of the Navy*, 1908, p. 424.

55. G. W. Patterson, "The Detection of Mercury in Explosives," and "Stability Tests of Smokeless Powders," filed with the cover letter, Jackson to Chief of the Bureau of Ordnance, 2 March 1909, RG 74, Entry 25, 22592/1, NARA; manuscript contents: In the second report, Patterson compared the various tests, establishing the first standards for the Navy.

56. Patterson, "Stability Tests of Smokeless Powders."

57. Report from Patterson, 14 April 1906, RG 74, Entry 25, 19265/6, NARA; Chief of the Bureau of Ordnance to Lt. Comdr. W. L. Howard, U.S. Naval Attache, Berlin, 11 September 1906, RG 74, Entry 25, 19340/4, NARA; Howard to Chief of the Bureau of Ordnance, 21 July 1906, RG 74, Entry 25, 19430, NARA. The last document was apparently incorrectly numbered and filed separately at 19430, although the packet dealing with the German equipment issue is numbered and filed at 19340.

58. Form prepared by C. S. Storm, 3 September 1907, RG 74, Entry 25, 21095/3, NARA. Service tests described in *Method of Investigation and Test of Smokeless Powder for Small Arms and Cannon* (Washington: GPO, 1910), p. 5; and in draft, chap. 4, *Ordnance Pamphlet No. 4* (1921), RG 181, Acc. 9959, 2108-1-2, PNAB, pp. 8–14.

59. Typical "Description of Manufacture" forms occur throughout the 1911, RG 74, Entry 25, NARA files; a sample, dated 9 September 1911, is at 21095/16.

60. Strauss to Chief of the Bureau of Ordnance, 4 December 1906, RG 74, Entry 25, 19265/17, NARA.

61. *Report of the Secretary of the Navy*, 1903, p. 744.
62. *Report of the Secretary of the Navy*, 1907, pp. 468–69.
63. *Report of the Secretary of the Navy*, 1909, pp. 405–6.
64. "Appropriation 'Experiments Bureau of Ordnance, 1910-1911-1912' ": Holden to Chief of the Bureau of Ordnance, 10 December 1911, RG 181, Acc. 9959, 2003-0-2, PNAB; *Report of the Secretary of the Navy*, 1911, p. 225.

CHAPTER 3. *Ordnance Technology in the World War I Era*

1. "Shot Near President," *Washington Post*, 8 July 1913; "President in Danger from 12-Inch Shell," *Virginian-Pilot and the Norfolk Landmark*, 8 July 1913; Grayson's personal relationship with Wilson: Arthur S. Link, *Woodrow Wilson: A Brief Biography* (Cleveland and New York: The World Publishing Company, 1963), p. 67; Jonathan Daniels, *The End of Innocence* (New York: Da Capo Press, 1972), p. 191. Wilson's original plans for the weekend included (1) a three-day cruise 1, 2, and 3 July from Hampton Roads up the Chesapeake and Potomac to Washington, (2) Friday, 4 July at Gettysburg, and (3) 5 and 6 July at Cornish, New Hampshire. As it turned out, Wilson and Grayson extended the New Hampshire stay by a few days to take advantage of a cool spell which set in on Monday; "Wilson to Attend, Will Speak to Gettysburg Veterans July Fourth," *Washington Post*, 29 June 1913; "Wilson Extends Visit," *Washington Post*, 8 July 1913. Wilson often made such *Mayflower* trips with Grayson, for brief vacations without the Secret Service along. Grayson describes one such yacht trip in detail in his memoir; evidence suggests it was the same trip recounted here: Cary T. Grayson, *Woodrow Wilson: An Intimate Memoir* (Washington, D.C.: Potomac Books, 1960), pp. 40–44.
2. Schuyler to Holden, 12 July 1913, Record Group (RG) 80, Entry 19, 6692/156, National Archives and Records Administration, Washington, D.C. (hereafter NARA).
3. "Shot Near President," *Washington Post*, 8 July 1913.
4. Ibid.; "President in Danger from 12-Inch Shell," *Virginian-Pilot and the Norfolk Landmark*, 8 July 1913; Dr. Grayson's prior service at Indian Head: *Report of the Secretary of the Navy* (Washington, D.C.: U.S. Government Printing Office 1904), p. 647 (hereafter *Report of the Secretary of the Navy*, for the year given).
5. "President in Danger from 12-Inch Shell," *Virginian-Pilot and the Norfolk Landmark*, 8 July 1913; "Shot Near President," *Washington Post*, 8 July 1913. Note that the Associated Press story derived from the *Virginian-Pilot*, which had paraphrased the conversation without attribution to source. Internal evidence suggests that Dr. Grayson was the source, as detailed in note 8, below.
6. McCulley to Josephus Daniels, 7 July 1913, RG 80, Entry 19, 6692/156, NARA. The yacht flew the "presidential flag" indicating the president was aboard. In the light of the risk of erratic shots, Schuyler's unilateral decision to conduct an impromptu demonstration without clearing through channels could be characterized by stronger language than McCulley's expression "manifest impropriety." Despite some small sensational treatment in the press, the incident never received front-page treatment. The incident with Woodrow Wilson did not become a blot on Schuyler's career; he was appointed officer in charge at Indian Head, 1931–32, under President Herbert Hoover.
7. Franklin Roosevelt to McCulley, 16 July 1913, RG 80, Entry 19, 6692/156, NARA.
8. McCulley to Daniels, 18 July 1913, RG 80, Entry 19, 6692/156, NARA. The story

of the fall of the rotating band splashing down near the yacht appeared only in the newspaper accounts, and in neither the McCulley nor Schuyler reports of the incident. Did the band really come off and splash down near the yacht? Such bands frequently did break off; the average newsman of the day would have been unlikely to have made up such a technically accurate detail, suggesting that the event occurred as reported and that the source was someone quite familiar with the Proving Ground. The band was described in the press as a 12-inch band; that was an error since the shell was a 14-inch shell. A decade earlier, in 1903, 12-inch guns had been tested but 14-inch had not. The particular error regarding the band diameter would most likely have been made by a former Proving Ground staff member rather than a current one. Grayson had served as Assistant Surgeon at the Proving Ground, August to November 1903. The *Pilot* and other papers quoted Wilson's remarks to Grayson when they were sitting alone and Wilson made no public statement, adding to the indication of Grayson as the probable source for the original news account. On the other hand, Grayson accompanied Wilson over the weekend and the next Monday to New Hampshire, when the story broke. If Grayson gave a story to the Virginia or Washington newspapers, he must have done so by phone from Gettysburg or New Hampshire. Regarding McCulley's denial that any "part of the shells fell within the vicinity of the ship," we need not assume that either he or the newspaper source was lying, since, technically, a rotating band was not "part of a shell" and a 300-foot distance could be regarded as beyond the yacht's "vicinity." An alternate explanation for the apparent discrepancy over the impact of the rotating band could be that the reported splashdown was not visible from McCulley's post but only from the afterdeck.

9. Holden to Daniels, 12 July 1913, RG 80, Entry 19, 6692/156, NARA.
10. Holden to Daniels, 19 July 1913, RG 80, Entry 19, 6692/156, NARA.
11. Clark (acting chief, Bureau of Ordnance) to Daniels, n.d., endorsed 24 July 1913, with a handwritten notation that the Assistant Secretary (Roosevelt) "says file and consider case settled," RG 80, Entry 19, 6692/156, NARA.
12. Typescript "Annual Report for Fiscal Year 1911," by Holden, RG 181, Acc. 9959, 2103-1, Philadelphia National Archives Branch, Philadelphia, Pa. (hereafter PNAB), pp. 17, 19.
13. Creighton to "War Department," 1 October 1908, RG 80, Entry 19, 6692/57, NARA.
14. Hull to von Meyer, 28 May 1909, RG 80, Entry 19, 6692/70, NARA; Secretary of the Navy to Congressman Hull, June 10, 1909, RG 80, Entry 19, 6692/70, NARA.
15. *Report of the Secretary of the Navy*, 1910, pp. 341–42.
16. Holden to Chief of the Bureau of Ordnance, 1 February 1911, RG 74, Entry 25, 24116/1, NARA.
17. Ibid.
18. Twining to Captain Winterhalter, 13 December 1912, RG 80, Entry 19, 6692/141, NARA.
19. House Committee on Naval Affairs, *Estimates Submitted by the Secretary of the Navy*, 62d Cong., 3d sess., 16 December 1912, pp. 225–26. The legend that a ricocheting shell had hit the Virginia side and caused a cow there to dry up, which was later purchased and kept at the Proving Ground, may be rooted in the incident referred to here. No primary documentation to support the legend has been uncovered, although from its repetition in public relations materials, it appeared rooted in uncited oral history sources.

20. The Whiskey Point juke was operated by a man named Mattox and his wife: 7 September 1910, RG 74, Entry 25, 20655/16, NARA.

21. *Federal Owned Real Estate under the Control of the Navy Department* (Washington: GPO, 1937), p. 192.

22. Holden to Chief of the Bureau of Ordnance, 6 January 1913, and Twining to Holden, 8 January 1913, RG 74, Entry 25, 2293/1, Washington National Records Center, Suitland, Md. (hereafter WNRC).

23. Holden to Chief of the Bureau of Ordnance, 13 March 1913, RG 74, Entry 25, 27793/3, NARA.

24. Typescript "Annual Report for Fiscal Year 1913," by Holden, RG 181, Acc. 9959, 2103-1-1, PNAB.

25. Typescript "Annual Report for Fiscal Year 1912," by Holden, RG 181, Acc. 9959, 2103-1-1, PNAB, p. 14.

26. Typescript "Annual Report for Fiscal Year 1916," by Hellweg, RG 181, Acc. 9959, 2103-1-2, PNAB, pp. 4–5.

27. Experimental work at Indian Head: typescript "Annual Report for Fiscal Year 1913," by Holden, RG 181, Acc. 9959, 2103-1-1, PNAB, pp. 5–6.

28. Typescript "Annual Report for Fiscal Year 1914," by Hellweg, RG 181, Acc. 9959, 2103-1-1, PNAB.

29. Typescript "Annual Report for Fiscal Year 1912," by Holden, RG 181, Acc. 9959, 2103-1-1, PNAB.

30. For changes in chemical laboratory: typescript annual reports for the fiscal years 1912–16, RG 181, Acc. 9959, 2103-1-1 and 2103-1-2, PNAB; Farnum's biographical information: undated public relations office materials, Indian Head (c. 1986), Naval Ordnance Station; Farnum to Twining, 5 March 1912, RG 74, Entry 26, Semi-Official Correspondence of the Bureau, Box 2, NARA.

31. Draft of chapter on smokeless powder (chap. 4) written in 1921 for "Ordnance Pamphlet No. 4," RG 181, Acc. 9959, 2108-1-2, PNAB; *Liberté* accident: *Report of the Secretary of the Navy*, 1912, p. 211; Ralph Earle, "The Destruction of the *Liberté*," *United States Naval Institute Proceedings* 37 (1911): 929–42. Hovgaard noted the explosions aboard the *Liberté* as well as another French ship, the *Jena*, in 1907, as affecting American concerns with powder: William Hovgaard, *Modern History of Warships* (Annapolis: U.S. Naval Institute, 1971), pp. 430–31.

32. Draft of chapter on smokeless powder (chap. 4) written in 1921 for "Ordnance Pamphlet No. 4," RG 181, Acc. 9959, 2108-1-2, PNAB. George von Meyer, the Secretary of the Navy, testified to Congress that French smokeless powder had similarities to U.S. powder, but the French "methods of manufacture, blending, and reworking are such as to cause the two powders to be radically different." He added that the French planned to replace their powder with "powder exactly like that used in the United States Navy," House Committee on Naval Affairs, *Estimates Submitted by the Secretary of the Navy*, 62d Cong., 3d sess., 13 January 1913, p. 649.

33. The D. Eng., granted by Worcester Polytechnic Institute in 1932, was identified in the records of WPI: "Class Directory," *The Journal of the Worcester Polytechnic Institute* 52 (September 1948): 8. The degree was granted at WPI by the president of the college, Admiral Ralph Earle, who had been officer in charge for three months at Indian Head in 1916, and bureau chief, 1916–20, before retiring and taking the academic post.

34. Typescript fiscal year reports 1912–15, RG 181, Acc. 9959, 2103-1-1 and 2103-1-2, PNAB; pp. 12–14 in 1912 report, p. 15 in 1913 report, p. 19 in 1914 report, and p. 17 in 1915 report.

35. Typescript "Annual Report for Fiscal Year 1915," by Hellweg, RG 181, Acc. 9959, 2103-1-2, PNAB, p. 16.

36. *Report of the Secretary of the Navy*, 1908, p. 423.

37. House Committee on Naval Affairs, *Estimates Submitted by the Secretary of the Navy*, 62d Cong., 3d sess., 13 January 1913, p. 649.

38. Daniels, *The End of Innocence*, pp. 113–14; annual price cited: *Report of the Secretary of the Navy*, 1916, p. 291.

39. Frank Freidel, *Franklin D. Roosevelt, the Apprenticeship* (Boston: Little, Brown and Company, 1952), pp. 218–19; *Report of the Secretary of the Navy*, 1913, p. 11.

40. Alfred Chandler and Stephen Salsbury, *Pierre S. du Pont and the Making of the Modern Corporation* (New York: Harper and Row, 1977); Mariann Jelinek, *Institutionalizing Innovation* (New York: Praeger, 1979). A contemporary defender: E. G. Buckner, "Is There a Powder Plot?," *Harpers*, 27 June 1914, p. 15; House Committee on Appropriations, *Hearings on H.R. 28186*, 62d Cong., 3d sess., 14, 16, 17 December 1912, pp. 207–66. A fuller exposition of the du Pont position can be found in: Smokeless Powder Department, E.I. du Pont de Nemours & Company, Inc., *A History of the du Pont Company's Relations with the United States Government, 1802–1927* (Wilmington: du Pont, 1928).

41. Freidel, *FDR, the Apprenticeship*, p. 218.

42. Typescript fiscal year reports 1913 and 1914, RG 181, Acc. 9959, 2103-1-1, and "Report of Action Taken During Recent Administration to Effect Economy and to Increase Efficiency," which gives a detailed explanation of the cost accounting system in place in 1911, RG 181, Acc. 9959, 2103-1, PNAB. The initial capital investment by 30 June 1914, had amounted to $1,406,696, as noted in House Committee on Naval Affairs, *Estimates Submitted by the Secretary of the Navy*, 66th Cong., 1st sess., 29 March 1916, vol. 3, p. 3423. In 1915 Daniels applied "interest on investment" rather than depreciation as a modern accountant would: *Report of the Secretary of the Navy*, 1915, pp. 56–57.

43. William S. Stevens, "The Powder Trust, 1872–1912," *Quarterly Journal of Economics* 27:444–81; William S. Stevens, "The Dissolution of the Powder Trust," *Quarterly Journal of Economics* 28 (1913): 202–7. The second article gives a report of the decision of the U.S. District Court of Delaware, 13 June 1912, ordering the breakup of the du Pont holdings. Hercules and Atlas were the two new firms eventually created out of the breakup, with Hercules eventually operating the Parlin, N.J., works.

44. *Hearings on H.R. 28186*, pp. 169–321.

45. Ibid.; *Report of the Secretary of the Navy*, 1913, p. 178; Holden to Twining, February 26, 1912, RG 181, Acc. 9959, 2900-1-1, PNAB.

46. *Report of the Secretary of the Navy*, 1914, p. 234.

47. Typescript "Annual Report for Fiscal Year 1914," by Hellweg, RG 181, Acc. 9959, 2103-1-1, PNAB, p. 14.

48. *Report of the Secretary of the Navy*, 1915, p. 301; *Report of the Secretary of the Navy*, 1916, pp. 289–90.

49. House Committee on Naval Affairs, *Armor Plant for U.S.*, 63d Cong., 3d sess., 24 November 1914.

50. U.S. Bureau of Ordnance, *Navy Ordnance Activities, World War 1917–1918* (Washington, D.C.: U.S. Government Printing Office, 1920), pp. 250–52; Powder Factory

report, 8 August 1919, RG 181, Acc. 9959, 2125-1-3, "Historical Data," PNAB; typescript "Annual Report for Fiscal Year 1917," by Lackey, RG 181, Acc. 9959, 2103-1-2, PNAB, p. 6.

51. Typescript "Annual Report for Fiscal Year 1917," by Lackey, RG 181, Acc. 9959, PNAB, p. 7.

52. Ibid., pp. 4–6; a short informative summary of Indian Head during World War I: *Navy Ordnance Activities*, pp. 250–54, 297, 299–300, 308.

53. "Powder Factory" report, 8 August 1919, RG 181, Acc. 9959, 2125-1-3, Historical Data, PNAB.

54. Ibid.; *Report of the Secretary of the Navy*, 1919, p. 503; the electrical air reduction process had been successfully worked out on the industrial scale in Norway and Germany by 1908: Charles Munroe, "The Nitrogen Question from the Military Standpoint," *United States Naval Institute Proceedings* 35 (1909): 715–27.

55. Explosive D: typescript "Annual Report for Fiscal Year 1917," by Lackey, RG 181, Acc. 9959, 2103-1-2, PNAB, p. 14; Explosive D characteristics: draft of "Ordnance Pamphlet 146," RG 181, Acc. 9959, 2108-1-1, PNAB; Cyclone fence and Explosive D production: typescript "Annual Report for the Fiscal Year 1918," by Lackey, RG 181, Acc. 9959, 2103-1-2, PNAB, pp. 7, 21; general lack of progress: "Powder Factory" report, 8 August 1919, RG 181, Acc. 9959, 2125-1-3, Historical Data, PNAB.

56. Typescript "Annual Report for Fiscal Year 1916," by Hellweg, RG 181, Acc. 9959, 2103-1-2, PNAB, pp. 30–31.

57. Typescript "Annual Report for Fiscal Year 1917," by Lackey, RG 181, Acc. 9959, 2103-1-2, PNAB.

58. Ibid., pp. 15, 23.

59. Ibid., p. 23.

60. Typescript "Annual Report for the Fiscal Year 1918," by Lackey, RG 181, Acc. 9959, 2103-1-2, PNAB, p. 3.

61. Typescript "Annual Report for the Fiscal Year 1919," by Lackey, RG 181, Acc. 9959, 2103-1-2, PNAB, p. 5.

62. Typescript "Annual Report for the Fiscal Year 1918," by Lackey, RG 181, Acc. 9959, 2103-1-2, PNAB, pp. 1–2.

63. Typescript "Annual Report for the Fiscal Year 1917," by Lackey, RG 181, Acc. 9959, 2103-1-2, PNAB, pp. 13–15.

64. Typescript annual reports, 1911–19, RG 181, Acc. 9959, 2103-1, PNAB.

65. *Navy Ordnance Activities*, p. 308.

66. Bureau of Yards and Docks, *Federal Owned Real Estate Under Control of the Navy Department* (Washington: GPO, 1937), pp. 359–60; Kenneth G. McCollum, ed., *Dahlgren* (Dahlgren, Va.: Naval Surface Weapons Center, June 1977), pp. 1–17.

CHAPTER 4. *Indian Head in Ordnance Research, 1921–1941*

1. *The Sun* (Baltimore, Md.), 3 August 1921; for details of incorporation: *Laws of the State of Maryland*, Maryland State Archives, Annapolis, Md., 1920, pp. 1177–91. Incorporation approved 16 April 1920.

2. For additional accounts reflecting the relative inactivity during the 1920s and 1930s see: *1890 Naval Proving Ground, Naval Powder Factory, Naval Propellant Plant* (Indian Head, Md., 1961), p. 43; Public Relations material, "Naval Propellant Plant Product History: 1900 to 1964," n.d., c. 1965, p. 1.

3. World of technology: Russell F. Weigley, *The American Way of War* (New York: Macmillan, 1973), pp. 249–53.

4. Peter Padfield, *Guns at Sea* (London: Hugh Evelyn, 1973), p. 282. In his major text on the warship, William Hovgaard, a naval architect, not an ordnance specialist, described the battle at length and then drew observations from it in connection with various elements of design. It is interesting to note that Hovgaard mentioned the fact that submarines were not present at the battle: William Hovgaard, *Modern History of Warships* (Annapolis: U.S. Naval Institute, 1971), pp. 231–35, 323.

5. Padfield, *Guns at Sea*, p. 278; Hovgaard, *Modern History of Warships*, p. 143.

6. Padfield, *Guns at Sea*, p. 279.

7. Ibid., p. 278; Hovgaard, *Modern History of Warships*, p. 436.

8. Padfield, *Guns at Sea*, p. 279; Hovgaard, *Modern History of Warships*, p. 438.

9. William B. Boyd and Buford Rowland, *U.S. Navy Bureau of Ordnance in World War II* (Washington, D.C.: U.S. Government Printing Office, n.d.), pp. 192–93.

10. Naval Ordnance Laboratory, *History of the Naval Ordnance Laboratory, 1918–1945: Scientific History*, vol. 3 (Washington, D.C.: United States Navy Yard, 1946), report no. 1000, narrative history no. 131c, p. 7 (hereafter *NOL Scientific History*).

11. Naval Ordnance Laboratory, *History of the Naval Ordnance Laboratory, 1918–1945: Administrative History*, vol. 1 (Washington, D.C.: U.S. Navy Yard, 1946), report no. 1000, narrative history no. 131a, pp. 4–5. According to well-established local legend, Dr. Patterson long opposed the establishment of the Naval Ordnance Station at Bryan's Road, Maryland, which would have been close to the Indian Head station, on the grounds that he did not want the added responsibility of the experimental work. Communication with the author from John Murrin.

12. *NOL Scientific History*, pp. 10–15. In 1927, for example, the EAU still continued to rely on the chemistry laboratory at Indian Head for investigation of such questions as the chemical deterioration of spotting projectiles.

13. Typescript "Annual Report for Fiscal Year 1920," by Greenslade, Record Group (RG) 181, Acc. 9959, 2103-1-3, Philadelphia National Archives Branch, Philadelphia, Pa. (hereafter PNAB), p. 18.

14. *NOL Scientific History*, pp. 10–15.

15. Ibid., p. 9.

16. Typescript "Annual Report for Fiscal Year 1924," by Pickens, RG 181, Acc. 9959, 2103-1-3, PNAB, p. 12.

17. *NOL Scientific History*, p. 15. It should be noted that Harry J. Nichols deserved personal credit for developing the fuze. According to a memorandum to the files, he at first did not patent the fuze in order to avoid the requirement that the precise nature of its design be publicized. Accordingly, he indicated, he expected a merit award to make up for his sacrifice of patent rights. In 1929 the Secretary of the Navy approved the award of ten thousand dollars. Then, since the details of the fuze design became more widely known, the Bureau approved a patent to Nichols. In any case, the fuze was never called the "Nichols Fuze," but rather it was designated VD7F, or Variable Delay, Mark 7 fuze. William D. Leahy to Bureau of Supplies and Accounts, 10 April 1929, RG 74, Entry 25, S78-1(26LD), Box S78-1 (26/BDF), Washington National Records Center, Suitland, Md. (hereafter WNRC).

18. *NOL Scientific History*, p. 16.

19. "Action of Navy Fuzes and Bursting Charges—Major Calibre, 1910–1929," n.d.,

RG 74, Entry 25, S78-1 (28/168), National Archives and Records Administration, Washington, D.C. (hereafter NARA), pp. 1–2.

20. Ibid., pp. 3–5.

21. Ibid., pp. 7–8.

22. Ibid., pp. 4–11.

23. Ibid.

24. Boyd and Rowland, *U.S. Navy Bureau of Ordnance in World War II*, p. 140; see also: "Review of Literature on Chemical Agents in Naval Armor-Piercing Shell (1917–1936)," 17 February 1936, RG 74, Entry 25, S78-1 (65/7), NARA, pp. 7–10. During the war, the American University unit had worked on armor-piercing shells which would release toxic gas after penetrating the ship; Indian Head provided the test station for some of the experimental shells. Later work indicated that recrystallization was not a solution: R. L. Beauregard, "History of Navy Use of Composition A-3 and Explosive D in Projectiles," 1 January 1971, NAVORD TR 71-1, Indian Head Technical Library, Naval Ordnance Station, Indian Head, Md. (hereafter IHTL).

25. H. R. Stark, "Diphenylamine—Analysis of Sample from DuPont Company," 23 June 1927, RG 74, Entry 25, S78-1 (51/13), NOS 3592, NARA; Robert H. Connery, *The Navy and the Industrial Mobilization in World War II* (New York: Da Capo Press, 1972), p. 37; Boyd and Rowland, *U.S. Navy Bureau of Ordnance in World War II*, p. 190.

26. Professional Notes, *United States Naval Institute Proceedings* 46 (1920): 1360–61.

27. Peter Hodges, *The Big Gun: Battleship Main Armament 1860–1945* (Annapolis: Naval Institute Press, 1980), p. 85.

28. Extensive test bombings of unarmed, dilapidated battleships by aircraft proved, to some, the effectiveness of bombing by air, while others remained skeptical. For a discussion of the arguments sparked by the Chesapeake tests, see Archibald D. Turnball and Clifford L. Lord, *History of United States Naval Aviation* (New Haven: Yale University Press, 1949), pp. 193–204.

29. Typescript "Annual Report for fiscal year 1920," by Greenslade, RG 181, Acc. 9959, 2103-1-3, PNAB; also, typescript "Annual Report for Fiscal Year 1924," by Pickens, RG 181, Acc. 9959, 2103-1-3, PNAB.

30. The early evolution of air power: Charles M. Melhorn, *Two-Block Fox: The Rise of the Aircraft Carrier, 1911–1929* (Annapolis: Naval Institute Press, 1974), pp. 21–38.

31. Typescript "Annual Report for Fiscal Year 1923," by Black, RG 181, Acc. 9959, 2103-1-3, PNAB, pp. 26–28.

32. Typescript "Annual Report for Fiscal Year 1924," by Pickens, RG 181, Acc. 9959, 2103-1-3, PNAB.

33. Details on the purification of nitrocellulose: "Purification of Nitrocellulose, Final Report," 14 October 1929, RG 74, Entry 25, NP8/S78-1 (51/38), NARA, pp. 1–2; information regarding du Pont's Climatic Trial Magazine stability test: "Development of New Stability Tests," 5 July 1918, RG 74, Entry 25, S78-1 (49/44), WNRC, pp. 1–6; also, "du Pont Stability Test at 80 degrees C. and 95% Relative Humidity," October 5, 1933, RG 74, Entry 25, S78-1 (49/145), NARA, unpaginated.

34. Commander H. F. Leary, "Military Characteristics and Ordnance Design," *United States Naval Institute Proceedings* 48 (July 1922): 1125.

35. Boyd and Rowland, *U.S. Navy Bureau of Ordnance in World War II*, p. 195.

36. Edward S. Farrow, *American Guns in the War with Germany* (New York: E. P. Dutton and Company, 1920), pp. 115–17.

37. Stark to Inspector of Ordnance in Charge, 27 January 1936, RG 181, Acc. 71A-7407, Box 21, Lead Azide, WNRC; Quinn to Farnum, 27 January 1938, RG 181, Acc. 71A-7407, Box 21, Lead Azide, WNRC; Woodbury to Farnum, 17 January 1938, RG 181, Acc. 71A-7407, Box 21, Lead Azide, WNRC.

38. "Specifications for Aluminum and Magnesium Powder for Pyrotechnic Compositions," 24 April 1934, RG 181, Acc. 71A-7407, Box 4, AA Tracers 65-64-0 #2, WNRC; Larimer to Inspector of Ordnance in Charge, 3 April 1933, RG 181, Acc. 71A-7407, Box 4, AA Tracers 65-64-0 #2, WNRC.

39. Inspector of Ordnance in Charge to Bureau of Ordnance, 15 October 1929, RG 74, Entry 25, NP8/S78-1 (51/37), WNRC; funds under National Industrial Recovery Act: Bureau of Ordnance to Inspector of Ordnance in Charge, 28 April 1937, RG 74, Entry 25, NP8/S78, WNRC.

40. Inspector of Ordnance in Charge to Bureau of Ordnance, 28 January 1938, RG 74, Entry 25, NP8/S78-1 (51), WNRC.

41. U.S. Naval Powder Factory Organization Personnel Pamphlet, 30 December 1939, RG 74, Entry 25, NP8/A3, WNRC. Exact personnel figures for each year through the interwar years have not been found in the course of research for this work. Later, retrospective comments and public relations materials indicated a work force in the range of three hundred as a low point and about eight hundred on the eve of World War II.

42. Walter W. Farnum, "Travel Orders," 12 August 1936, RG 74, Entry 25, NP8/P16 (281), NARA, pp. 1–2, 6–7.

43. Coster-Parran Collection, Southern Maryland Studies Center, Charles County Community College, La Plata, Maryland.

CHAPTER 5. *Indian Head: The Navy and the Community, 1890–1940*

1. The scanty records of the period are imprecise as to exact numbers. The estimate is derived from early photographs found in the Coster-Parran Collection, Southern Maryland Studies Center, Charles County Community College; Jack Brown et al., *Charles County Maryland: A History* (So. Hackensack, N.J.: Custombook, Inc., 1976), p. 72.

2. Dept. of the Interior, Census Office, *Population of the United States at the Eleventh Census: 1890*, pt. 1 (Washington, D.C.: U.S. Government Printing Office, 1895), p. 24.

3. Maryland's slow agricultural decline: *Maryland: A History 1632–1974*, ed. Richard Walsh and William Lloyd Fox (Baltimore: Maryland Historical Society, 1974), pp. 397–98; for tobacco farming: Sharon L. Camp, "Modernization: Threat to Community Politics" (Ph.D. diss., Johns Hopkins University, 1976), p. 50.

4. *The Sun* (Baltimore, Md.), 29 April 1910; *The Sun*, 27 April 1910; Camp, "Modernization," pp. 50–51, 55–56.

5. U.S. Department of the Navy, *Report of the Secretary of the Navy* (Washington, D.C.: U.S. Government Printing Office, 1901), p. 705 (hereafter *Report of the Secretary of the Navy*, for the year given).

6. Mason and the doctor: *Report of the Secretary of the Navy*, 1893, p. 307.

7. *The Sun*, 23 August 1974; Mattingly on first Board of Trustees: The Charles County Retired Teachers Association, *A Legacy: One- and Two-Room Schools in Charles County*

(La Plata, Md.: Dick Wildes Printing Company, 1984), p. 29; for first commissioners see, "Minutes: Indian Head Town Commissioners, 9 June 1920, to 7 June 1961," 9 June 1920, Indian Head Town Hall.

8. Retired Teachers Association, *One- and Two-Room Schools*, pp. 23–24.

9. Actions of the Board: "Board Minutes," 28 February 1898, Charles County Board of Education, Administrative Office; school stories: Retired Teachers Association, *One- and Two-Room Schools*, pp. 23–24, and Dorothy B. Artes, "Birth of Indian Head and Her Schools," the collection of Dorothy B. Artes, Indian Head, Md.

10. *Proving Ground Fragments*, 5 December 1919, collection of Dorothy B. Artes, Indian Head, Md.

11. Charles County storm on inauguration day: *The Sun*, 29 April 1910; Jansen at Indian Head: *Proving Ground Fragments*, 5 December 1919, collection of Dorothy B. Artes, Indian Head, Md.

12. *Proving Ground Fragments*, 5 July 1919; *Proving Ground Fragments*, 5 December 1919, both in collection of Dorothy B. Artes, Indian Head, Md.

13. Information on Mr. and Mrs. Evans and the histories of the churches in the collection of Dorothy B. Artes, Indian Head, Md.

14. William B. Rainsford, "A Short Historical Sketch of Events Leading up to the Establishment of Perseverance Lodge No. 208, A. F. & A. M.," 1953, collection of Dorothy B. Artes, Indian Head, Md.

15. Twining to Irvine, 14 September 1911, Record Group (RG) 74, Entry 26, Box 2, 26 May to 29 December 1911, National Archives and Records Administration, Washington, D.C. (hereafter NARA).

16. Strauss to Olmstead, 2 February 1915, RG 74, Entry 26, Box 4, NARA.

17. James B. Allen, *The Company Town in the American West* (Norman: University of Oklahoma Press, 1966), pp. 3–6, 71.

18. Farnum to Twining, 5 March 1912, RG 74, Entry 26, Box 2, NARA; Twining to Farnum, 7 March 1912, RG 74, Entry 26, Box 2, NARA; Indian Head to Twining, 26 February 1912, RG 181, Acc. 9959, 2900-1-1, Philadelphia National Archives Branch, Philadelphia, Pa. (hereafter PNAB).

19. Emma McWilliams, "My Memoirs," p. 25. In her memoir, Mrs. McWilliams tells of her move to Indian Head from a Charles County farm, her experiences as a housekeeper, and her years as a naval station employee. This memoir, which spans seven decades, provides important material regarding the history of Indian Head. We thank John McWilliams, her grandson, for the use of her memoir.

20. *The Sun*, 27 April 1922.

21. Hellweg to Chief of the Bureau, 14 December 1913, RG 181, Acc. 9959, 2900-1-1, PNAB.

22. Naval Proving Ground to Secretary of the Navy, 15 October 1914, RG 181, Acc. 9959, 2605-0-1, PNAB; the beginnings of banking in Indian Head: *The Sun*, 28 September 1916.

23. The system of crossing Mattawoman Creek was related in a conversation between Rodney Carlisle and Bernard Cox; see also *The Sun*, 5 May 1916.

24. Mudd's case for the footbridge: *The Sun*, 5 May 1916; approval for the bridge: *The Sun*, 23 May 1916; opening of the footbridge: photograph, Historian, Naval Ordnance Station, Indian Head, Md.

25. Assigning of enlisted men: U.S. Naval Proving Ground, RG 181, Acc. 9959, 2125-1-3, Historical Data, PNAB; Lackey's observations: Rear Adm. H. E. Lackey,

Retired, "Duty at Indian Head," 20 February 1944, collection of Dorothy Artes, Indian Head, Md.

26. McWilliams, "My Memoirs," p. 28.

27. *Eleventh Census*, Pt. 1, p. 177; Dept. of the Interior, Census Office, *Fourteenth Census of the United States Taken in the Year 1920: Population 1920: Number and Distribution of Inhabitants*, vol. 1 (Washington: GPO, 1921), p. 454. The *Fourteenth Census* lists the population by district in Charles County for the preceding two decades as well as for 1920.

28. Lackey, "Duty at Indian Head," collection of Dorothy Artes, Indian Head, Md.

29. Lackey, "Duty at Indian Head," collection of Dorothy Artes, Indian Head, Md.; transfer of funds to Glasva school: Artes, "Birth of Indian Head."

30. *Proving Ground Fragments*, 5 September 1919, collection of Dorothy B. Artes, Indian Head, Md.

31. Naming school after Lackey: *Proving Ground Fragments*, October 1920, collection of Dorothy B. Artes, Indian Head, Md.; Lackey's return: Lackey, "Duty at Indian Head," collection of Dorothy B. Artes, Indian Head, Md.

32. *Proving Ground Fragments*, 5 August 1919; discontent: "Annual Report for Fiscal Year 1923," RG 181, Acc. 9959, 2103-1-3, PNAB, p. 1.

33. Inspection of Indian Head: *The Sun*, 30 July 1921; larger guns tested at Dahlgren during the war: *The Sun*, 3 August 1921.

34. *The Sun*, 3 August 1921.

35. *The Sun*, 17 May 1922.

36. *The Sun*, 10 March 1922.

37. *The Sun*, 27 April 1922.

38. *The Sun*, 11 June 1922; restriction of work force and output: Indian Head Annual Reports 1920–24, RG 181, Acc. 9959, 2103-1-3, PNAB.

39. First meeting of the commissioners: "Minutes: Indian Head Town Commissioners, 9 June 1920, to 7 June 1961," 9 June 1920, Indian Head Town Hall; *Laws of the State of Maryland* (Annapolis: George T. Melvin, State Printer, 1920), chap. 590.

40. Annual Report for Fiscal Year 1921, RG 181, Acc. 9959, 2103-1-3, PNAB.

41. "Minutes: Indian Head Town Commissioners, 9 June 1920, to 7 June 1961," 30 November 1920, Indian Head Town Hall; electricity to Indian Head according to notes taken from "Old town notebook kept by Mr. Charles E. Wright," collection of Dorothy B. Artes, Indian Head, Md.

42. Johnson to Bureau of Supplies and Accounts, 1 December 1933, RG 74, Entry 25, NP8/L11-4 (5/5), Washington National Records Center, Suitland, Md. (hereafter WNRC), and Wilson to Assistant Secretary of the Navy, 21 June 1937, RG 74, Entry 25, NP8/L11-4 (5/15), WNRC; petition: President of the Potomac and Chesapeake Steamboat Company to Josephus Daniels, 26 October 1914, RG 80, Entry 19, 6692/177, NARA.

43. *The Sun*, 12 September 1922; for jousting: interview with Vince Hungerford; activities at the fairs: *Smokeless Flashes*, 20 December 1957, and *Proving Ground Fragments*, October 1920 and October 1921, both in collection of Dorothy B. Artes, Indian Head, Md.

44. Church cornerstone: *The Sun*, 3 November 1923; school renovation: *The Sun*, 18 May 1928.

45. *Report of the Secretary of the Navy*, 1925, pp. 244–45.

46. Boarded-up houses: *Report of the Secretary of the Navy*, 1926, p. 224; marines on station: Brady and Bertha Bledsoe, interview with James Gilchrist, Indian Head,

Md., 24 January 1989; extra horse for girlfriend: Brady and Bertha Bledsoe, conversation with James Gilchrist, 24 January 1989; Dorothy Beecher Artes, interview by Lois Rand, Indian Head, Md., 21 March 1978, collection of Dorothy B. Artes, Indian Head, Md.

47. McWilliams, "My Memoirs," pp. 35–36.
48. McWilliams, "My Memoirs," p. 40.
49. *The Sun*, 29 October 1933; a full collection of the work of the Civilian Conservation Corps at Indian Head: RG 95, Entry 145, Boxes 128–133, NARA.
50. President Roosevelt on the Potomac river: Haines to Chief of the Bureau of Yards and Docks, 12 July 1940, RG 74, Entry 25, NP8/N1 (24), NARA; President following up on erosion problem: Chief of the Bureau of Ordnance to Inspector of Ordnance in Charge, Naval Powder Factory, 1 November 1940, RG 74, Entry 25, NP8/N1, NARA.
51. "Minutes: Indian Head Town Commissioners, 16 September 1929, to 8 June 1931," Indian Head Town Hall.
52. "Minutes: Indian Head Town Commissioners," 10 September 1931, Indian Head Town Hall; Artes, interview with Rand, 21 March 1978.
53. Condition of Route 224: Haines to Chief of Bureau of Yards and Docks, 11 July 1940, RG 74, Entry 25, NP8/N2 (1), WNRC; condition of Route 224 and shopping trips on boats: "War Time History of the U.S. Naval Powder Factory," pp. 50–53, collection of Dorothy B. Artes, Indian Head, Md.
54. Dr. Frank Susan, interview with James Gilchrist, Indian Head, Md., 24 January 1989.
55. *The Sun*, 26 January 1936.
56. "Minutes: Indian Head Town Commissioners," 21 August 1935, and 23 November 1935.
57. "Minutes: Indian Head Town Commissioners," 21 August 1935.
58. Camp, "Modernization," pp. 56–57.
59. Opening of hospital: *The Sun*, 7 May 1939; new houses: "War Time History of the U.S. Naval Powder Factory," pp. 51, 54–57, and *The Sun*, 26 November 1940, and 26 March 1941; press release: Federal Works Agency, 2 December 1941, Physicians' Memorial Hospital, La Plata, Vertical File, Maryland Department, Enoch Pratt Library, Baltimore, Md.
60. Jay Carsey, interview with Rodney P. Carlisle, Washington, D.C., March 3, 1989.
61. Susan interview, 24 January 1989.

CHAPTER 6. *World War II*

1. Weekly Progress Report, #44, Indian Head Construction Project 6086, 15 November 1941, Record Group (RG) 74, Entry 25, NP8/A1 (164), National Archives and Records Administration, Washington, D.C. (hereafter NARA); Glennon to Chief of the Bureau of Ordnance, 28 January 1942, RG 74, Entry 25, NP7/8A-1 (199), NARA; Project Order 11114-Ord, 3 April 1941, RG 74, Entry 25, NP8/A1-1, NARA; Hersey to Chief of the Bureau of Ordnance, 28 April 1942, RG 74, Entry 25, NP8/S78-1 (4), NARA.
2. Aerojet Corporation founded by California Institute of Technology professors: Clayton R. Koppes, *JPL and the American Space Program: A History of the Jet Propulsion Laboratory* (New Haven: Yale University Press, 1982), pp. 16–17.
3. Albert B. Christman, *Sailors, Scientists, and Rockets* (Washington, D.C.: Naval His-

tory Division, 1971), p. 124; carpet rolls: Blandy to the Chief of Ordnance, War Department (Army), 3 December 1943, RG 74, Entry 25, NP8/S78, NARA.

4. Trip reports: J. B. Nichols and Thomas F. Dixon, 6 August 1942, RG 74, Entry 25, NP8/N5-28, Photographs of California Institute of Technology, NARA; Hersey to Chief of the Bureau of Ordnance, 14 August 1942, RG 74, Entry 25, NP8/S76, NARA; Hussey to Glennon, 20 July 1944, and 5 August 1944, RG 74, Entry 25, NP8/S78, NARA.

5. Christman, *Sailors*, p. 126; trip reports: Nichols and Dixon, 6 August 1942, RG 74, Entry 25, NP8/N5-28, Photographs of California Institute of Technology, NARA.

6. Outgoing Confirmation of Telephone Message, Ensign T. F. Dixon and Lt. C. H. Brooks, 26 November 1942, RG 74, Entry 25, NP8/N5-28, NARA.

7. Hersey to Chief of the Bureau of Ordnance, 10 August 1943, RG 74, Entry 25, N8/P16, NARA.

8. Blandy to Commanding Officer, Powder Factory, 25 November 1943, RG 74, Entry 25, NP8/S78, NARA.

9. Chief of the Bureau of Ordnance to Commanding Officer, 25 March 1944, RG 74, Entry 25, NP8/S78, NARA; Commanding Officer to Chief of the Bureau of Ordnance, 16 November 1944, RG 74, Entry 25, NP8/S78, NARA; Chief of the Bureau of Ordnance to Commanding Officer, 5 September 1944, RG 181, Acc. 71A-7407, Box 29, Conf. 1943, #3, Washington National Records Center, Suitland, Md. (hereafter WNRC); carpet rolls: Chief of the Bureau of Ordnance to Chief of Ordnance (Army), 3 December 1943, RG 74, Entry 25, NP8/S78, NARA.

10. Chief of the Bureau of Ordnance to Commanding Officer, 19 November 1944, and 26 December 1944, RG 74, Entry 25, NP8/S78, NARA.

11. Chief of the Bureau of Ordnance to Commanding Officer, 7 June 1944, RG 74, Entry 25, NP8/S78, NARA.

12. Chief of the Bureau of Ordnance to Commanding Officer, 20 July 1944, and 5 August 1944, RG 74, Entry 25, NP8/S78, NARA.

13. Travel Report: Thames to Commanding Officer, 7 February 1944, RG 181, Acc. 71A-7407, Box 20, Chemist-Powder Expert and Final Records, WNRC.

14. Ibid.; Thames to Chief Chemist, 28 October 1943, RG 181, Acc. 71A-7407, Box 20, Chemist-Powder Expert Problems in Progress, WNRC.

15. Thames to Commanding Officer, trip report, 7 February 1944, RG 181, Acc. 71A-7407, Box 20, Chemist-Powder Expert and Final Records, WNRC; Stern to Farnum, 24 January 1944, RG 181, Acc. 71A-7407, Box 9, Ballistic Lab C60-18-2 General Information, WNRC.

16. Glennon to Chief of the Bureau of Ordnance, 14 January 1944, RG 74, Entry 25, NP8/P16, NARA.

17. Chief of the Bureau of Ordnance to Chief of Naval Operations, 25 April 1944; Chief of the Bureau of Ordnance to Commanding Officer, 2 June 1944, and 31 July 1944; all in RG 74, Entry 25, NP8/S78, NARA.

18. Thames to Chief Chemist, 24 January 1946, RG 181, Acc. 71A-7407, Box 20, Memo to Chief Chemist, WNRC.

19. Glennon to Chief of the Bureau of Ordnance, 7 February 1944, RG 74, Entry 25, NP8/P16/00, NARA.

20. Photoflash rocket: Hussey to Commanding Officer, 21 February 1944; rocket flare: Hussey to Commanding Officer, 21 March 1944; "window" rockets: Hussey to Commanding Officer, 24 January 1944; Jet Propulsion Laboratory, Memorandum Report, "Development of Window Rockets," 10 April 1944; rocket

flare: Glennon to Chief of the Bureau of Ordnance, 11 August 1944; 1,500 pound rocket bomb: Glennon to Chief of the Bureau of Ordnance, 23 October 1944; Jet Propulsion Laboratory, Memorandum Report, 16 June 1944; Jet Propulsion Laboratory, "Development of Illuminating Rockets for Motor Torpedo Boat Use," 28 April 1944; all filed chronologically in RG 74, Entry 25, NP8/S78, NARA.

21. Appleton, Memorandum Report, "Monthly Progress Report, 1 July 1944, Jet Propulsion Laboratory, Indian Head," 4 July 1944, RG 74, Entry 25, NP8/A9, NARA. On 23 October 1944, responding to an 13 October request, Glennon noted that all files connected with the JPL were shipped to Inyokern "at the time this activity was transferred to that station." Last JPL document from Indian Head found in research was 11 August 1944. The actual date of transfer is thus narrowed to August or September 1944.

22. Blandy to Inspector in Charge, 6 June 1942, RG 74, Entry 25, NP8/A1-1, NARA; Hersey to Chief of the Bureau of Ordnance, 10 July 1942, RG 74, Entry 25, NP8/A1-1 S76-1/2, NARA.

23. Hersey to Chief of the Bureau of Ordnance, 10 July 1942, RG 74, Entry 25, NP8/A1-1 S76-1/2, NARA.

24. Blandy to Officer in Charge, Explosives Investigation Laboratory, 30 July 1942, and Ordnance Investigation Memorandum No. 2, "Development of technique of cutting mine and bomb cases by means of special explosives," 28 July 1942, both in RG 74, Entry 25, NP8/A1-1, NARA.

25. Chief of the Bureau of Ordnance to the Chief of Naval Personnel, 24 August 1942, RG 74, Entry 25, NP8/A1-1, NARA. In the cited "references" of this document appointing Klein, the date of the relationship between the EIL and Indian Head becoming "effected" was noted as 25 July 1942; the document establishing the EIL was dated 17 July 1942, which should be regarded as the anniversary of the institution.

26. Hersey to Chief of the Bureau of Ordnance, 10 July 1942, RG 74, Entry 25, NP8/A1-1 S76-1/2, NARA.

27. Variety of memoranda in RG 74, Entry 25, NP8/A1-1, NARA, dated 28 July 1942, 18 January 1944, and 18 November 1944, filed chronologically.

28. EIL progress report on "OIM #125, Ignition of Oil Film on Water," 31 January 1944, RG 74, Entry 25, NP8/A1-1, NARA; EIL Report No. 6, "Trial of Special Shaped Charges," 31 May 1943, RG 74, Entry 25, NP8/S78, NARA.

29. Monthly reports summarized the activities. "Monthly Report of Progress on Experimental and Radiographic Projects Assigned to this Activity," Commanding Officer (EIL) to Chief of the Bureau of Ordnance, 18 July 1944, and 16 November 1944, RG 74, Entry 25, NP8/A9, NARA. By the time of these reports, the CO of the EIL was Lt. T. F. Darrah.

30. Chief of the Bureau of Ordnance to Inspector of Ordnance, 11 December 1942, RG 74, Entry 25, NP8/A-1, NARA.

31. Commanding Officer to Chief of the Bureau of Ordnance, 13 June 1944, RG 74, Entry 25, NP8/A9, NARA; Chief of the Bureau of Ordnance to Commanding Officer, 25 November 1944, RG 74, Entry 25, NP8/A1-1, NARA; "Monthly Report of Progress," 16 November 1944, RG 74, Entry 25, NP8/A9, NARA; Commanding Officer to Commanding Officer of Ordnance Investigation Laboratory, RG 181, Acc. 71A-7407, Box 29, Confidential-1945, WNRC. Note that neither the British nor the Russian ordnances were designated "CEE." Instead, an "FE" number

was assigned to Russian materials, which might have stood for "Friendly" or "Foreign Equipment." The British materials were simply not numbered.

32. "Monthly Report of Progress," 16 November 1944, RG 74, Entry 25, NP8/A9, NARA; Commanding Officer to Commanding Officer of Ordnance Investigation Laboratory, 18 May 1945, RG 181, Acc. 71A-7407, Box 29, Confidential-1945, WNRC.

33. Darrah to Chief of the Bureau of Ordnance, 18 July 1944, RG 74, Entry 25, NP8/A9, NARA.

34. EIL Report No. 6, "Trial of Special Shaped-Charges," 31 May 1943, RG 74, Entry 25, NP8/S78, NARA.

35. Chief of the Bureau of Ordnance to Officer in Charge, 7 March 1952, RG 74, Acc. 0010681, Box 102, NP8, 3-1-52, WNRC.

36. Haines to Assistant Secretary of the Navy, 23 January 1940, RG 74, Entry 25, NP8/P16 (422), NARA.

37. Farnum to Inspector, 25 March 1941, and Chief of the Bureau of Ordnance to Blandy, 20 June 1941, both in RG 74, Entry 25, NP8/S78, NARA.

38. The Tennessee Powder Company: Inspector of Ordnance in Charge to Chief of the Bureau of Ordnance, 11 September 1942, RG 74, Entry 25, NP8/S78-1 (51), NARA; Maumelle Ordnance Works: Inspector of Ordnance in Charge to Chief of the Bureau of Ordnance, 19 June 1942, RG 74, Entry 25, NP8/S78-1 (4), NARA; Powder EX-5021: Glennon to Chief of the Bureau of Ordnance, 15 June 1944, RG 74, Entry 25, NP8/S78 Re2, NARA; Powder EX-5020: Glennon to Chief of the Bureau of Ordnance, 27 April 1944, RG 74, Entry 25, NP8/S78 Re2, NARA.

39. Analyzing propellants: Glennon to Chief of the Bureau of Ordnance, 15 June 1944, RG 74, Entry 25, NP8/S78 Re2, NARA; Kray's trip: Glennon to Chief of the Bureau of Ordnance, 19 July 1944, RG 74, Entry 25, NP8/S78 Re2, NARA.

40. Analysis of pyrotechnic chemicals: Commanding Officer to Commanding Officer, U.S. Naval Ordnance Plant, 7 August 1944, RG 74, Entry 25, NP8/S70, NARA; luting: Commanding Officer to Chief of the Bureau of Ordnance, 10 July 1944, RG 74, Entry 25, NP8/S78, NARA; shellacs: Hersey to Chief of the Bureau of Ordnance, 4 October 1943, RG 74, Entry 25, NP8/A8-3, NARA; revision of JAN specifications: Commanding Officer to Chief of the Bureau of Ordnance, 13 October 1944, RG 74, Entry 25, NP8/S78-1 (4), NARA; cartridge cases and technical reports: Glennon to Chief of the Bureau of Ordnance, 4 October 1944, RG 74, Entry 25, NP8/A8-3, NARA.

41. Hersey to Chief of the Bureau, 20 October 1943, RG 74, Entry 25, NP8/N19 AD3, NARA.

42. Work Report, Chemical Laboratory, 1 December 1945, RG 181, Acc. 71A-7407, Box 24, Experiments: Work Reports, WNRC.

CHAPTER 7. *Postwar Patterns*

1. Hussey to Commanding Officer, 23 June 1944, Record Group (RG) 74, Entry 25, NP8/S78, National Archives and Records Administration, Washington, D.C. (hereafter NARA).

2. Ibid. GALCIT stood for Guggenheim Aeronautical Laboratory, California Institute of Technology. The history of the JATO "GALCIT 61-C" formula is given in Clayton R. Koppes, *JPL and the American Space Program: A History of the Jet Propulsion Laboratory* (New Haven: Yale University Press, 1982), pp. 12–13.

3. EIL report on OIM #218, "X-Ray Examination of Jet Assist Take-Off Units," 24 October 1944, RG 74, Entry 25, NP8/S78, NARA.

4. Bureau of Ordnance letter of 12 March 1945, RG 181, Acc. 71A-7407, Box 12, NPF-19 Re2d 05-1 (426-7), Washington National Records Center, Suitland, Md. (hereafter WNRC).

5. Projects for the Ballistics Laboratory: Moore to Thames, 10 September 1946, RG 181, Acc. 71A-7407, Box 6, Physical Development (general), WNRC, and Moore to Thames, 3 December 1946, RG 181, Acc. 71A-7407, Box 6, Operating Engineering, WNRC; motor tests per month: Hanlon to Chief of the Bureau of Ordnance, 18 March 1947, RG 74, Acc. 5595, Box 119, NP8, 4-1-47, WNRC.

6. JATO inspection: Chief of the Bureau of Ordnance to Commanding Officer, 10 September 1946, RG 181, Acc. 71A-7407, Box 6, Experiments, WNRC.

7. Definitions of JATO units: Chief of the Bureau of Ordnance to Commanding Officer, 5 October 1948, and Chief of the Bureau of Ordnance to Commander U.S. Naval Ordnance Test Station, 26 November 1948, both in RG 181, Acc. 71A-7407, Box 10, Confidential File 1948, WNRC; sample of JATO designations for JATO units from Allegany Ballistics Laboratory: "Current Nomenclature for ABL JATO Units," March 1952, RG 181, Acc. 71A-7407, Box 4, JATO Records, WNRC.

8. Hussey to Commanding Officer, 15 November 1945, RG 181, Acc. 71A-7407, Box 6, Experiments, WNRC.

9. Monthly Progress Report No. 1: Propellant Research and Development Problems, 15 August 1946, RG 181, Acc. 71A-7407, Box 2, Monthly Progress Rept. Misc., WNRC.

10. Ibid.

11. Chief of the Bureau to Commanding Officer, 3 July 1947, RG 74, Acc. 5595, NP8, 1947, WNRC. This document, "The Propellant Research and Development Program at the Naval Powder Factory," prepared by J. S. Brady for Bureau Chief Hussey, was drafted on 30 June 1947, and dated 3 July.

12. Hanlon to Chief of the Bureau, 12 December 1947, RG 74, Acc. 5595, NP8/Re2-L5, 1947, WNRC.

13. "Quarterly Report No. 1" by F. C. Thames, 1 December 1947–29 February 1948, RG 181, Acc. 71A-7407, Box 6, loose material, WNRC.

14. Monthly Progress Report No. 10, 15 June 1947, Propellant Research and Development Problems, RG 181, Acc. 71A-7407, Box 2, WNRC.

15. Internal memorandum by Schoeffel, 26 February 1947, RG 74, Acc. 5595, Box 119, NP8/ES-MHD, 2-16-47 to 2-28-47, WNRC.

16. Ibid.

17. Letter to Sasscer: Hussey to Sasscer, 26 March 1947, RG 74, Acc. 5595, Box 119, NP8, 4-1-47 to 4-30-47, WNRC; phone conversation with Tydings: "Memorandum of Telephone Conversation," Hussey, 27 February 1947, RG 74, Acc. 5595, Box 119, NP8, 3-1-47 to 3-15-47, WNRC; reduction in number of civilian employees: Hussey to Sasscer, 5 March 1947, RG 74, Acc. 5595, Box 119, NP8, 3-1-47 to 3-15-47, WNRC.

18. Internal memorandum,   April 1947, RG 181, Acc. 71A-7407, Box 24, Experiment 551-B, WNRC.

19. Forrestal to Tydings, 12 June 1947, RG 74, Acc. 5595, NP8, 1947, WNRC.

20. Walsh to "A" (internal memorandum), 1 April 1947, RG 181, Acc. 71A-7407, Box 24, Experiment 551-B, WNRC.

21. Pilot plant planning: Hussey to Secretary of the Navy, 4 February 1947, RG 74,

Acc. 5595, Box 119, NP8, 2-1-47 to 2-15-47, WNRC, and Hussey to Secretary of the Navy, 17 March 1947, RG 181, Acc. 71A-7407, Box 24, Experiment 551-B, WNRC; authorization of pilot plant: Hussey to Commanding Officer, 3 June 1947, RG 74, Acc. 5595, NP8, 1947, WNRC.

22. Recommendation to establish another pilot plant: Hanlon to Chief of the Bureau, 18 March 1947, RG 74, Acc. 5595, Box 119, NP8/A MS, 4-1-47 to 4-30-47, WNRC; priorities for plant construction set by Capt. Voegeli: "Quarterly Report No. 1" by Thames, 1 December 1947–29 February 1948, RG 181, Acc. 71A-7407, Box 6, loose material, WNRC; recommendation to set up a production plant for fine-grade nitroguanidine: internal memorandum by Hogarth, 10 November 1948, RG 181, Acc. 71A-7407, Box 6, Nitroguanidine, WNRC.

23. Selection of Mayer for visit: Thames to Farnum, 4 June 1947, and Farnum to Commanding Officer, 3 June 1947, in RG 181, Acc. 71A-7407, Box 13, Nitroglycerine, WNRC; Mayer's travel orders, 1 July 1947, and Mayer's report, 30 September 1947, both in RG 181, Acc. 71A-7407, Box 13, Nitroglycerine, WNRC; essential to establish batch process: Chief of Bureau to Commanding Officer, 14 October 1947, Acc. 71A-7407, Box 13, Nitroglycerine, WNRC. Mayer's arguments might have been more convincing had he gone to Switzerland and spoken with Biazzi himself; as it was, he visited plants following both the Schmidt and Biazzi plans and combinations of them in Britain and Germany.

24. Order to proceed with a batch process plan: Noble to Commanding Officer, 14 October 1947, RG 181, Acc. 71A-7407, Box 6, loose material, WNRC; suspension and progression of the different plants: "Quarterly Report No. 2" by Thames, 1 March 1948-31 May 1948, RG 181, Acc. 71A-7407, Box 6, loose material, WNRC; scheduled plant completion dates: "Quarterly Report No. 1" by Thames, 1 December 1947–29 February 1948, RG 181, Acc. 71A-7407, Box 6, loose material, WNRC; nitroglycerin batches: Thames to Director of Research and Development, 11 December 1951, RG 181, Acc. 71A-7407, Box 8, WG for Production Dept., WNRC.

25. Incorporation of pH meters: Murrell to Resident Officer in Charge of Construction, 17 July 1952, RG 74, Acc. 0010681, Box 102, NP8/MA3, 7-1-52 to 7-31-52, WNRC; drowning tanks: Schoeffel to Chief of Bureau of Yards and Docks, 30 September 1952, RG 74, Acc. 0010681, Box 102, loose material, WNRC.

26. Chief of the Bureau of Ordnance to Commanding Officer, 19 March 1951, RG 181, Acc. 71A-7407, Box 13, Nitroglycerine, WNRC.

27. Chief of the Bureau of Ordnance to Chief of the Bureau of Yards and Docks, 8 April 1952, RG 181, Acc. 71A-7407, Box 13, Nitroglycerine, WNRC. There are two drafts of the letter, both with the same date. Both address funding for the Biazzi plant and for completion of the cast plant. An undated review of Biazzi production, c. August 1954: Biazzi Nitroglycerine Plant, Current Status, RG 181, Acc. 71A-7407, Box 13, Nitroglycerine, WNRC.

28. Chief of the Bureau of Ordnance to Commanding Officer, 13 May 1949, RG 74, Acc. 80040, NP8, WNRC.

29. Chief of the Bureau of Ordnance to Secretary of the Navy, 2 April 1952, RG 74, Acc. 10682, NP8, 1 April 1952–30 June 1952, WNRC. The 2.75-inch rocket was known as the Mighty Mouse. Warren Bowie, interview with Rodney Carlisle, Indian Head, Md., 30 March 1989.

30. Craig to Chief of the Bureau of Ordnance et al., 1 June 1948, RG 74, Acc. 6829, WNRC.

31. Scanland's report on progress on the recommendations: Scanland to Chief of the Bureau of Ordnance, 21 January 1952, RG 74, Acc. 10681, Box 102, NP8, 1 January 1952, through 31 January 1952, WNRC.
32. Commanding Officer to Chief of the Bureau of Ordnance, 23 September 1952, RG 74, Acc. 0010682, NP8, 1 July 1952, through 30 September 1952, WNRC.
33. Joe Browning, interview with Rodney Carlisle, Miami, Fla., 17 February 1989.
34. Ibid.; Project Plan on Weapon A: Benson to Chief of the Bureau, 17 June 1954, RG 181, Acc. 71A-7407, Box 13, Weapon A, WNRC.
35. Ibid.
36. Ibid.
37. Ibid.
38. Ibid.
39. Ibid.
40. Ibid.
41. Scanland to Chief of the Bureau of Ordnance, 7 April 1952, RG 181, Acc. 71A-7407, Box 29, Misc. #2, WNRC.

CHAPTER 8. *Years of Transition, 1955–1963*

1. Jack Raymond, *Power at the Pentagon* (New York: Harper & Row, 1964), gives a readable treatment of the issues of Defense Department management.
2. "Burning Surface Inhibitors for Double-Base Propellants," typescript, 1959, Record Group (RG) 181, Acc. 71A-7407, Box 45, Correspondence June 1959, Washington National Records Center, Suitland, Md. (hereafter WNRC), p. 2.
3. The correlation between Sputnik and missile development is a commonplace of the era; it is documented, for example: House Subcommittee for Special Investigations of the Committee on Armed Services, *Utilization of Naval Powder Factory, Indian Head, Md.*, 85th Cong., 2d sess., 10 and 11 July 1958, p. 595. An excellent recent treatment of the impact of Soviet advances is found in Walter A. McDougall, *. . . the Heavens and the Earth: A Political History of the Space Age* (New York: Basic Books, 1985).
4. Discussions of SJPs: Anciet J. Perk, interview with Rodney Carlisle, Indian Head, Md., 27 April 1989, and Al Camp, interview with Rodney Carlisle, Indian Head, Md., 13 April 1989.
5. Conclusion of a fatal accident investigation: typescript, "Annual Report," by Holden, 30 June 1911, RG 181, Acc. 9959, 2103-1, Philadelphia National Archives Branch, Philadelphia, Pa., p. 17.
6. Scanland to Chief of the Bureau of Ordnance, 7 April 1952, RG 181, Acc. 71A-7407, Box 29, Misc. #2, WNRC.
7. Atkins to Chief of the Bureau of Ordnance, 15 October 1958, RG 181, Acc. 71A-7407, Box 45, no file name, WNRC.
8. Jay Carsey, interview with Rodney Carlisle, Washington, D.C., 3 March 1989, and Joe Browning, interview with Rodney Carlisle, Miami, Fla., 17 February 1989.
9. *Utilization of Naval Powder Factory*, p. 539.
10. Ibid.
11. Browning, interview with Carlisle, 17 February 1989.
12. Corcoran to Benson, 8 July 1954, RG 181, Acc. 71A-7407, Box 49, Quality Control Program, WNRC.

13. Skolnik to Corcoran, 2 December 1954, RG 181, Acc. 71A-7407, Box 49, Quality Control Program, WNRC.

14. Skolnik to Benson, 18 November 1954, RG 181, Acc. 71A-7407, Box 45, Gimlet Rocket, WNRC.

15. Browning, interview with Carlisle, 17 February 1989, and Carsey, interview with Carlisle, March 3, 1989.

16. Task Group 97: W. B. Foster, "Introduction to Panel of 8th Institute on Research Administration at the American University in April 1963," Records of the Office of the Director of Naval Laboratories, R.C. 3-1, Series 4, Acc. #82-18, Box 14, Task 97: Enhancement of DOD In-House R & D, David Taylor Research Center, Carderock, Md.; Bell report: Senate, *Report to the President on Government Contracting for Research and Development*, 87th Cong., 2d sess., 1962, S. Doc. 94, Bureau of the Budget (Washington, D.C.: U.S. Government Printing Office, 1962), pp. 21–22.

17. John Murrin, interview with Rodney Carlisle, Indian Head, Md., 17 January 1989. Some commanding officers also moved into the rocket industry. King went to work for Aerojet: *Utilization of Naval Powder Factory*, pp. 606–7; Atkins went to work for Thiokol: Opal Willis, interview with Rodney Carlisle, Indian Head, Md., 30 March 1989.

18. Murrin, interview with Carlisle, 17 January 1989, and Browning, interview with Carlisle, 17 February 1989.

19. King to Chief of the Bureau of Ordnance, 2 May 1956, RG 181, Acc 71A-7407, Box 49, Correspondence from R&D, WNRC.

20. King to Chief of the Bureau of Ordnance, 19 November 1956, RG 181, Acc. 71A-7407, Box 46, Answer to Pink Route Sheet, WNRC; H. N. Sternberg and E. Roberts, "Recent Naval Powder Factory Developments in the Field of Catapult Propellants," 1956, RG 181, Acc. 71A-7407, Box 38, JAN Paper, WNRC.

21. Murrin, interview with Carlisle, 17 January 1989.

22. Terrier booster grains: "NPF Project Plan (Proposed) FY 59," RG 181, Acc. 71A-7407, Box 48, 510-525/56003/44058, WNRC; type-life accelerated aging programs: Atkins to Chief of the Bureau of Ordnance, 4 August 1958, RG 181, Acc. 71A-7407, Box 48, Terrier Booster, WNRC.

23. Atkins to Chief of the Bureau of Ordnance, 4 August 1958, RG 181, Acc. 71A-7407, Box 48, Terrier Booster, WNRC.

24. "Report of Meeting on ABL Weapon 'A' (950 Yard) Propellant Grain," 15 April 1954, RG 181, Acc. 71A-7407, Box 13, Weapon "A," WNRC.

25. Weapon "A" Meeting, Minutes, 26 October 1955, RG 181, Acc. 71A-7407, Box 49, Jan. 1956 Ltrs. to BuOrd, WNRC.

26. King to Chief of the Bureau of Ordnance, 1 February 1956, RG 181, Acc. 71A-7407, Box 49, Correspondence from R&D, WNRC; Browning, interview by Carlisle, 17 February 1989.

27. Meeting on the malfunction of the 2.75" FFAR: King to Chief of the Bureau of Ordnance, 7 March 1956, RG 181, Acc. 71A-7407, Box 49, Correspondence from R&D, WNRC.

28. The fall and rise of the 2.75" FFAR: John F. Judge, "Navy Meeting Need for 2.75-In. Rocket," reprint from *Missiles and Rockets: The Weekly of Advanced Technology*, 22 November 1965, Indian Head Technical Library, Naval Ordnance Station, Indian Head, Md. (hereafter IHTL); problems with inhibitor wrapping: Camp, interview with Carlisle, 13 April 1989.

29. March 1956 conferences: R. W. Sears, "12.75 Rocket Conference Notes of 13 Mar 1956," RG 181, Acc. 71A-7407, Box 48, Weapon A, WNRC, and "Minutes of Weapon A Meeting Held at the Research and Development Department, Naval Powder Factory, Indian Head, Maryland," 21 March 1956, RG 181, Acc. 71A-7407, Box 49, Jan. 1956 Ltrs. to BuOrd, WNRC; trip report: Worcester to Director, R&D Dept., 25 April 1956, RG 181, Acc. 71A-7407, Box 48, Weapon A, WNRC.

30. Roberts to King, 29 May 1956, RG 181, Acc. 71A-7407, Box 49, Correspondence from R&D, WNRC.

31. F. J. Worcester, "Naval Powder Factory Presentation," August 20–22, 1957, RG 181, Acc. 71A-7407, Box 31, Long Range Planning, WNRC.

32. Ibid.

33. Ibid.

34. Ibid.

35. Ibid.

36. Browning, interview with Carlisle, February 17, 1989.

37. Ibid.

38. Ibid.

39. Ibid.

40. Establishment of the board: Scanland to Chief of the Bureau of Ordnance, 9 May 1952, RG 74, Acc. 0010681, NP8, Box 102, 4-1-52 to 4-30-52, WNRC; report of Advisory Board meeting: Advisory Board to Chief of the Bureau of Ordnance, 27 September 1957, RG 181, Acc. 71A-7407, Box 45, Advisory Committee, WNRC, and King to Chief of the Bureau of Ordnance, RG 181, Acc. 71A-7407, Box 45, Advisory Committee, WNRC.

41. Collections of the minutes of the board: RG 181, Acc. 71A-7407, Box 45, Advisory Committee, WNRC, and Office of the Commanding Officer, Advisory Board Correspondence, Naval Ordnance Station, Indian Head, Md. (hereafter NOS); Wesche and Browning corresponding about agenda items: Browning to Wesche, 13 September 1960, RG 181, Acc. 71A-7407, Box 45, Advisory Committee, WNRC.

42. Emphasis on foundational research: Skolnik to King, 7 January 1958, RG 181, Acc. 71A-7407, Box 45, Advisory Committee, WNRC; Skolnik's remarks about minutes: Skolnik to Director of Production and Engineering, 27 May 1958, RG 181, Acc. 71A-7407, Box 45, Advisory Committee, WNRC.

43. King to Chief of the Bureau of Ordnance, 24 June 1958, RG 181, Acc. 71A-7407, Box 45, Advisory Committee, WNRC.

44. Ibid.

45. Browning, interview with Carlisle, 17 February 1989.

46. Atkins to Chief of the Bureau of Ordnance, 27 September 1958, RG 181, Acc. 71A-7407, Box 45, Advisory Committee, WNRC.

47. Browning, interview with Carlisle, February 17, 1989.

48. Ibid.

49. Ibid.

50. Wesche to Chief of the Bureau of Naval Weapons, 19 October 1960, RG 181, Acc. 71A-7407, Box 45, Advisory Committee, WNRC.

51. Browning's effort to turn over personnel: Browning, interview with Carlisle, 17 February 1989; Advisory Board's comments: Advisory Board to Chief of the Bureau of Ordnance, 29 September 1959, RG 181, Acc. 71A-7407, Box 45, Advisory Committee, WNRC.

52. Browning, interview with Carlisle, 17 February 1989, and Bill Jenkins, interview with Rodney Carlisle, Indian Head, Md., 26 July 1988. The tendency of technical directors of R and D institutions to hire young researchers was widespread in government laboratories: Hans Mark and Arnold Levine, *The Management of Research Institutions: A Look at Government Laboratories* (Washington, D.C.: National Aeronautics and Space Administration, 1984), p. 145.

53. Browning to Chief of the Bureau of Naval Weapons, 14 December 1959, RG 181, Acc. 71A-7407, Box 48, Terrier Booster, WNRC.

54. Browning, interview with Carlisle, 17 February 1989.

55. Ibid.

56. Ibid.

57. Ibid. Browning's memory is confirmed by a letter Wesche wrote to one of the board members, stating in part, "In order to give the new Technical Director complete freedom in establishing a new board, I now consider the current board disestablished." Wesche to Ball, 20 February 1962, Advisory Board Correspondence, Office of Commanding Officer, NOS.

58. Preliminary design studies: Skolnik to Director, Special Projects Office, 5 February 1958, RG 181, Acc. 71A-7407, Box 45, Correspondence, WNRC; proposed water quenching system for the Polaris missile: Skolnik to Director, Special Projects Office, 12 March 1958, RG 181, Acc. 71A-7407, Box 45, Correspondence, WNRC; work with 2-2-DNPOH: "Polaris Propellant Development," Allotment 30610, RG 181, Acc. 71A-7407, Box 16, Answer Pink Route Sheets Jan 1958–June 1958, WNRC.

59. "Cost Summary of Pilot Processing and Process Development for Polaris Program," written in August or September 1958, RG 181, Acc. 71A-7407, Box 46, Polaris Proposal, WNRC.

60. Watt to R, 3 January 1959, RG 181, Acc. 71A-7407, Box 34, Polaris Jan. 1959, WNRC.

61. Ibid.; description of Polaris Project Management Office: "Biweekly Report 1, Polaris B," 6 February 1959, RG 181, Acc. 71A-7407, Box 34, Polaris January 1959, WNRC; Watt to R and P, 9 January 1959, RG 181, Acc. 71A-7407, Box 34, Polaris Jan. 1959, WNRC.

62. Nichols to Watt, 15 January 1959, RG 181, Acc. 71A-7407, Box 34, Polaris Jan. 1959, WNRC.

63. Milestones for Polaris Project: Dodgen to Director, Special Projects Office, 15 January 1959, RG 181, Acc. 71A-7407, Box 34, Polaris Jan. 1959, WNRC; Line of Balance meeting: Watt to Commanding Officer, 23 January 1959, RG 181, Acc. 71A-7407, Box 34, Polaris Jan. 1959, WNRC.

64. Watt to Distribution, 26 October 1959, RG 181, Acc. 71A-7407, Box 34, Polaris July 1959, WNRC.

65. "Status of the Naval Propellant Plant Programs," 17 November 1959, RG 181, Acc. 71A-7407, Box 34, Polaris Dec. 1959, WNRC.

66. Tsao to K1, "Polaris Biweekly Report for Period 2–17 November 1959," November 1959, and "Biweekly Polaris Report," 9 November 1959, both in RG 181, Acc. 71A-7407, Box 34, Polaris Dec. 1959, WNRC.

67. Primary responsibility of Indian Head and Allegany: Smith to Belmore, 6 May 1960, RG 181, Acc. 71A-7407, Box 35, Polaris May 1960, WNRC; scale model motor work: "Agenda Items for Seventh Polaris Second-Stage Alternate Motor Coordination Committee Meeting," 17 May 1960, RG 181, Acc. 71A-7407, Box 41,

Polaris Weekly, WNRC; U.S. Naval Propellant Plant, "Mission and Activities, December 1961," 1961, Box 6, IHTL.

68. J. E. Wilson, "Polaris Progress Summary, Week Ending April 1, 1961," RG 181, Acc. 71A-7407, Box 41, Polaris Weekly, WNRC.

69. Joe L. Browning, "Polaris Reliability Programs Monthly Status Report, Month of June 1962," n.d., RG 181, Acc. 71A-7407, Box 40, no file name, WNRC.

70. Examples of distribution lists: D. J. Quagliarello, "Polaris Progress Summary, Week Ending 29 September 1962," n.d., RG 181, Acc. 71A-7407, Box 41, Ballistics Division Monthly Progress Report, WNRC, and O. F. Dreyer and Joe L. Browning, "Polaris Program Progress Report, 1 July–31 December 1963," 10 April 1964, RG 181, Acc. 71A-7407, Box 40, Polaris Program, WNRC.

71. Polaris funds wind down: Murrin, interview with Carlisle, 17 January 1989; work continuing at NPP: "Polaris Program Progress Report, 1 July–31 December 1963," 10 April 1964, RG 181, Acc. 71A-7407, Box 40, Polaris Program, WNRC, and Joe L. Browning, "Casting Powder Status Report, for the Week Ending 3 April 1965," RG 181, Acc. 71A-7407, Box 33, Career, WNRC; HMX discussion: Perk, interview with Carlisle, 27 April 1989.

CHAPTER 9. *From Naval Propellant Plant to Naval Ordnance Station*

1. Joe Browning, interview with Rodney Carlisle, Miami, Fla., 17 February 1989; mission defined by tasks accrued: John F. Kincaid, "Report of the U.S. Naval Propellant Plant Advisory Board (4th Meeting) 29–30 April 1964," Advisory Board Reports, Office of Commanding Officer, Naval Ordnance Station, Indian Head, Md. (hereafter NOS).

2. Browning, interview with Carlisle, 17 February 1989.

3. Sol Skolnik, "A Case for a Solid Propellant Standards Laboratory," *The Generator*, December 1958, Indian Head Technical Library, NOS (hereafter IHTL), pp. 4–7.

4. Ibid.

5. Ibid. Skolnik ended his article without much of a claim for doing the work at Indian Head. While he started strong and suggested that Indian Head could be the bureau, the references to the early days of the Powder Factory were isolated. Documentary evidence suggests that the material regarding the Powder Factory was inserted at the commanding officer's suggestion; in other words, it was not Skolnik's original intention to argue that the solid propellant standards laboratory be Indian Head but simply that there needed to be one. The article fades off at the end, without a clinch. CO's suggestion: King to Chief of the Bureau of Ordnance, 22 May 1958, RG 181, Acc. 71A-7407, Box 45, Correspondence, Washington National Records Center, Suitland, Md. (hereafter WNRC).

6. Browning, interview with Carlisle, 17 February 1989.

7. Ibid.; Jay Carsey, interview with Rodney Carlisle, 3 March 1989.

8. Naval Ordnance Station, "Naval Ordnance Station, Indian Head, Maryland: Its Role in Technical Support to the Naval Ordnance Systems Command and Chief of Naval Material," ca. 1971, IHTL, pp. 11–14.

9. John Murrin, interview with Rodney Carlisle, Indian Head, Md., 17 January 1989; Advisory Board Reports for 1966, Office of Commanding Officer, NOS.

10. Browning, interview with Carlisle, 17 February 1989.

11. O. A. Wesche and J. E. Dodgen, "Sidewinder 1A: Procedures for the Production

of the Propellant Grain Mk 29 Mod 1 and the Auxiliary Power Unit Mk 1 Mod 0," September 1960, IHTL.

12. Ibid., pp. 5–6.

13. Ibid., p. 11.

14. Ibid., passim.

15. Naval Propellant Plant, "Production of Plastisol Nitrocellulose," *The Generator*, December 1958, IHTL, pp. 1–3.

16. Indian Head product lines: Naval Propellant Plant, "Flow Charts of Products," ca. 1960, IHTL. By 1959 the acid production was dropped and supplies were purchased commercially: Naval Propellant Plant Advisory Board to Chief of the Bureau of Ordnance, 29 September 1959, Advisory Board Reports, Office of Commanding Officer, NOS.

17. Naval Propellant Plant, "Tour Guide," ca. 1962, IHTL. This work is undated, but an internal map was produced in 1961, and the document refers to Browning as technical director. He received that appointment in June 1962.

18. Ibid., p. 5.

19. Ibid., p. 6.

20. Ibid., p. 7.

21. Ibid., pp. 8–9.

22. Ibid., p. 10.

23. Ibid., p. 11.

24. Ibid., pp. 13–14.

25. Ibid., p. 15.

26. Naval Propellant Plant, "NPP Facility Requirements: Technical Briefs," 1 July 1962, and "NPP Facility Requirements Analysis," 1 July 1962, IHTL. Comparison of the facility requirements plans in July 1962 with the plans in January 1963 suggests that a good deal of the Polaris line materials were installed in this period. Plans for 1963: Naval Propellant Plant, "NPP Facility Requirements: Analysis," 1 January 1963, and "NPP Facility Requirements: Technical Briefs," 1 January 1963, IHTL.

27. James Edward Wilson, Jr., "The Research and Development Department" (master's thesis, George Washington University, February 1963), p. 2.

28. Ibid., pp. 65–69.

29. Consulting contract: O. F. Dreyer and Joe L. Browning, "Scientific Services Offered at the National Aeronautics and Space Administration Langley Research Center, Langley, Virginia," 15 March 1964, IHTL; X motors: O. F. Dreyer and Joe L. Browning, "Technical Capability: U.S. Naval Propellant Plant," 25 June 1965, IHTL, pp. 4–5, and Dr. John F. Kincaid et al., "Report of the U.S. Naval Propellant Plant Advisory Board," 3–4 April 1963, Advisory Board Reports, Office of Commanding Officer, NOS, p. 7.

30. O. F. Dreyer and Joe L. Browning, "Design Criteria Guideline for Inert-Diluent Process Production Plant," September 16, 1963, IHTL, pp. 1, 3.

31. Dreyer and Browning, "Design Criteria Guideline," pp. iii, 2.

32. Ibid., passim.

33. Naval Propellant Plant, "Inert Diluent Process Presentation," ca. 1966, IHTL.

34. Ibid. Construction on the Inert Diluent Plant commenced in July 1966 with an expected completion date of December 1967. The contractor was John C. Grimberg Co. of Rockville: "Navy to Build Safer Plant for Propellants," *Evening Star*, 8 July 1966.

35. Murrin, interview with Carlisle, 17 January 1989.
36. Ibid.
37. Ibid.
38. Ibid.
39. Ibid.
40. Ibid.; patent information: King to Commander, U.S. Naval Ordnance Laboratory, 2 January 1958, RG 181, Acc. 71A-7407, Box 16, Answer Pink Route-Jan. 1958–June 1958, WNRC.
41. Murrin, interview with Carlisle, 17 January 1989.
42. C. L. Adams et al., "Hazard Evaluation of Otto Fuel II," 27 July 1966, IHTL.
43. Naval Ordnance Systems Command, "Manufacture of Otto Fuel II by the Biazzi Process," 15 July 1967, NAVORD OD 30980, IHTL, p. 2.
44. Manufacture and handling of Otto Fuel: ibid.; Naval Ordnance Systems Command, "Safety and Handling Instructions, Otto Fuel II, Third Revision," 1 September 1967, NAVORD OP 3368, IHTL; Naval Ordnance Systems Command, "Otto Fuel II: Safety, Storage, and Handling, Fourth Revision," 1 November 1969, NAVORD OP 3368, IHTL; Grainger's adventures: Lu Grainger, interview with Rodney Carlisle, Ft. Washington, Md., 16 May 1989.
45. E. L. Wickens and B. D. Biddix, "Torpedo Mark 46 Mod 1 Igniter/Booster Qualification Program: Final Report," 30 November 1967, IHTL.
46. The minutes for the Advisory Board for these years are in the Advisory Board Reports file in the Office of the Commanding Officer at NOS. A recommendation of name change: O. F. Dreyer to Chief of the Bureau of Naval Weapons, 17 June 1965, Office of Commanding Officer, Advisory Board Reports, NOS.
47. Dreyer and Browning, "Technical Capability: U.S. Naval Propellant Plant," p. 3.
48. Ibid., passim.
49. Ibid., p. 6.
50. Ibid.
51. Ibid., pp. 11–15; in addition, the "capability" report summarized technical capabilities in surveillance, in the NASA Hazards program, in Quality Control, Solid Propellant Surveillance, Rocket Motor Rework, Field Services, Static-Testing Services, Safety, Documentation, Value Engineering, and Computer Services.
52. Ibid., Exhibit 27.
53. Ibid., pp. 17–21.
54. Manpower figures: Wesche to Chief of the Bureau of Naval Weapons, 30 April 1963, and John F. Kincaid, "Report of the U.S. Naval Propellant Plant Advisory Board (4th Meeting), 29–30 April 1964," both in Advisory Board Reports, Office of the Commanding Officer, NOS.
55. John F. Judge, "Navy Meeting Need for 2.75-In. Rocket," *Missiles and Rockets: The Weekly of Advanced Technology*, 22 November 1965, IHTL.
56. David W. Creason and Paul H. Raftery, "2.75-Inch FFAR Cost Reduction Study Project 1: Automatic Recycle of ELBA Solvent," 30 May 1969, IHTR 292, IHTL, pp. 4–6.
57. Leslie R. Olsen and Joe Browning, "Proposal for Testing High-Energy Propellants in Zuni Motors," 13 October 1966, IHTL.
58. Naval Ordnance Systems Command, "Rocket, Line-Throwing, 2.75-Inch, EX 48 Mod 0 Launcher, Line-Throwing, EX 131 Mod 0: Description, Operation, and Maintenance," 15 December 1966, NAVORD OP 3519, IHTL.

59. Jay Thornburg, interview with Rodney Carlisle, Indian Head, Md., 9 February 1989, and Browning, interview with Carlisle, 17 February 1989; both men independently remembered this event and offered it as illustrative of the positive mood in the mid-sixties.

60. "Narrative History of the Naval Ordnance Station, Indian Head, and of the Gun Systems Division," ca. 1973, unpaged, public relations material, IHTL; Carsey, interview with Carlisle, 3 March 1989; Browning, interview with Carlisle, 17 February 1989.

61. "Narrative History, Gun Systems Division," IHTL, passim; Carsey, interview with Carlisle, 3 March 1989; Browning, interview with Carlisle, 17 February 1989.

62. Jeffrey S. Kornblith and J. L. Swedlow, "Design of an Optimum Sabot for Long Range Bombardment Ammunition," 9 July 1971, IHTR 339, IHTL.

63. M. A. Henderson and D. W. Knutson, Sr., CDR, U.S. Navy, "Joint Final Report on the Operational Firings of the 8-Inch Gunfighter Long Range Bombardment Ammunition," 26 October 1970, IHTR 328, IHTL.

64. Ibid.; Browning, interview with Carlisle, 17 February 1989.

65. Kornblith and Swedlow, "Design of an Optimum Sabot," passim.

66. Hans Mark and Arnold Levine, *The Management of Research Institutions: A Look at Government Laboratories* (Washington, D.C.: NASA, Scientific and Technical Information Branch, 1984), pp. 62, 70.

67. Naval Ordnance Station, "Role in Technical Support," passim.

68. Ibid., pp. 2–4.

69. B. W. Frese, Jr., and Joe Browning, "Development and Technology Technical Programs Review, Fiscal Year 1970," 15 January 1971, IHSP 71-68, IHTL.

70. Naval Ordnance Station, "Role in Technical Support," p. 10.

71. Ibid., passim.

72. Frese and Browning, "Development and Technology Technical Programs," p. 1.

73. Ibid., passim.

74. Ibid.

75. Ibid.

76. Ibid.

77. Naval Ordnance Station, "Long Range Planning Task Force Team for Gun Systems: Historical Summary," 30 May 1975, IHSP 75-121, IHTL.

78. Ibid.; Carsey, interview with Carlisle, 3 March 1989.

79. Naval Ordnance Station, "Gun Systems: Historical Summary."

80. Ibid.

81. "Narrative History of the Naval Ordnance Station."

82. Ibid.

83. Stephen E. Mitchell, "Development and Evaluation of NOSOL-363 Gun Propellant," 30 October 1975, IHTR 437, IHTL.

84. Ibid.

85. Grainger, interview with Carlisle, 16 May 1989.

86. Dominic Monetta, communication with the author, August 1989.

87. Browning, interview with Carlisle, 17 February 1989.

CHAPTER 10. *The Post-Vietnam Years*

1. *Maryland Independent* (Waldorf, Md.), 29 October 1975.

2. Through late 1975, after the appointment of Lee, this forum became even less

utilized, and after the reorganization which Lee implemented by January 1976 this meeting structure and its minutes were abandoned: Management Information Notes from August 1972, Box 12, Indian Head Technical Library, Naval Ordnance Station, Indian Head, Md. (hereafter IHTL).

3. Date of Lee's appointment: *Citizen News*, 28 October 1975.

4. Lee's reorganization plan: Naval Ordnance Station, *Profile*, November 1975, Box 6, Pollution, IHTL.

5. Ibid.

6. Lee's file of others' speeches: "Background Speeches," Box 12, IHTL; Lee also kept a file of his predecessors' speeches: "Collected Speeches of JLB, 1959–1974," Box 12, IHTL; Lee prepared a speech for Rear Adm. W. E. Meyer: "Dedication of Morton Thiokol, Inc., Wasatch Div., Brigham City, Utah 3/16/84," Box 13, IHTL; David Lee, "State of the Station Address—1978 by D. E. Lee," and "1981-TD Gives State-of-Station Message," both in Box 13, IHTL.

7. Indian Head exhibits: folder "Reports from the 'TD' D. Lee," Box 13, IHTL; Dave Lee's suggestions and directives: Boxes 12 and 13, IHTL.

8. Jay Carsey, interview with Rodney Carlisle, Washington, D.C., 3 March 1989. For a fuller description of Carsey's relationship with Indian Head, see Jonathan Coleman, *Exit the Rainmaker* (New York: Atheneum, 1989).

9. David Lee, "1976: Technical Director's First Annual Report," 1976, Box 12, IHTL; for more on the public relations effort: Carsey, interview with Carlisle, March 3, 1989.

10. D. E. Lee, "Report from the TD," Box 12, IHTL. The reports in this collection range from April 1976 to March 1981.

11. Bob Spotz, communication with author.

12. Naval Ordnance Station, "Naval Ordnance Station, Indian Head, Maryland," 20 June 1977, Box 12, IHTL.

13. Naval Air Systems Command, "Technical Manual: Description, Preparation for Use, and Handling Instructions: Rocket Catapults and Rocket Motors for Aircrew Escape Systems," 15 December 1972, NAVAIR 11-85-1, IHTL.

14. CAD transfer to NOS: Naval Ordnance Station, "Ordnance Devices: Strategic Planning Conference, Charles County Community College, June 1983" (7 Booklets), Box 12, IHTL; new CAD building: Lee, "Report from the TD," 15 April 1976; Frankford Arsenal: Lee, "Report from the TD," 22 April 1976; B-1 system: Lee, "Report from the TD," 3 June 1976; delay cutter development program: Lee, "Report from the TD," 9 June 1977.

15. Representative at General Dynamics: Lee, "Report from the TD," 29 April 1976; West German aircraft: Lee, "Report from the TD," 3 February 1977; purchase order for ejection system: Lee, "Report from the TD," 10 March 1977; "can-do attitude:" Lee, "Report from the TD," 17 November 1977.

16. Accident investigations: Lee, "Report from the TD," 14 April 1977, and 28 April 1977.

17. Martin Marietta Aerospace: Rolf K. Broderson and Vito O. Bravo, "Fluidic Sequencer Development, Phase II: Improved Accuracy," January 1977, Rept. N00174-76-C-0083, IHTL; more stringent performance "envelopes": Vincent P. Marchese, "Development of a Steel Resonance Tube for the Flueric Cartridge Initiator," October 1977, IHTL; "category 6.3": Lee, "Report from the TD," 10 November 1977. There was some hope for the fluidic system which was recommended for 6.2 funding for FY 79: Lee, "Report from the TD," 28 April 1977.

18. Standard Missile: Lee, "Report from the TD," 5 August 1976, 12 August 1976, and 16 December 1976; booster batch: Lee, "Report from the TD," November 4, 1976; ordnance lots: Lee, "Report from the TD," 26 May 1977; units to rejections: Lee, "Report from the TD," 21 July 1977.
19. Training shapes contract: Lee, "Report from the TD," 27 July 1977; follow-on work with Tomahawk: Lee, "Report from the TD," 3 November 1977; Lu Grainger, interview with Rodney Carlisle, Ft. Washington, Md., May 16, 1989; Vince Hungerford, interview with Rodney Carlisle, Indian Head, Md., 17 January 1989.
20. Lee, "Report from the TD," 3 February 1977.
21. Money to improve JATOs: Lee, "Report from the TD," 10 March 1977; batches: Lee, "Report from the TD," 9 June 1977; Operation Deep Freeze: Lee, "Report from the TD," 8 December 1977.
22. LANCE and Bullpup: Lee, "Report from the TD," 3 February 1977; new Harpoon: Lee, "Report from the TD," 3 March 1977.
23. Lee, "Report of the TD," 24 February 1977.
24. Funding for Sea Gnat: Lee, "Report from the TD," 5 May 1977; design changes: Lee, "Report from the TD," 30 June 1977.
25. Lee, "Report from the TD," 7 April 1977.
26. Lee, "Report from the TD," 12 May 1977.
27. Naval Ordnance Station, "Naval Ordnance Station, Indian Head, Maryland," 20 June 1977, Box 12, IHTL.
28. Ibid.
29. Lee, "State of the Station Address-1978 by D. E. Lee."
30. Ibid., pp. 1–3.
31. Ibid., pp. 5, 8.
32. Ibid., p. 7.
33. Ibid., p. 7.
34. Lee's handwritten notes are in the margins: ibid.; Carsey, interview with Carlisle, 3 March 1989; Grainger, interview with Carlisle, 16 May 1989; Anciet J. Perk, Jr., interview with Rodney Carlisle, Indian Head, Md., 27 April 1989; Hungerford, interview with Carlisle, 17 January 1989.
35. Lee, "State of the Station Address-1978 by D. E. Lee."
36. Ibid., pp. 12–13.
37. David Lee, "Technical Directors Talk with Industry," 1981, and Office of the Technical Director, "NOS/Industry Interchange," 1981, both in Box 13, NOS/Industry Interchange-TD, IHTL.
38. Lee, "Technical Directors Talk with Industry."
39. Attrition problems: Lee, "State of the Station Address-1978 by D. E. Lee," pp. 15–16, view-graphs #14, #15; recruiting efforts: "TD Raps w/Class of 1978, Newly Recruited Engineers," Box 13, IHTL; recruiting trips: Lee, "Report from the TD," 24 November 1976.
40. Otto Fuel production: Lee, "Report from the TD," 30 March 1978, and 24 August 1978; Standard Missile production: Lee, "Report from the TD," 23 March 1978, 24 August 1978, 9 February 1978, and 31 August 1978.
41. Study series: John Fedor and Nick Lakis, "A Continuation of the Gun Weapon System Replacement Program Coordination Effort Study," September 1979, Pub. 1661-01-1-2010, Archive Copy AD-A074295, IHTL; N. Lakis, "Technical and Management Support for the Gun Weapon System Replacement Program," August 1980, Pub. 1665-01-1-2254, IHTL; N. Lakis and M. Hallahan, "Final Report: Plan-

ning Support for Maintenance and Overhaul of Gun Weapon Systems," May 1981, Pub. 1685-01-1-2459, IHTL.

42. N. Lakis, "Technical and Management Support," p. G-3.

43. Ibid., passim.

44. Maintaining armament posture: ibid., G-3; information flow: J. Fedor and H. Mashaw, "A Continuation of the Gun Weapon System Replacement Program Coordination Effort Study," October 1978, Publication 1655-02-2-1818, Archive Copy AD-A061631, IHTL; Fedor and Lakis, "Coordination Effort Study"; Lakis, "Technical and Management Support"; Lakis and Hallahan, "Planning Support." The reports by Fedor and Mashaw and by Fedor and Lakis are summarized in the Fedor and Lakis report on page 7–2. Buying spare parts: N. Lakis, "Technical and Management Support," p. G-7.

45. Lakis, "Technical and Management Support," pp. 3-5 through 3-7, and 3-10 through 3-12. The report incorrectly identifies Newport as "Mayport."

46. OEA, Inc., "Interior Ballistics Computer Model 5-Inch/54 Caliber Navy Gun Using Consolidated NACO Propellant," 1 October 1977, AD B059650, IHTL; fluid-pressure sequencing devices: Marchese, "Flueric Cartridge Initiator," passim; testing methods and devices: E. B. Fisher, "Investigation of Ignition Effects on Closed Bomb Test Results," March 1977, IHTL, pp. i, 7, 21. Fisher's work commented on CPIA M2 Propellant Manual, June 1964.

47. A. Michael Varney and Paul W. Morgan, "Final Report: An Overview of the Status of Flame Spreading Models in Porous Gun Propellants to Identify Areas for Experimental Investigation," 23 February 1978, IHTL. Information on the various laboratories is on pp. 1–2, and the "referee" for laboratory-to-laboratory testing is on pp. 36–37, and in illustration 38.

48. Naval Ordnance Station, "A Briefing for Capt. R. D. Buchwald," July 1979, Box 12, IHTL.

49. Richard Nixon, Executive Order 11507, 4 February 1970, Box 6, Pollution, IHTL; pollution reduction work at Indian Head: F. W. Bewely et al., "Propellant Disposal Facility Phase I: Summary Report," 15 June 1972, Box 6, Pollution, IHTL; K. L. Wagaman and T. J. Sullivan, "Industrial Preparedness Measure: Propellant Disposal/Reclamation Facility Design," 28 September 1973, Box 6, Pollution & Antipollution, IHTL; Naval Ordnance Station, Profile, May 1970, Box 6, Pollution, IHTL; Naval Ordnance Station, Profile, February 1971 and April 1971, both in Box 6, Pollution & Antipollution, IHTL.

50. Support groups: Naval Ordnance Station, "A Briefing for Capt. R. D. Buchwald," passim; CAD/PAD work: S. P. Gary, "1975 Command History: OPNAV Report 5750-1," Box 6, IHTL, p. 7; "cognizant field activity": Naval Ordnance Station, "Briefing for Admiral Steven A. White, Chief of Naval Material, 8/23/84, NOS," Box 12, IHTL.

51. J. Fedor, "Final Summary Report: Depot Acquisition Management and Engineering Support for Close-In Weapon System (CIWS), Phalanx MK 15," Pub. 1695-01-1-2543, IHTL.

52. Commander, Naval Sea Systems Command, to Chief of Naval Operations, 16 August 1977, Record Group (RG) 181, Acc. 85-0075, Merger CENO/CASDO, Washington National Records Center, Suitland, Md. (hereafter WNRC), and Bill Jenkins, interview with Rodney Carlisle, Indian Head, Md., 26 July 1988.

53. Naval Ordnance Station, "Strategic Planning Conference," June 1983 (7 Booklets), Box 12, IHTL.

54. Ibid.; Hungerford, interview with Carlisle, 17 January 1989; Grainger, interview with Carlisle, 16 May 1989.

55. Naval Ordnance Station, "Ordnance Devices: Strategic Planning Conference, Charles County Community College, June 1984" (14 Booklets), Box 13, IHTL; efficacy despite doubts: D. S. Taylor to A, 25 February 1989, D. E. L. Policy Papers, Box 12, IHTL; "Corporate Plan": John R. Sweetman to PMS309, 24 September 1982, RG 181, Acc. 85-0075, Box 9, no file name, WNRC.

56. Naval Ordnance Station, "1985 Corporate Planning Conference," Box 13, IHTL.

57. Ibid., for milestones see the "Surface/Underwater Weapons and Explosives," same file.

58. Naval Ordnance Station, Box 12, D. E. L. Policy Papers, IHTL.

59. Ibid. Evidence that Lee modelled his papers on one by Meyer is found in the fact that he received a Meyer "Guidance and Policy Paper" (#84-31) dated November 1984, which he kept in his file apparently as a model. Memoranda to department heads before that date followed a conventional format, simply headed "To all Department Heads." Memoranda to department heads after that date followed the Meyer format in layout, title, and phrasing.

60. D. E. Lee to All Operating Department Heads, 25 March 1983, D. E. L. Policy Papers, Box 12, IHTL.

61. Ibid.; Perk, interview with Carlisle, 27 April 1989.

62. Establishment of test readiness reviews: D. E. Lee to Distribution, 2 January 1985, Box 13, D. E. Lee Policy Papers, IHTL; Perk, interview with Carlisle, 27 April 1989.

63. Rotation of young engineers: Lee, "Report from the TD," 19 August 1976; AMB-PDC: Naval Ordnance Station, "PDC-AMB Conference: PDC-AMB Membership Directory," 29 April 1988, IHTL, pp. 8, 29.

64. D. Lee to Distribution, memorandum entitled "Technical Director's Guidance and Policy Paper—Engineers and Scientists Development," 3 December 1984, and D. Lee to Distribution, memorandum entitled "Technical Director's Guidance and Policy Paper—Selective Rotation Program," 15 March 1985, both in D. E. L. Policy Papers, Box 12, IHTL.

65. D. S. Taylor to A, 25 February 1986, Box 12, D. E. L. Policy Papers, IHTL.

66. Resume of Dominic J. Monetta; Dominic Monetta, interviews with Rodney Carlisle, Washington, D.C., 15 April and 10 June 1989.

67. Although Monetta did not cite specifically from the work, the terms he used to describe the roles gained currency from Roger Porter, *Presidential Decision Making* (New York: Cambridge University Press, 1980). In 1989 the concept underwent something of a revival with the conscious adoption of the concept in the George Bush administration: *Time*, 21 August 1989, pp. 18.

68. In Porter's version of this concept, the Executive Office of the White House plays the role of "Honest Broker," while the cabinet officers, in the "multiple advocacy" mode, represent the "Robber Barons." Monetta appeared to want to create two degrees of brokerage, calling those closer to him—and presumably more objective—his "Praetorian Guard," yet working to reduce the special-interest role of the department heads. Monetta, interviews with Carlisle, 15 April and 10 June 1989.

CHAPTER 11. *The Second Century Begins*

1. Rodney Carlisle, *Navy RDT&E in an Age of Transition: A Survey Guide to Contemporary Literature* (Washington D.C.: Naval Historical Center, 1997).

2. For complete histories of these facilities, see Jeffery Dorwart with Jean Wolf, *Philadelphia Navy Yard: From the Birth of the U.S. Navy to the Nuclear Age* (Philadelphia: University of Pennsylvania Press, 1999); William Anspacher, Betty H. Gay, Donald E. Marlowe, Paul B. Morgan, and Samuel J. Raff, *The Legacy of the White Oak Laboratory: Accomplishments of NOL/NSWC* (Dahlgren, VA: Naval Surface Warfare Center, Dahlgren Division, 2000); Rodney Carlisle, *Where the Fleet Begins: A History of the David Taylor Research Center* (Washington, D.C.: Naval Historical Center, 1998).

3. *Profile*, 12 January 1990, p. 3.

4. Ibid.

5. *Flash Point*, 11 January 1991, p. 4. The monthly base newspaper changed its name in 1991 from *Profile* to *Flash Point*. An occasional insert, devoted to social events and personal news, the *Arrowhead*, allowed *Flash Point* to focus primarily on technical, policy, and managerial issues, becoming an excellent source for the historical record through the 1990s.

6. *Flash Point*, July 1991, p. 3, 12.

7. *Flash Point*, October, 1991 p. 3.

8. *Flash Point* November and December, 1991 p. 3; "Nerd" editorial: *Flash Point*, December 1992, p.3.

9. *Flash Point*, January 1992, p. 3.

10. *Flash Point*, March 1992, p. 3.

11. Indian Head Division Corporate Operations Office, Economic Impact Report FY 1994, p. 5; the accretion of personnel at Carderock from both its Annapolis field activity and from White Oak are discussed in Rodney Carlisle, *Where the Fleet Begins: A History of the David Taylor Research Center* (Washington, D.C.: Naval Historical Center, 1998), pp. 451-453.

12. Indian Head Division Corporate Operations Office, Command History, 1998, p.20.

13. Trick, interview with Carlisle, 16 November 2000.

14. Ibid.

15. Lacey, interview with Carlisle, 8 May 2000.

16. Indian Head Division, Office of Public Affairs, Press Release, 1 February 1999.

17. U.S. Department of Commerce, Bureau of Export Administration, Office of Strategic Industries and Economic Security, "National Security Assessment of the Cartridge & Propellant Actuated Device Industry," October 1995, p. iv.

18. U.S. Department of Commerce, Bureau of Export Administration, Office of Strategic Industries and Economic Security, "National Security Assessment of the Cartridge & Propellant Actuated Device Industry—An Update" December 2000, p. ii-iii.

19. Indian Head Division, Office of Public Affairs, Press Release, 20 July 1998.

20. Indian Head Division, Marketing Plan, October 1999, p. 18.

21. Meaders, interview with Carlisle, 19 April 2001; IHDIV Marketing Plan, 1999, p. 29.

22. IHDIV Marketing Plan, October 1999, p. 18-23.

23. IHDIV Marketing Plan, October 1999, p. 15.

24. Informal discussions, Captain Seidband with Carlisle, May 2001.

25. Lacey interviews with Carlisle, 8 May 2001; 24 May 2001.

26. Lacey interviews with Carlisle, 8 May 2001, 24 May 2001; Nanos (NAVSEA) and Lockard (NAVAIR) to Assistant Secretary of the Navy for Research, Development, and Acquisition, 23 March 1999. The approach of consolidation emerged from the Senior Navy Oversight Group at a 17 December 1998 briefing on a proposed Navy internal plan for the RDT&E infrastructure, authorized by Section 912C of the 1998 National Defense Authorization Act.

27. *Maryland Independent*, 22 July 1998, and 29 July 1998.

28. *Flash Point*, January 1999, p.9.

29. Lacey interviews with Carlisle, 8 May 2001; 24 May 2001.

30. Program, CECD signing ceremony, 14 December 1998, University of Maryland, College Park.

31. Award Ceremony, 24 May 2001.

32. Edward A. Lustig, Jr., "Design and Development of Improved Insensitive Munitions Warhead for M72A6 Light Anti-Armor Weapon (LAW) System," IHTR 2161, 2 April 1999.

33. Allyn C. Buzzell, "APOBS to 'Fly' with New IHDIV- developed explosive;" "The IM Concept ," (sidebar story), *Flash Point* May 2001, p.1, 9.

34. ONR Press Release, dated 16 May 2000. For a full history of the MAST program, see Rodney Carlisle and William Ellsworth, *Shaping the Invisible: Development of the Advanced Enclosed Mast/Sensor System, 1992-1999* (Washington, D.C.: Office of Naval Research, 2001).

35. Larine Barr, "NOS takes over Production of Mine Clearing Device," *Flash Point*, 11 January 1991, p. 1; comment to author by Harvey Camp.

36. Mitchell, interview with Carlisle, 9 February 2001.

37. IHTR 2051, pp. 3-4, 9-10.

38. Diane Nell, et al., "Explosive Cratering and Channeling of Shallow Water Bottoms," IHTR 1924, 30 September 1996, pp. 1-3.

39. Conversation, Carlisle with John Brough, July 26, 2001.

40. "Continuous Process Scale Up Facility," press release, 26 May 1999, Public Affairs Office, IHDIV.

41. Paul J. Smith, et al., "Application of Microelectromechanical Systems (MEMS) Technology to Safety and Arming (S&A) and Fuzing Systems," IHTR 1842, 1 September 1996, p. 2-3.

42. Memorandum for Acting Assistant Secretary of the Navy for RD&A from David Oliver, Principal Deputy Under Secretary of Defense, 20 April 2001.

43. Indian Head Division Program Management Review, 16 November 2000.

44. Jean A Goertner and Delbert L. Lehto, "Summary of Technical Support for *Seawolf* Shock Test: Potential Impact on Marine-Mammal Hearing," IHTR 1872, 15 May 1996, pp. i , 3-7.

45. Mark A. Hancock, "Navy Pollution Abatement Ashore Program: Ordnance and Energetic Material Environmental Quality Requirements of the Navy and Marine Corps Shoreside Facilities," IHTR 2313, 27 November 2000, pp. viii, and passim.

46. Tim Dunn, et al., "Processing and Combustion of Propellants and Explosives," IHTR 2213, 24 September 1999, pp. v-vi, 1-4.

47. Randall J. Cramer, "Green Energetics: Report to the Strategic Environmental Research and Development Program," IHTR 2321, 12 January 2001, p. v.

48. Susan T. Peters, "Get the Lead Out," IHTR 1863, 29 1996, pp. 1-6.

Glossary of Technical Terms

Amatol: An explosive mixture of ammonium nitrate and TNT.

Ammonium nitrate: NH_4NO_3. A colorless, crystalline salt which serves as a very stable and insensitive high explosive; also used in fertilizer.

Ammonium perchlorate: NH_4ClO_4. A salt that forms colorless or white rhombic and regular crystals; it decomposes at 150° C., and the reaction is explosive at higher temperatures.

Anti-radar missile: (Abbreviated ARM). Missile designed to home in and destroy enemy radar stations; examples include the Standard ARM and the Shrike.

Anti-submarine rocket: (Abbreviated ASROC). Torpedo designed to destroy fast, deep-running submarines. It is a solid-propellant, rocket-driven torpedo that is surface-launched.

Augmented rotating band: See rotating band.

Ammonium picrate: See Explosive D.

Ballistite: A smokeless propellant containing nitrocellulose and nitroglycerin; used in some rocket, mortar, and small-arms ammunition.

Biazzi process: A continuous flow process of producing nitroglycerin and Otto Fuel. It is one of the safest methods known, as very little of the raw product is present in any one place at any one time.

Black powder: A low explosive consisting of an intimate mixture of potassium nitrate or sodium nitrate, charcoal, and sulfur.

Boar: Early U.S. Navy air-to-surface rocket boosted bomb.

Brown powder: A modified form of black powder which uses less sulfur and brown charcoal from partially burned light wood or straw. It burns more slowly than black powder and improves gun ballistics. Also called cocoa powder.

Bullpup: A Navy solid propellant air-to-surface guided missile first produced in the late 1950s with a range of seven to ten miles; it could carry either conventional or nuclear warheads.

C-3: Composite-3, a powerful type of plastic explosive made with tetryl.

CAD: See cartridge actuated device.

Caliber: The diameter of a projectile or the diameter of the bore of a gun or launching tube; for example, a caliber .22 cartridge has a diameter of approximately 0.22 inch (5.6 millimeters).

Cartridge actuated device: (Abbreviated CAD). A small explosive device providing energy and time delay used for a variety of purposes, including ejection of personnel and arms from aircraft.

Casting: A method of producing propellants by mixing the components in a mold and heating and curing the filled mold until the components form a single, solid mass.

Cathode-ray oscilloscope: A test instrument which uses a beam of electrons to make rapidly changing electrical quantities visible on a fluorescent screen.

Cellulose: $(C_6H_{10}O_5)_n$. The main polysaccharide in living plants, forming the skeletal structure of the plant cell wall; a polymer of B-D-glucose units linked together, with the elimination of water, to form chains comprising 2,000–4,000 units.

Cellulose acetate: An acetic acid ester of cellulose; a tough, flexible, slow-burning, and long-lasting thermoplastic material used for a variety of purposes.

Centrifuge: Rotating device used to separate liquids with different densities or to separate particles suspended in a liquid.

Chromatography: A method of separating and analyzing chemical mixtures by determining how rapidly the substances are absorbed by a known solid material such as activated carbon or silica gel.

Cocoa powder: See brown powder.

Cordite: Type of nitrocellulose prepared by treating cotton fiber or purified wood pulp with a mixture of nitric and sulfuric acids to produce an explosive, smokeless powder. Also known as pyrocellulose.

Cordite-N: A derivative of cordite made with nitroglycerin and nitroguanidine.

Cosgrove process: A method of hardening armor plate by welding a hard nickel-steel face to a softer backing of the same material.

Cure: To change the properties of a resin material by chemical polycondensation or additional reactions.

Cyclonite: (Abbreviated RDX). $(CH_2)_3N_3(NO_2)_3$. Highly explosive white crystalline powder. Also known as cyclo tri-methylenenitramine.

DBX: Depth Bomb Explosive. A grey solid explosive developed during World War II to replace torpex. It is a mixture of TNT, cyclonite, ammonium nitrate, and aluminum.

Deflagration: A chemical reaction accompanied by vigorous evolution of heat, flame, sparks, or spattering of burning particles.

Delay action fuze: A fuze designed to explode several seconds after contact, enabling the charge to penetrate through armor.

Detonator: A device, such as a blasting cap, which uses an initial primer to set off a high-explosive charge.

Dielectric heating: Heating of a nominally electrical insulating material due to its own electrical losses when the material is placed in a varying electrostatic field.

Diethylphthalate: $C_6H_4(CO_2C_2H_5)_2$. Clear, colorless, odorless liquid with bitter taste; used as a cellulosic solvent, wetting agent, alcohol denaturant, and mosquito repellant, and in perfumes.

Di-n-butyl-sebacate: $C_4H_9OCO(CH_2)_8OCOC_4H_9$. Colorless liquid used as a plasticizer; component of Otto Fuel II.

2-2-dinitroproponal: (Abbreviated DNPOH). $H_3CC(NO)_2CH_2OH$. Compound used to synthesize the nitropolymers and nitroplasticizers used in Polaris propellants.

Diphenylamine: $(C_6H_5)_2NH$. Colorless leaflets; used as an additive in propellants to increase the storage life by neutralizing the acid products formed when nitrocellulose decomposes. Also known as phenylaniline.

DNPOH: See 2-2-dinitroproponal.

Double-base rocket propellant: A solid rocket propellant containing two main explosive ingredients, such as nitrocellulose and nitroglycerin.

Electric firing mechanism: Firing mechanism using a firing magneto, battery, or alternating-current power in circuit with an electric primer; one side of the line is connected by an insulated wire to the primer, and the other side is grounded to the frame of the weapon.

Electrolysis: A chemical reaction where an electric current is passed through an electrolyte solution or through a molten salt.

ELBA: See ethyl lactate-butyl acetate.

Emulsion: A stable mixture of two or more liquids held in suspension by small amounts of a substance called an "emulsifier"; examples include milk and paint.

Ethyl cellulose: $(C_8H_{14}O_5)_n$. The ethyl ester of cellulose; it has film-forming properties and is inert to alkalies and dilute acids; used in adhesives, lacquers, coatings, and as an inhibitor in ballistite-based rocket propellants.

Ethyl centralite: $C_2H_5(C_6H_5)NCON(C_6H_5)C_2H_5$. White crystalline solid used as a stabilizer in smokeless powder and rocket propellant; also called sym-diethyldiphenylurea.

Ethyl lactate-butyl acetate: (Abbreviated ELBA). A mixture of two compounds: ethyl lactate, CH_3CHOH, and butyl acetate, $CH_3COOCH(CH_3)C(C_2H_5)$. Flammable solvent used to dissolve nitrocellulose.

Explosion: A chemical reaction or change which takes place in a very short space of time and generates high temperatures and a large quantity of gas.

Explosive D: $NH_4C_6H_2O(NO_2)_3$. Compound with stable yellow and forms of orthorhombic crystals; used as a military explosive for armor-piercing shells. Also known as ammonium trinitrophenolate and ammonium picrate.

Extrusion: A process in which a hot or cold semisoft solid material, such as metal or plastic, is forced through the orifice of a die to produce a continuously formed piece in the shape of the desired product.

Flechette: A small fin-stabilized missile, a large number of which can be loaded in an artillery canister or in a warhead. Also a metal dart designed to be dropped from aircraft.

Fluoroscope: A fluorescent screen designed for use with an x-ray tube to permit direct visual observation of x-ray shadow images of objects interposed between the x-ray tube and the screen.

Folding-fin aircraft rocket: (Abbreviated FFAR). A rocket whose control surfaces fold back and conform to the shape of the body, enabling the weapon to be inserted into a smaller launching tube; examples include the 4- to 5-inch Zuni, the Gimlet, and the 2.75-inch Mighty Mouse.

Gemini program: Series of manned U.S. space flights between 1965 and 1966; included first U.S. space walk and first manned docking of two spacecraft in orbit.

Gimlet: A modified 2.75-inch folding-fin, air-to-surface missile.

Grain: A single, elongated molding or extrusion of solid rocket propellant, regardless of size or shape.

Graphite powder: A crystalline form of carbon with numerous uses, including as a lubricant and a component of paints.

Guanidine picrate: $C_{13}H_{11}N_9O_{14}$. An explosive crystalline salt.

Guncotton: Any of various nitrocellulose explosives of high nitration (13.35–13.4 percent nitrogen) made by treating cotton with nitric and sulfuric acids; used principally in the manufacture of single-based and double-based propellants.

Harpoon: A long-range, antiship missile guided by radar; can be launched by air, ships, or submarine.

Harvey process: A method of hardening armor plate by increasing the carbon content of the surface steel.

Heptane: $CH_3(CH_2)_5CH_3$. A hydrocarbon; water soluble, flammable, colorless liquid; used as an anesthetic, solvent, and chemical intermediate, and in standard octane-rating tests.

High-bulk density nitroguanidine: (Abbreviated HBNQ). A modified form of nitroguanidine which is packed closely together in a rhombic crystal structure, making a stable high explosive.

High capacity shell: A projectile with thin walls and high explosive loading for use where no special penetrative qualities are required.

HMX: Cyclotetramethylenetetranitramine. $(CH_2)_4N_4(NO_2)_4$. A white crystalline powder used as a high energy oxidizer for several propellants and as an additive in several explosives.

Hydrolysis: Chemical reaction of a substance with water; for example, when a salt dissolves. In aqueous solutions of electrolytes, the reactions of cations with water to produce a weak base or of anions to produce a weak acid.

Ignition: The process of starting a fuel mixture burning or the means of starting this process.

Incendiary bomb: A bomb designed to be dropped from an aircraft to destroy a target by the effects of combustion.

Inert diluent process: A continuous automated process of manufacturing solid propellants. The process uses a liquid carrier to carry and mix the propellant ingredients.

Inhibitor: A inert substance which is capable of stopping or retarding a chemical reaction; used in propellant grains to slow down the reaction and stop the grain from burning through its casing.

Jet assist takeoff unit: (Abbreviated JATO). An auxiliary jet-producing unit or units, usually rockets, added to an engine to provide additional thrust.

Krupp process: A method of hardening armor plate similar to the Harvey process, except that chromium is added and the carbon content of the surface steel is greater.

LANCE: A surface-to-surface missile system designed to provide artillery support for infantry units.

Lead azide: $Pb(N_3)_2$. Unstable, colorless needles that explode at 350° C.; used as a detonator for high explosives.

Low vulnerability ammunition: (Abbreviated LOVA). Ordnance developed to reduce the high barrel wear rates common with ordinary shells; it uses cyclonite combined with an inert binder/plasticizer instead of nitrocellulose or nitroguanidine.

Mercuric chloride: $HgCl_2$. An extremely toxic compound that forms white, rhombic crystals; used for the manufacture of other mercuric compounds, as a fungicide, and in medicine and photography.

Mercury fulminate: $Hg(CNO)_2$. A grey crystalline powder which explodes at its melting point; used for explosive caps and detonators. Also called mercuric cyanate.

Metallurgy: The science and technology of metals and alloys.

Metriol trinitrate: $CH_3C(CH_2NO_3)_3$. A viscous explosive oil developed before World War II. It is more stable than nitroglycerin. Also called 1,1,1 trimethylol ethane.

Micro-crystallography: Study of the molecular geometry of crystals and their internal arrangement.

Mighty Mouse: A 2.75-inch solid propellant, folding-fin aircraft rocket intended for use against ground targets.

Minol: An explosive mixture of aluminum and ammonium nitrate.

Monopropellant: A type of propellant used in rocket engines in which the fuel and oxidizer are part of a single mixture.

Munroe Effect: The result of shaping an explosive charge so that the explosive force can be controlled and channeled in a specific direction.

Nickel steel: Carbon steel containing up to 9 percent nickel as a major alloying element. It is harder than ordinary steel, though not as hard as Harveyized or Krupp steel.

NG: See nitroglycerin.

Nitrasol: A solid propellant containing plastisol nitrocellulose and liquid nitrate esters. Used in large missile engines.

Nitrate of soda: $NaNO_3$. Very soluble crystalline solid. Also known as sodium nitrate.

Nitric acid: HNO_3. A colorless or yellowish liquid; used for chemical synthesis, explosives, and fertilizer manufacture, and in metallurgy, etching, engraving, and ore flotation. Also known as aqua fortis.

Nitrocellulose: (Abbreviated NC). Any of several esters of nitric acid, produced by treating cotton or some other form of cellulose with a mixture of nitric and sulfuric acids; used as an explosive and propellant.

Nitroglycerin: (Abbreviated NG). $CH_2NO_3CHNO_3CH_2NO_3$. Highly unstable, explosive, flammable pale-yellow liquid; used as an explosive, to make dynamite, and in medicine.

Nitroguanidine: (Abbreviated NQ). $H_2NC(NH)NHNO_2$. Explosive white, needle-shaped solid; used in explosives and smokeless powder.

Nitromethane: CH_3NO_2. An oily and colorless liquid nitroparaffin compound; used as a monopropellant for rockets, in chemical synthesis, and as an industrial solvent, in resins, waxes, fats, and dye stuffs.

NQ: See nitroguanidine.

Otto Fuel: A stable monopropellant composed of nitrate ester in solution with a desensitizing and stabilizing agent. Otto Fuel I contains PGDN and triacetin; Otto Fuel II contains PGDN and di-n-butyl-sebacate.

Oxidizer: Reactive compound which gains electrons during an oxidation-reduction chemical reaction; the propellant ingredient which provides oxygen for the burning process.

PAD: See propellant actuated device.

PBX: Plastic Bonded Explosive. An improved explosive made from a mixture of cyclonite, HMX, PETN, and a plastic binder. It has a high mechanical strength and excellent chemical stability, and is impervious to shock.

Pentaerythritol tetranitrate: (Abbreviated PETN). $C(CH_2ONO_2)_4$. Shock-sensitive material used in explosives, blasting caps, and in the preparation of some monopropellants.

Pentolite: Combination of pentaerythritol tetranitrate and TNT.

Percussion firing mechanism: Any firing mechanism which fires the primer by percussion.

PETN: See pentaerythritol tetranitrate.

PGDN: See propylene glycol dinitrate.

Phalanx: A seaborne weapons system using radar-guided 20mm Gatling guns to shoot down enemy missiles.

Phenylaniline: See diphenylamine.

Picric acid: $C_6H_2(NO_2)_3OH$. Poisonous, explosive, and highly reactive yellow crystals used in explosives, dyes, and matches. Also known as carbazotic acid and picronatic acid.

Pig lead: A high carbon lead made by reducing lead ore in a blast furnace.

Plasticizer: Organic compound added to a polymer either to facilitate processing or add flexibility and toughness.

Plastisol: A vinyl resin dissolved in a plasticizer to make a pourable liquid; useful for moldings and castings.

Polaris missile: A U.S. Navy surface-to-surface intermediate-range ballistic missile designed to be launched from submarines and surface ships for accurate bombardment of small target areas with conventional or nuclear warheads at ranges up to 2,500 nautical miles (4,600 kilometers).

Polymer: A macromolecule formed by bonding five or more identical units together in a chain; examples include cellulose and most synthetic plastics.

Potassium sulfate: K_2SO_4. Colorless crystals with bitter taste; used as an analytical reagent, as fertilizer, and in aluminum and glass manufacture. Also known as salt of Lemery.

Potting resin: Resin used to encapsulate or cover other materials.

Primer: A small, sensitive explosive component which initiates the explosive reaction without scattering the explosive charge; classed according to the method of initiation, for example, percussion primer, electric primer, or friction primer.

Propellant: A combustible substance which produces large quantities of heat and ejection particles.

Propellant actuated device: (Abbreviated PAD). A small rocket device used for a variety of purposes, including ejection of personnel and arms from aircraft.

Propylene glycol dinitrate: (Abbreviated PGDN). $CH_3CHNO_3CH_2NO_3$. Principal component of Otto Fuel. An explosive liquid. Also known as 1,2-dinitro[xy]propane.

Propyl nitrate: $CH_2CH_2CH_2NO_3$. A white, slightly explosive liquid used in some rocket propellants.

RDX: See cyclonite.

RAPEC: (Rocket Assisted Pilot Ejection Catapult). A rocket attached to a pilot's seat; used for emergency ejections.

Rosaniline: $C_{20}H_{19}N_3$. Brownish-red crystals; used as a dye in chemical tests.

Rotating band: A thin strip of metal placed around a shell in order to improve its range and muzzle velocity. An augmented rotating band is a slightly thicker strip of metal used when a gun barrel has become worn down with use.

Sabot: Lightweight carrier in which a subcaliber projectile is centered to permit firing the projectile in the larger-caliber weapon; the sabot diameter fills the bore of the weapon from which the projectile is fired.

SAE number: A classification of motor, transmission, and differential lubricants to indicate viscosities, standardized by the Society of Automotive Engineers. SAE numbers do not connote quality of the lubricant.

SCAMP: An antipersonnel missile which scatters hundreds of small darts.

Semple plunger fuze: A centrifugal plunger which operates to maintain a fuze in a safe condition until centrifugal force unlocks and moves the firing pin into the armed position.

Shell-destroying tracer: A tracer the includes an explosive element designed to destroy the projectile after it has passed the target point to avoid impact in friendly territory.

Shrike: Navy antiradar missile developed in the early 1960s; it had a range of ten miles.

Sidewinder: A U.S. Navy air-to-air missile with an infrared guidance system and a speed of over Mach 2; the missile seeks enemy planes by homing in on the heat emitted by the target.

Signal pyrotechnics: Method of communicating using rockets or flares.

Single-base powder: An explosive or propellant powder in which nitrocellulose is the only active ingredient.

Smokeless powder: Nitrocellulose containing 13.1 percent nitrogen with small amounts of stabilizers (amines) and plasticizers usually present, as well as various modifying agents (nitrotoluene and nitroglycerin salts); used in ammunition.

Snark: The U.S. armed forces' first intercontinental missile; it had a length of 69 feet, weighed 60,000 pounds, and had a range of 5,500 nautical miles. First deployed in 1959.

Sparrow: A U.S. Navy air-to-air guided missile having a speed of over 1,500 miles (2,400 kilometers) per hour; it is guided to its target by a radar beam transmitted by the launching airplane.

Specific impulse: A measure of a propellant's performance equal to the thrust in pounds divided by weight flow rate per second. Also known as specific thrust, it is a way of expressing the "push" of a rocket fuel.

Spectrophotometry: The process of measuring the wavelength range of radiant energy absorbed by a sample under analysis; this measurement can then be used to determine the sample's chemical composition.

Spectroscopy: The branch of physics concerned with the production, measurement, and interpretation of electromagnetic spectra arising from either emission or absorption of radiant energy by various substances.

Standard missile: Standardized follow-up to the Navy's Tartar and Terrier ship-to-air guided missiles.

Star shell: Projectile, with a time fuze, that releases a parachute flare at any desired height; used for lighting up an area. Also known as an illuminating projectile.

Static firing: The test firing of a rocket motor which is bolted to a test stand.

Sulfuric acid: H_2SO_4. A toxic, corrosive, and strongly acidic, colorless liquid that is miscible with water and dissolves most metals and melts; used in industry in the manufacture of chemicals, fertilizers, and explosives, and in petroleum refining. Also known as dipping acid.

Talos: A U.S. Navy surface-to-air guided missile having a speed of about Mach 3, a range of about twenty-five miles (forty kilometers), and beam-rider guidance; one version can carry nuclear warheads, and it can also be used against ships and shore bombardment targets.

Tartar: A U.S. Navy surface-to-air guided missile intended primarily for use on destroyers; it is a smaller version of Terrier and has about the same range.

Terrier: A U.S. Navy surface-to-air guided missile using beam-rider guidance and having a range of about ten miles (sixteen kilometers).

Tetryl: $(NO_2)_3C_6H_2N(NO_2)CH_3$. A yellow, crystalline compound used in explosives and ammunition; also called N-methyl-N,2,4,6-tetranitroaniline or tetralite.

Tetrytol: A highly explosive mixture of tetryl and TNT.

Theodolite: An optical instrument consisting of a telescope mounted so that it can swing freely; used to follow trajectories.

TNT: See trinitrotoluene.

Tomahawk: A liquid-fueled subsonic missile with a rocket booster.

Torpex: An explosive mixture of cyclonite, TNT, aluminum powder, and wax.

Tracer shell: A shell containing a pyrotechnic mixture to make the flight of the projectile visible by day and night.

Triacetin: $C_3H_5(CO_2(H_3))_3$. A colorless, combustible oil found in cod liver and butter; used as a plasticizer and as a solvent.

Triethylene glycol dinitrate: $CH_3CH(OC_2H_4NO_3)_2$. An explosive liquid. Also called 1,2-bis(2-hydroxyethoxy)ethane dinitrate.

Trinitrotoluene: (Abbreviated TNT). $CH_3C_6H_2(NO_2)_3$. Highly explosive compound used in shells, bombs, and missiles.

Turbojet: Form of jet engine with a compressor mounted in front which forces the air flow back through the burning fuel, producing a high velocity stream of air.

Type-life surveillance: A series of controlled experiments to see how a piece of apparatus will react when exposed to the different conditions it will experience during the course of its life.

Very signals: Bright flares used for signalling or illuminating.

Viscosity: The resistance that a gaseous or liquid system offers to flow when it is subjected to a shear stress. Also known as flow resistance.

Weapon A: A Navy anti-submarine weapon developed during World War II; it was a multibarreled depth charge projector with a range of about 900 yards.

Window: A radar countermeasure designed to confuse enemy radars by scattering strips of metal chaff, wire, or bars; the pieces of metal give resonance at expected enemy radar frequencies.

Wood pulp: The cellulosic material produced by reducing wood mechanically or chemically and used in making paper and cellulose products.

Zuni: A folding-fin, air-to-surface unguided rocket with solid propellant; can be armed with various types of heads, including flares, fragmentation, and armor-piercing.

Bibliography

Bibliographic Note

Archival materials cited in the endnotes were collected from the holdings of the Indian Head Technical Library, the National Archives and Records Center holdings in Philadelphia, Washington, D.C., and Suitland, Maryland, and in other repositories as noted. The endnotes direct the reader to file name, date or box, collection, record group, and repository. Technical publications, which were not widely distributed, although in print, have been treated as primary sources when collected from the Indian Head Technical Library. Newspaper items are cited as primary sources in the endnotes only. The following list of secondary sources represents works directly utilized or serving as background sources. The Navy Department Library, at the Navy Yard in Washington, D.C., often served to supplement the collection of the Library of Congress, especially in the search for older naval and ordnance works.

Secondary Sources

Alden, John D. *The American Steel Navy.* Annapolis: Naval Institute Press, 1972.

Alger, Philip R. "The High Explosives in Naval Warfare." *United States Naval Institute Proceedings* 26 (1900): 245-78.

_____. "Ordnance and Armor." *United States Naval Institute Proceedings* 27 (1901): 529-50.

Allen, James B. *The Company Town in the American West.* Norman: University of Oklahoma Press, 1966.

Allison, David K., Ben G. Keppel, and C. Elizabeth Nowicke. *D. W. Taylor.* Washington, D.C.: David Taylor Naval Ship Research and Development Center, 1987.

Anspacher, William, et al. *The Legacy of the White Oak Laboratory: Accomplishments of NOL/NSWC.* Dahlgren, Va.: NSWC, Dahlgren, 2000.

Artes, Dorothy B. "Birth of Indian Head and Her Schools." Indian Head, Md.: Collection of Dorothy B. Artes, n.d. Mimeographed.

Beauregard, R. L. *History of the Navy Use of Composition A-3 and Explosive D in Projectiles.* NAVORD TR-70-1. Indian Head, Md.: Naval Ordnance Station, 1970.

Bernadou, Jean Baptiste. *The Development of Smokeless Powder.* New York and Annapolis: n.p., 1897-1898.

Bernadou, Jean Baptiste, trans. "Pyro-Collodion Smokeless Powder." *United States Naval Institute Proceedings* 28 (1897): 644-54, and continued in *United States Naval Institute Proceedings* 29 (1898): 605-16.

Boyd, William B., and Buford Rowland. *U.S. Navy Bureau of Ordnance in World War II.* Washington, D.C.: U.S. Government Printing Office, n.d.

Brown, Jack D., William A. Diggs, Gladys S. Jenkins, J. Karpiak, Elwood M. Leviner, Mary Clare Matthews, Janie MacInnis, Rona R. Schaepman, and Frederick Tilp. *Charles County, Maryland: A History.* So. Hackensack, N.J.: Custombook, 1976.

Buckner, E. G. "Is There a Gunpowder Plot?" *Harper's Weekly,* 27 June 1914, p. 15.

Camp, Sharon L. "Modernization: Threat to Community Politics." Ph.D. diss., The Johns Hopkins University, 1976.

Carlisle, Rodney. *Management of the U.S. Navy Research and Development Centers During the Cold War: A Survey Guide to Reports*. Washington, D.C.: Naval Historical Center, 1996.

_____. *Navy RDT&E Planning in an Age of Transition: A Survey Guide to Contemporary Literature*. Washington, D.C.: Naval Historical Center, 1997.

_____. *The Relationship of Science and Technology: A Bibliographic Guide*. Washington, D.C.: Naval Historical Center, 1997.

_____. *Where the Fleet Begins: A History of the David Taylor Research Center, 1898-1998*. Washington, D.C.: Naval Historical Center, 1998.

Carlisle, Rodney, and August W. Giebelhaus. *Bartlesville Energy Research Center: The Federal Government in Petroleum Research, 1918-1983*. Washington, D.C.: U.S. Dept. of Energy, Office of Scientific and Technical Information, 1985.

Carlisle, Rodney, with Joan Zenzen. *Supplying the Nuclear Arsenal: American Production Reactors, 1942-1992*. Baltimore: Johns Hopkins University Press, 1996.

Carlisle, Rodney, and William Ellsworth. *Shaping the Invisible: Development of the Advanced Enclosed Mast/Sensor System, 1992-1999*. Washington, D.C.: Office of Naval Research, 2001.

Cathcart, William Ledyard. "George Wallace Melville." *Journal of the American Society of Naval Engineers* 24 (1912): 477-511.

Chandler, Alfred, and Steven Salsbury. *Pierre S. du Pont and the Making of the Modern Corporation*. New York: Harper & Row, 1977.

The Charles County Retired Teachers Association. *A Legacy: One- and Two-Room Schools in Charles County*. La Plata, Md.: Dick Wildes Printing Company, 1984.

Christman, Albert B. *Sailors, Scientists, and Rockets*. Washington, D.C.: Naval History Division, 1971.

"Class Directory." *Journal of the Worcester Polytechnic Institute* 52, no. 1 (September 1948): 1-134.

Cleland, David I., and William R. King. *Systems, Organizations, Analysis, Management: A Book of Readings*. New York: McGraw-Hill, 1969.

Coleman, Jonathan. *Exit the Rainmaker*. New York: Atheneum, 1989.

Coletta, Paolo. *Admiral Bradley A. Fiske and the American Navy*. Lawrence: University Press of Kansas, 1979.

Connery, Robert H. *The Navy and the Industrial Mobilization in World War II*. Washington, D.C.: U.S. Government Printing Office, n.d.

Constant, Edward W. *The Origins of the Turbojet Revolution*. Baltimore, Md.: The Johns Hopkins University Press, 1980.

Cooling, Benjamin F. *Grey Steel and Blue Water Navy*. Hamden, Conn.: Archon, 1979.

Daniels, Jonathan. *The End of Innocence*. New York: Da Capo Press, 1972.

Deming, W. E. *Out of Crisis*. Cambridge, Mass.: MIT Center for Advanced Engineering Study, 1986.

Dorwart, Jeffery, with Jean Wolf. *Philadelphia Navy Yard from the Birth of the U.S. Navy to the Nuclear Age*. Philadelphia: University of Pennsylvania Press, 1999.

Duncan, Francis. *Rickover and the Nuclear Navy: The Discipline of Technology*. Annapolis, Md.: Naval Institute Press, 1990.

Drucker, Peter F. *The Practice of Management*. New York: Harper & Row, 1954.

Earle, Ralph. "The Destruction of the *Liberté*." *United States Naval Institute Proceedings* 37 (1911): 929-42.

———. "The Development of Our Navy's Smokeless Powder." *United States Naval Institute Proceedings* 40 (1914): 1041-57.

Farrow, Edward S. *American Guns in the War with Germany*. New York: E. P. Dutton and Company, 1920.

Federal Advisory Commission. "Federal Advisory Commission on Consolidation and Conversion of Defense Research and Development Laboratories," (Adolph Report). Washington, D.C.: The Commission, 1991.

Fiske, Bradley A. *From Midshipman to Rear Admiral*. New York: Century Company, 1919.

Freidel, Frank. *Franklin D. Roosevelt, the Apprenticeship*. Boston: Little, Brown and Company, 1952.

Grayson, Cary T. *Woodrow Wilson: An Intimate Memoir*. Washington: Potomac Books, 1960.

Guttman, Oscar. "Twenty Years Progress in Explosives." In *Annual Report of the Board of Regents of the Smithsonian Institution for the Year Ending June 30, 1908*, pp. 263-300. Washington, D.C.: U.S. Government Printing Office, 1909.

Howe, Thomas C. *Power and Change: The History of the Office of the Chief of Naval Operations*. Washington, D.C.: Naval Historical Center, 1989.

Henderson, R. W. "The Evolution of Smokeless Powder." *United States Naval Institute Proceedings* 30 (1904): 353-372.

———. "The Naval Torpedo Station." *United States Naval Institute Proceedings* 29 (1903): 193-201.

Hodges, Peter. *The Big Gun: Battleship Main Armament* 1860-1945. Annapolis: Naval Institute Press, 1980.

Hovgaard, William. *Modern History of Warships*. Annapolis: U.S. Naval Institute, 1971.

Ingersoll, R. R. *Text-Book of Ordnance and Gunnery—Compiled and Arranged for the Use of Naval Cadets, U.S. Naval Academy*. 4th ed. Annapolis: U.S. Naval Institute, 1899.

Jelinek, Mariann. *Institutionalizing Innovation*. New York: Praeger, 1979.

Judge, John F. "Navy Meeting Need for 2.75-inch Rocket." *Missiles and Rockets: The Weekly of Advanced Technology*, 11 November 1965, no pagination.

King, Randolph W., Admiral, ed. *Naval Engineering and American Sea Power*. Baltimore: Naval and Aeronautical Press, 1989.

Koontz, Harold, and Cyril O'Donnell. *Management: A Book of Readings*. 2d ed. New York: McGraw-Hill, 1968.

Koppes, Clayton R. *JPL and the American Space Program: A History of the Jet Propulsion Laboratory*. New Haven: Yale University Press, 1982.

Lackey, Henry E. "Duty at Indian Head." Indian Head, Md.: Collection of Dorothy B. Artes, n.d. Mimeographed.

Leary, H. F. "Military Characteristics and Ordnance Design." *United States Naval Institute Proceedings* 48 (1922): 1125-37.

Lord, Clifford L. *History of the United States Naval Aviation*. New Haven: Yale University Press, 1980.

Mark, Hans, and Arnold Levine. *The Management of Research Institutions: A Look at Government Laboratories*. Washington, D.C.: National Aeronautics and Space Administration, 1984.

Maryland General Assembly. *Laws of the State of Maryland.* Annapolis: State Printer, 1920.

Maxim, Hiram. *My Life.* New York: McBride, Nast & Company, 1915.

McCollum, Kenneth G., ed. *Dahlgren.* Dahlgren, Va.: Naval Surface Weapons Center, 1977.

McCormick, Charles. *The Power of People.* Baltimore, Md.: McCormick, 1948.

McDougall, Walter A. *. . . the Heavens and the Earth: A Political History of the Space Age.* New York: Basic Books, 1985.

McGregor, Douglas. *The Human Side of Enterprise.* New York: McGraw-Hill, 1960.

McWilliams, Emma. "My Memoirs." Indian Head, Md.: Collection of John McWilliams, 1973. Mimeographed.

Melhorn, Charles M. *Two-Block Fox: The Rise of the Aircraft Carrier, 1911-1929.* Annapolis: Naval Institute Press, 1974.

Method of Investigation and Test of Smokeless Powder for Small Arms and Cannon. Washington, D.C.: U.S. Government Printing Office, 1910.

Monetta, Dominic J. "PAM: A Research and Development Project Appraisal Methodology." Ph.D. diss., University of Southern California, 1983.

Munroe, Charles. "The Nitrogen Question from the Military Standpoint." *United States Naval Institute Proceedings* 35 (1909): 715-27.

Naval Ordnance Laboratory. *History of the Naval Ordnance Laboratory, 1918-1945: Administrative History.* Washington, D.C.: United States Navy Yard, 1946.

Naval Ordnance Laboratory. *History of the Naval Ordnance Laboratory, 1918-1945: Scientific History.* Washington, D.C.: United States Navy Yard, 1946.

Naval Ordnance Station. "A Narrative History of the Naval Ordnance Station, Indian Head, and of the Gun Systems Division." Indian Head, Md.: Naval Ordnance Station, c. 1975. Typescript.

Naval Ordnance Station, Public Relations Material. *1890 Naval Proving Ground, Naval Powder Factory, Naval Propellant Plant.* Indian Head, Md.: Naval Ordnance Station, 1961.

Naval Ordnance Station, Public Relations Material. "Naval Propellant Plant Product History: 1900 to 1964." Indian Head, Md.: Naval Ordnance Station, c. 1965.

"Old Town Notebook Kept by Mr. Charles E. Wright." Indian Head, Md.: Collection of Dorothy B. Artes.

Padfield, Peter. *Guns at Sea.* London: Hugh Evelyn, 1973.

Polmar, Norman. *Ships and Aircraft of the U.S. Fleet.* Annapolis, Md.: Naval Institute Press, 1987.

Porter, Roger. *Presidential Decision Making.* London: Cambridge University Press, 1980.

"Professional Notes." *United States Naval Institute Proceedings* 25 (1899): 236-37.

"Professional Notes." *United States Naval Institute Proceedings* 26 (1900): 226-30.

"Professional Notes." *United States Naval Institute Proceedings* 28 (1902): 415-26.

"Professional Notes." *United States Naval Institute Proceedings* 46 (1920): 1360-61.

Rainsford, William B. "A Short Historical Sketch of Events Leading up to the Establishment of Perseverance Lodge No. 208, A. F. & A. M." Indian Head, Md.: Collection of Dorothy B. Artes, 1953. Mimeographed.

Raymond, Jack. *Power at the Pentagon.* New York: Harper & Row, 1964.

Reynolds, Clark G. *Famous American Admirals.* New York: Van Nostrand Reinhold, 1978.

Robison, Samuel S. *A History of Naval Tactics from 1530 to 1930: The Evolution of Tactical Maxims.* Annapolis: U.S. Naval Institute, 1942.

Rowland, Buford, and William B. Boyd. *U.S. Navy Bureau of Ordnance in World War II.* Washington, D.C.: U.S. Government Printing Office, n.d.

Smokeless Powder Department, E. I. du Pont de Nemours & Company, Inc. *A History of the du Pont Company's Relations with the United States Government, 1802-1927.* Wilmington, Del.: du Pont Company, 1928.

Sontag, Sherry, and Christopher Drew. *Blind Man's Bluff: The Untold Story of American Submarine Espionage.* New York: Public Affairs, 1998.

Spector, Ron. *Admiral of the New Empire: The Life and Career of George Dewey.* Columbia: University of South Carolina Press, 1974.

Stevens, William S. "The Powder Trust, 1872-1912." *Quarterly Journal of Economics* 27 (1912): 444-81.

_____. "The Dissolution of the Powder Trust." *Quarterly Journal of Economics* 28 (1913): 202-7.

Strauss, Joseph A. "Smokeless Powder." *United States Naval Institute Proceedings* 27 (1901): 733-38.

_____. "The Stability of Smokeless Powder." *United States Naval Institute Proceedings* 36 (1910): 929-42.

Turnbull, Archibald D., and Clifford L. Lord. *History of United States Naval Aviation.* New Haven: Yale University Press, 1949.

U.S. Congress. House. Committee on Appropriations. *Hearings on H.R. 28186.* 62d Cong., 3d sess., 14, 16, and 17 December 1912.

_____. Committee on Naval Affairs. *Armor Plant for U.S.* 63d Cong., 3d sess., 24 November 1914.

_____. *Estimates Submitted by the Secretary of the Navy.* 62d Cong., 3d sess., 16 December 1912, and 13 January 1913.

_____. *Estimates Submitted by the Secretary of the Navy.* 66th Cong., 1st sess., 2 March 1916.

_____. Subcommittee for Special Investigations of the Committee on Armed Services. *Utilization of Naval Powder Factory, Indian Head, Maryland.* 85th Cong., 2d sess., 10 and 11 July 1958.

_____. Senate. *Report of the Secretary of the Navy Relative to the Cost of Armor Plate for Vessels of the U.S. Navy.* 53d Cong., 3d sess., 1894-1895, S. Doc. 1453.

_____. Senate. *Report to the President: Government Contracting for Research and Development.* ["Bell Report"]. 87th Cong., 2d sess., 1962, S. Doc. 94, Bureau of the Budget.

U.S. Department of Commerce, Bureau of the Census. *Fourteenth Census of the United States Taken in the Year 1920. Population 1920: Number and Distribution of Inhabitants.* Vol. 1. Washington, D.C.: U.S. Government Printing Office, 1921.

_____.*Population of the United States at the Eleventh Census: 1890.* Pt. 1. Washington, D.C.: U.S. Government Printing Office, 1895.

U.S. Department of the Navy. *Annual Report of the Secretary of the Navy.* Washington, D.C.: U.S. Government Printing Office, 1890-1920.

_____. Bureau of Ordnance. *Navy Ordnance Activities, World War, 1917-1918.* Washington, D.C.: U.S. Government Printing Office, 1920.

_____. Bureau of Yards and Docks. *Federal Owned Real Estate Under the Control of the Navy Department.* Washington, D.C.: U.S. Government Printing Office, 1937.

Vincenti, Walter G. *What Engineers Know and How they Know it: Analytic Studies from Aeronautic History*. Baltimore: Johns Hopkins University Press, 1990.

Walsh, Richard, and William Lloyd Fox, eds. *Maryland: A History, 1632-1974*. Baltimore, Md.: Schneidereith & Sons, 1974.

Weigley, Russell F. *The American Way of War*. New York: Macmillan, 1973.

Weir, Gary. *Forged in War: The Naval Industrial Complex and American Submarine Construction, 1940-1961*. Washington, D.C.: Naval Historical Center, 1993.

Wenk, Edward Jr. *Making Waves: Engineering Politics and the Social Management of Technology*. Urbana: University of Illinois Press, 1995.

White, Leonard. *Introduction to the Study of Public Administration*. 4th ed. New York: MacMillan, 1955.

Wilson, James E., Jr. "The Research and Development Department." Master's thesis, George Washington University, 1963.

Index

A

Accidents, 22, 45, 140, 171, 242
Advanced Mine School, 115
Advisory Board, 152–154, 155
Aerojet General, Inc.,156, 163
Aircraft, 65, 76
Alger, Philip R., 11, 13, 17
Allegany Ballistics Laboratory (ABL), 106, 123, 135
Annapolis Proving Ground, 4–5
Anti-radar missile (ARM), 192, 197, 198
Anti-submarine rocket (ASROC), 166
Armor-testing, 10–16; Proving Ground, 35–37
Assistant Management Board, 155, 163, 207, 217, 241
Atkins, Griswold T.,144, 147, 153
Automatic Dynamic Analysis of Mechancial Systems (ADAMS), 229

B

Ballistite, 78, 107, 109
Base Realignment and Closure (BRAC), 214–215, 220–222, 226
Benson, William H., 133
Bethlehem Steel Company, 3, 13, 14, 51, 57, 77, 237
Biazzi, 129–131, 167, 172, 240
Bisson, Arthur E. (Prize), 229
Black powder, 2–3, 21, 36, 74
Bloch, Claude C., 28, 66
Booz, Allen, and Hamilton; evaluation of Naval Propellant Plant, 174
Brough, John, 231
Brown powder, 2, 3, 21
Browning, Joseph L. (Joe), 133, 134–135, 137–138, 140, 151–156, 161–164, 169, 173–174, 176, 178, 180, 182–183, 187, 200, 204, 216–217, 240–241, 242

Bullpup, 166, 170, 193
Bureau of Ordnance, 1–2, 4, 5, 11, 15, 27, 30, 34, 41, 50, 76
Bureau of Weapons, 163

C

C-3, plastic explosive, 175
Cagle, W. C., 119
California Institute of Technology, 106, 107, 108, 110, 111, 119, 240
Captured enemy equipment, examination of, 114–118
Carnegie Steel Company, 3, 13, 14, 51, 53, 57, 237
Cartridge actuated device (CAD), 46–149, 190–192, 200, 201, 223–224, 232, 244
Carsey, Jay, 104, 180, 189
Cast Propellant Plant, 136, 141, 167, 229
Catapults, 133
Cellulose, 35
Central Naval Ordnance Management Information System (CENO), 202–203
Chappell, Dennis, 224
Charles County, Maryland, 33, 64, 89, 90, 185–186, 189, 214
Chemical Biological Incident Response Force (CBIRF), 226
Cocoa powder, 2, 19
Computer Applications Support and Development Office (CASDO), 202–203
Corcoran, William J., 140, 144
Cordite, 2, 21, 78
Cordite-N plant, 130, 134–135
Cosgrove process, 36
Couden, Albert R., 20, 22, 24, 29
Customer Advisory Board, 224–225
Cyclonite (RDX), 119, 210

317

D

Dahlgren, Virginia, 44–45, 47, 62, 64, 66, 68, 69, 75, 239

Daniels, Josephus, 44–45, 51–57, 189

Dashiell, Robert Brooke, 1–18, 22, 24, 29, 65, 186, 214, 237

Davis Torpedo Gun, 59

David Taylor Research Center, 177–178

Dayton, J. H., 4–5

Delay action fuze, 70–73

Dennison, Wayne, 182

Desert Storm, 215, 229

Desk F, 66, 73, 76, 239

Desk Q, 66, 73, 239

Dewey, George, 8, 19

Dick, F. F., 69, 81, 152, 239

Dieffenbach, A. C., 26, 31, 32, 36, 39

Di-n-butyl-sebacate, 171

Dinitropropanol (DNPOH), 156, 157, 165

Diphenylamine, 50, 82, 238

Dodgen, James, 153, 156

Double-base powder, 128

Double-base rocket propellant, 129

du Pont Company, 20, 21, 22, 33, 38, 53, 54, 55, 56, 75, 84, 94, 163, 238

E

Earle, Ralph, 28, 66

Eaton Canyon, 107–108, 110, 111

Energetics Manufacturing Technology Center (MANTECH), 230–231

Environmental concerns, 187, 189, 232–234, 242

Equal opportunity, 187, 189

Ethyl centralite, 109, 165, 181

Ethyl lactate-butyl acetate (ELBA), 175

Experimental Ammunition Unit (EAU), 66, 69–72, 77, 143, 239

Explosive D (ammonium picrate), 49, 59, 74, 75, 78, 81, 239

Explosives, 17, 51, 62

Explosives Investigation Laboratory, 105, 112, 114–118, 240

Explosive Ordnance Disposal Technical Center (EODTC), 113, 166, 191, 226, 240, 241

Extrusion plant, 66, 239

F

Farnum, Walter, 33, 50, 55, 56, 59, 60, 61, 81, 83–86, 94, 100, 116, 118–119

Fiske, Bradley A., 4, 66

Flashless powder, 68

Flechette, 181

Folger, William M., 5, 6, 7, 8, 9, 11, 12, 13, 22, 213

Folding-fin aircraft rocket (FFAR), 131, 148

Frankford Arsenal, 191

Frese, Bernard W., Jr., 178, 182

Furlong, William R., 101

Fuzes, 17, 40, 48, 51, 61, 70, 73

G

Gary, Stanley P., 182, 188

Gas Research Institute (GRI), 207

Gering, J. J, 31–32

Gimlet, 144

Goldwater-Nichols Act, 215

Grainger, Lu, 172, 192, 194, 197

Greenslade, John W., 71

Guanidine picrate, 75

Gulf War, 243 (see also Desert Storm)

Guncotton, 23, 24, 56

H

Haines, Preston B., 118

Hanlon, Byron H., 126, 129

Harpoon missile, 192–193, 237

Harveyized steel, 13–14, 15, 16, 36, 53, 237

Hebert, F. E., 142–143

Hellweg, Julius F., 49, 50, 59–60, 94–95

Hercules Powder Company, 106, 127, 158, 159, 163, 172, 238

Hersey, Mark L., 109, 115, 119–120, 125

High altitude research projectile (HARP), 176

High-bulk density nitroguanidine (HBNQ), 165, 210

High capacity shells (HC shells), 82–83

Holden, Jonas H., 33, 40, 44, 46, 48–49, 55

Hoyer, Steny, 220

Hungerford, Vincent C., 182, 192, 195, 203, 209

Hussey, George F., 124, 128

I

I. G. Farben Co., 85
Illuminating shells, 68
Indian Head, Maryland;
    development of town
        churches, 92–93
        housing, 34, 60
        incorporation, 64, 91, 99
        living conditions, 41, 60
        medical care, 103
        schools, 60, 91–92, 96–97, 100
        social life, 100
Inert diluent plant, 169–170
Insensitive Munitions Advanced
    Development-High Explosives
    Program (IMAD-HE), 228
Internet, 231–232
Inyokern, California, 105, 106, 114, 118

J

Jet Assist Takeoff Units (JATOs), 122–
    123, 165, 166, 167, 193, 201, 216, 217
Jet Propulsion Laboratory, 105, 113–
    114, 240
JP powder, 109, 122
Jutland, Battle of (1 June 1916), 67–72, 76

K

King, George E., 146, 147, 152
Kirk, A. G., 58
Klein, D., 115, 116, 122
Korean War, 131, 142
Kruppized steel, 36

L

Labor;
    shortages, 33, 45, 59, 61, 96
    skilled, 33
    unskilled, 33
    professional, 197
    union, 227
Lacey, Mary, 216, 226–227, 243, 244
Lackey, Henry E., 28, 57–62, 96, 97
LANCE missile, 193
Land acquisitions, 8–9, 35, 47, 60
Lauritsen, Charles, 107–108
Lead azide, 83

Leary, Herbert F., 28, 82
Lee, David, 186–190, 192–200, 203–206,
    208, 211, 216, 217, 229, 241
Long Range Bombardment Ammuni-
    tion (LRBA), 176–177, 218
Long Range Planning Task Force Team
    for Gun Systems (LORAP TAFT
    FOGS), 180–181
Low-Vulnerability Ammunition
    (LOVA), 210, 219
Lower Station, 44, 62, 239 (see also
    Dahlgren, Virginia)

M

Mason, Newton E., 15–16, 17, 22, 29,
    39, 40, 42
Mattawoman Creek, 2, 46, 47, 95, 187
Maxim, Hiram, 21–22
Maxwell, David G., 220
Melville, George, 4, 39
Mendeleev, Dmitri, 24
Mercuric chloride, 37
Metriol trinitrate, 181
Midvale Steel Company, 51, 57
Mighty Mouse (2.75 Inch rocket), 219,
    228, 241
Microelectromechanical Systems
    (MEMS), 231–232
Milkulski, Barbara, 220
Mitchell, Billy, 63, 66, 76, 79, 80
Moffett, William A., 66
Monetta, Dominic J., 177, 182, 194,
    206–211, 216–217, 220, 224, 242
Monopropellant, 170–171
Mueller, Kurt, 221
Munroe, Charles, 7, 21, 22, 23, 115
Murrin, John W., 146, 171–172

N

Nanos, George P., 226
National Aeronautics and Space
    Administration (NASA), 169, 242
National Defense Research Council
    (NDRC), 106
Naval Air Systems Command
    (NAVAIR), 163, 189, 191–193, 200,
    201, 227

Naval Gun Factory, 4, 8–11
Naval Ordnance Systems Command (NAVORD), 163, 175, 199, 202
Naval Ordnance Laboratory, 66, 73, 215, 217
Naval Ordnance Station, 162
Naval Ordnance Test Station (NOTS), 123, 127, 163
Naval Powder Factory, 24–27, 60, 122, 240
Naval Powder Factory Explosives Research Laboratory, 120
Naval Propellant Plant, 141, 156, 167–170, 240
Naval Proving Ground, 8–10, 29, 35–37, 238
Naval Research Laboratory, 113, 196
Naval Sea Systems Command (NAVSEA), 66, 189, 194, 198, 200, 201, 202, 204, 210, 222, 225, 226, 243
Naval Ship Systems Command (NAVSHIP), 189, 199
Naval Surface Warfare Center, 216
Newport Torpedo Factory, 5, 7, 21, 23, 25
Newton, Wayne J., 220
Nichols, Harry J., 74–75, 215
Nicholson, Edwin P., 210, 218, 242
Nickel-steel, 36, 53, 57
Nitrate of soda, 59
Nitric acid, 21, 26, 86, 165, 168
Nitrocellulose, 21, 23, 24, 35, 38, 82, 107, 108, 165
Nitroglycerin (NG), 21, 22, 107, 108, 165, 167
Nitroguanidine (NQ), 119, 135, 165
Nitromethane, 165
No-Solvent Propellant (NOSOL), 179, 181, 242
Nobel, Alfred, 21, 22, 74

O
Occupational Safety and Health Administration (OSHA), 211
Office of Naval Research (ONR), 222, 224, 225, 229
Ordnance Environmental Support Office, 201
Otto Fuel, 170–172, 190, 198, 210

P
Patterson, George W., 23, 24, 31–42, 45, 48, 50–53, 55–56, 60–61, 69, 70, 78, 81–85, 186, 213, 238
Perk, Anciet J. (A. J.), 209
Personnel, 131–136, 139–141
Phalanx Close-In Weapon System, 202
Pilot Ejection Catapult Cartridge (PECC), 146
Pilot plants, 125, 127–131, 167, 240
Plastic Bonded Explosives (PBX), 228–229
Polaris, 138, 151, 153, 156–159, 240
Pollution, 187, 201
Potomac River, 1–2, 36, 43, 62, 213
Powder, 2, 3, 20–24, 37; standardization of, 47–50; production of, 55–56
Production Department, 134, 135
Professional Development Council (PDC), 205, 241
Program Evaluation Review Technique (PERT), 151, 173
Program Management Office, 225
Profile, 189
Propellant actuated device (PAD), 159, 190–191, 201
Propellant grain, 146, 167
Propellants, 167
Propyl nitrate, 170–171
Propylene glycol dinitrate (PGDN), 171–172
Proving Ground, 35–37; standardization of, 47–50
Public Law, 313–145

Q
Quality control, 166
Quality Surveillance Department, 144–145

R
RDX, see cyclonite
Recruitment, 59–60, 90, 139–140, 154, 198, 240
Reitlinger, Otto, 170–171

Research and Development Department, 124–125, 131, 144, 166, 182, 239–240
Roberts, Eugene, 133, 146
Rocket-Assisted Pilot Ejection Catapult (RAPEC), 166, 191
Rocket-Assisted-Torpedo (RAT), 147
Rocketdyne Company, 163, 169
Rocket grains, 107, 166
Rockets, 49, 64;
    2.75-inch, 131, 148, 175
    modifications: Mark, Mod, 65, 164, 239
Roosevelt, Franklin D., 44, 57
Rosaniline, 50

S
Sarbanes, Paul, 220
SCAMP, 181
Scanland, Francis W., Jr., 132, 135, 141
Schoeffel, Malcolm F., 127, 128, 130
Schuyler, Garret L., 28, 29, 43–44, 46
Sea Automated Data Systems Activity (SEAADSA), 202–203
Sea Automatic Data Support Office (SEAADSO), 202
Seabury breech, patent suit, 7–8
Semple plunger fuze, 73
Severn River, 4
Shells-testing, 10–16, 40, 45;
Shrike, 173, 196, 198
Sidewinder Propulsion Unit (SPU), 147, 150, 152, 196
Sims, William S., 66
Skolnik, Sol, 132–134, 144, 147, 152–153, 162
Smith, Roger, 216, 217–219, 221, 223, 224, 242–243
Smokeless powder, 2–3, 7, 16, 20–24, 34, 36, 38–39, 41, 45, 47, 48, 50–51
Solid Propellant Advisory Group, 127
Spanish-American War, 19–20, 238
Sparrow, 173–174
Special Projects Office, 138, 151, 174, 216
Spotz, Robert H., 190–191
Standard Job Procedure (SJP), 140
Standard Missile, 195, 203

Standard Operating Procedure (SOP), 165
Stark, Harold R., 29, 66
Strauss, Joseph, 20, 24, 26–30, 35, 36, 42, 213
Stump Neck, Maryland, 35, 45–47, 60, 114, 166
Sulfuric acid plant, 34, 60, 168

T
Tadlock, James D., 209
Talos missile, 152
Tartar missile, 173
Taylor, David W., 4, 39
Taylor, Frederick, 25
Technical Director, 153
Terrier missile, 146–149, 152
Thames, F. C., 83, 110, 111–112, 127, 129, 132, 147, 164
Thiokol Corporation, 163, 172
Thornburg, Jay, 175–176, 182, 191, 209
Tomahawk cruise missile, 192–193, 203
Torpedoes, 170–171
Tracers, 49
Tracy, Benjamin, 2, 3, 5, 9, 12, 13, 14, 22, 53, 237
Training programs, 28–30, 117, 205
Triacetin, 167, 171
Trick, John, 221
Trinitrotoluene (TNT), 76
Twining, Nathan C., 47, 55, 93
Tydings, Millard, 102, 127–129

V
Variable delay fuze, 73–75
Very signals, 71
Vietnam War, 148, 162, 174–175, 177, 182
Voegeli, Clarence E., 127, 129, 130

W
Walsh, John J., 220
Washington Naval Conference, 76
Washington Navy Yard, 1, 4, 69, 239
Wassman, Bill, 221
Weapon A, 134, 146–149, 150, 152;
    (SOP), 165

Weapons Quality Engineering Center
(WQEC), 193–194
Weapons Systems Simulation pro-
gram, 203
Wendt, George F., 209, 210, 242
Wesche, Otis A., 153–154, 155
Whiskey Point, 47
Whitney, William C., 3, 4, 237
Wiegand, James, 182
Wilson, James, 166, 168–169
Wilson, Woodrow, 43–45, 47, 51, 53, 57

Worcester, F. J., 149
Worcester report, 149–150
World Trade Center (attack on), 243
World War I, 45, 49f, 57–61, 74, 238,
243
World War II, 69, 105–120, 239–240,
Wynne, Stephen, 194

Z
Zihlman, Fred, 145–146
Zuni, 152, 175, 180, 228, 241